THE BASTARD BRIGADE

The True Story of the Renegade
Scientists and Spies Who Sabotaged
the Nazi Atomic Bomb

SAM KEAN

Back Bay Books
Little, Brown and Company
New York   Boston   London

Back Bay Books / Little, Brown and Company
Hachette Book Group
1290 Avenue of the Americas, New York, NY 10104
littlebrown.com

Originally published in hardcover by Little, Brown and Company, July 2019
First Back Bay trade paperback edition, July 2020

Back Bay Books is an imprint of Little, Brown and Company, a division of Hachette Book Group, Inc. The Back Bay Books name and logo are trademarks of Hachette Book Group, Inc.

The publisher is not responsible for websites (or their content) that are not owned by the publisher.

The Hachette Speakers Bureau provides a wide range of authors for speaking events. To find out more, go to hachettespeakersbureau.com or call (866) 376-6591.

Illustrations by Kevin Cannon

ISBN 9780316381680 (hc) / 9780316381673 (tpb)
Library of Congress Control Number: 2019939370

10 9 8 7 6 5 4 3 2 1

LSC-C

Printed in the United States of America

Praise for Sam Kean's

THE BASTARD BRIGADE

The True Story of the Renegade Scientists and Spies Who Sabotaged the Nazi Atomic Bomb

"*The Bastard Brigade* is as entertaining as it is fascinating. Kean's colloquial expressions and metaphors provide levity to the gritty history of a world at war, with the survival of freedom, and possibly humanity, hanging in the balance. He never lets the reader forget what was at stake…Kean's page-turner about a still-too-little-understood chapter in history deserves a prominent place in WWII collections." —*Booklist* (starred review)

"Kean is on the verge of the transcendent throughout as he chronicles the lives of a bunch of renegade scientists and spies willing to do almost anything to stop the Nazi threat."
—Elaine Margolin, *The Jerusalem Post*

"A thrilling tale of wartime derring-do meets a richly researched story of postwar intellectual exploitation…Perfect as a first foray into this period, and I defy any reader not to be drawn into the world of unlikely spies and Nazi Nobel Prize winners that Kean paints so vividly and infuses with such energy." —Charlie Hall, *Science*

"An exciting read for fans of World War II history, espionage tales, and the development of nuclear weapons."
—*Library Journal* (starred review)

"An exciting history of the battle for atomic supremacy during World War II...Throughout, Kean eschews erudite fastidiousness for consistent action and brio. Beginning with the title, the narrative is an engrossing cinematic drama...Vivid derring-do moves swiftly through a carefully constructed espionage thriller."

—*Kirkus Reviews*

THE BASTARD
BRIGADE

Also by Sam Kean

Caesar's Last Breath
The Tale of the Dueling Neurosurgeons
The Violinist's Thumb
The Disappearing Spoon

Strange things may seem reasonable to men who know only enough to fear the worst.

—Thomas Powers

Contents

Contents

Part III: 1942

Part IV: 1943

Contents

Part V: 1944

Contents

Part VI: 1945

Author's Note

I'm often asked, after talks or readings, why I've never written a book about physics. After all, I majored in physics in college, and I still think it's the most romantic of the sciences. No other field has such incredible scope, taking as its domain everything from the structure of subatomic particles to the fate of the cosmos, not to mention all the human-sized things in between. Know physics, know the universe.

But in my previous four books, I've more or less ignored physics, focusing instead on chemistry, genetics, neuroscience, and the atmosphere. Why? The short answer is that I also had to be true to my second major in college—English literature. That is, what I really love doing is telling *stories,* and when I'm planning a book, I look for rip-roarin' stories first and foremost. I want heroes and villains, conflict and drama, plot twists and redemption. And frankly, I just haven't found a physics topic that captured my imagination enough to write a whole book about it.

Until now. *The Bastard Brigade* is just the sort of physics-adventure tale I always wanted to tell—about the epic quest to stop the Nazi atomic bomb. Science drives this story, no question, but the heart of it is the extraordinary men and women who took on this duty and who were willing to use any means necessary—espionage, sabotage, subterfuge, even murder—to achieve it. No matter what type of story we're talking about, it's the characters that draw us in, and there are pirates and Nobel Prize–winners here, heads of state and Hollywood starlets, people of great strength and people of contemptible weakness.

Above all they're human beings—people thrown into situations that reveal them at their best and worst.

The Bastard Brigade is also something of a departure for me, a new challenge as a writer. In all my other books, I took one central topic (the periodic table, the human brain, et cetera) and spun out a few dozen tales. As a result, the chapters were largely independent and could stand alone, like a collection of short stories. This book is more unified, more of a novel. Because while there are several threads to the plot, the book really tells one larger story overall, and the truth emerges only in the collective actions of the characters.

And because the characters are central to this adventure, I thought it might help to include a list of them as a reference, on page 425. (I've tried not to spoil anything.) If you need a reminder of who's who, you can always flip back there and peek.

Above all, I hope you enjoy the book. I love physics so much that I wanted to be careful about my first foray into it, and this story is absolutely worth the wait.

—spk

THE BASTARD BRIGADE

Prologue: Summer of '44

As the soldiers darted out of the cottage, the doorframe near their skulls exploded in splinters. This wasn't the first time someone had shot at Boris Pash that day, and it wouldn't be the last. An hour earlier Pash and a lieutenant had crept into the booby-trapped forest surrounding this seaside cottage in northern France. Seven brave resistance fighters had already died in these woods, but Pash had a swashbuckling—some said reckless—streak and had plunged ahead anyway. His mission: to capture a local scientist. As for why he needed to capture him, Pash was keeping mum. But echoing through his mind that day were the last words he'd heard from his bosses in Washington a few weeks earlier: "Any slight delay in reaching your targets might cost us tremendous losses, or even the war."

This wasn't an exaggeration. Pash led a team of scientific commandos called the Alsos Unit, who roamed around Europe collecting secrets about the most dreadful threat they could imagine: the Nazi atomic bomb project. Because Alsos ("all-soss") worked independently, unattached to any larger military group, people called it the Bastard Unit. But the nickname was equally apt for Pash himself, a hard-charging World War I veteran whose unruliness behind enemy lines gave his minders back in Washington gastric ulcers.

At the same time, the desk jockeys needed a bastard like Pash: he took on missions no one else could or would. Like hunting down a scientist in a seaside village in France that was still under Nazi control. The man in question was a Nobel Prize–winning physicist rumored to be collaborating with the Germans on nuclear research.

3

His capture could therefore disrupt the entire Nazi bomb project and keep atomic weapons out of Adolf Hitler's hands.

But after slinking past all the pressure mines and tripwires in the forest, Pash and his sidekick had arrived at the cottage to find something sickening: nothing. The door was ajar and the cottage abandoned, stripped bare and full of debris. They searched everywhere, but there were no documents, no equipment, and certainly no nuclear scientist. Washington had feared that even a "slight delay" in finding the target could cost the Allies the war. Now the target had vanished. A dejected Pash and his lieutenant made ready to leave. At which point bullets splintered the doorframe near their heads. Then came the machine-gun fire.

Both men dived to the dirt outside and began belly-crawling into the cover of the woods. Given the secret nature of his mission, Pash had told very few people what they were up to that day. He therefore had no idea who was firing at them or why—Nazis, Americans, French renegades of dubious loyalty. Whoever it was had one clear objective: to make Pash and his sidekick the eighth and ninth casualties in the hunt for the French nuclear physicist.

Meanwhile, as Boris Pash was dodging gunfire, the Bastard Unit's new scientific chief was weathering a calamity of his own. Samuel Goudsmit, a soft and somewhat dandyish nuclear physicist, had arrived in London shortly after D-Day, just in time to see the first V-1 rockets smash down. In the dead of night, people in the city would hear a buzzing noise in the dark above them, until the rocket's motor cut and it began to plunge. Several seconds of dreadful silence followed; many held their breaths until the *boom*. Afterward, there might be another second or two of silence, until the screams began— at which point there would be no more silence that night.

The next morning, Goudsmit ("Gowd-smit") had the uncomfortable job of inspecting the V-1 craters with a Geiger counter. Military officials would drag him from disaster to disaster, all but pushing him down the smoldering slopes to listen for the telltale clicks of radioactivity. The Nazi high command was furious about the D-Day invasion, and the Allies feared that they'd retaliate by lobbing nuclear weapons across the English Channel. The V-rockets seemed an ideal delivery vehicle, and it fell to Goudsmit to scour the pits they left behind.

Although he didn't detect any radioactivity, that didn't mean Goudsmit could relax. To the contrary, he soon received orders that were far more hazardous—to invade the dragon's lair of the Reich and hunt for nukes in mainland Europe. Even the checklist to help him pack for the mission looked menacing. It recommended finding a wool stocking hat "for use with helmet." Who would be shooting at him? And good Lord, a gas mask? Most ominous of all, the checklist recommended he update his will and pay up his life insurance. He might as well call his wife right now and tell her he was a goner. It turned out that no American insurance firm would cover a member of the Bastard Unit anyway. *Let's get this straight. You're going to infiltrate Nazi territory to hunt down an atomic superweapon, and you want life insurance? We'll pass.* Whereas Boris Pash saw the nuclear commando work as an adventure, Goudsmit foresaw only danger and the certainty of his own death.

Indeed, Goudsmit likely would have skipped the war and stayed home in comfort if greater forces hadn't compelled him. As a European Jew, born in Holland, he was determined to fight back against Hitler. His status as one of the few Allied nuclear scientists not working on the Manhattan Project put him in a unique position as well: he had the general knowledge to interrogate Nazi scientists about fission research, but not enough specific knowledge of bombs to give away any secrets if he was (*gulp*) captured and tortured. Moreover, he spoke several European languages, and counted many top German physicists as friends.

Or at least he used to. After years of war, he'd come to hate some of them. He'd been particularly close with the legendary quantum physicist Werner Heisenberg, even letting Heisenberg stay at his home on occasion. But Goudsmit's affection had crumbled into ash after Heisenberg joined the German nuclear bomb program. Goudsmit felt betrayed, and it pushed his mind to sinister places. At one point he suggested, in complete seriousness, deploying a black-ops team into Germany to kidnap his erstwhile friend. And as rumors about the Germans intensified, Goudsmit found himself participating in even darker deeds—including a plot to send a former Major League Baseball player into Switzerland with a gun and a cyanide pill, to assassinate Heisenberg at a scientific meeting.

But more than anything else, beyond even his obsession with Heisenberg, Samuel Goudsmit was joining the war in Europe on a personal mission. Hitler's machinations had trapped his family in Holland, and his elderly mother and father had been rounded up and arrested. The last letter he'd received from them was postmarked from a concentration camp, and he'd been sick with worry ever since. Goudsmit was joining the Bastard Unit to fight Hitler, certainly, and to stop the Nazi atomic bomb. He also needed to find his parents.

The V-1 craters that Samuel Goudsmit inspected in London were terrifying enough, but scientific spies across Europe had already heard rumors of even deadlier V-weapons to come—the V-2s and mysterious V-3s, missiles that promised greater range, greater speed, greater destruction. All of which was fine with Joe Kennedy. The greater the danger, the greater the glory for him.

In August 1944, Joseph Kennedy Jr. was stationed in England, and he whiled away his days writing letters home to his little brother John, future president of the United States. Like every pilot—he flew for the navy—Joe wrote salacious things about girls in the letters and complained of boredom and hardship in the countryside. In reality, his status as a Kennedy gave him privileges that most grunts could only dream about—fresh eggs, white silk scarves, a Victrola, a humidor, a bicycle to pedal to church. He could even commandeer planes to London sometimes to pick up cases of scotch and Pabst Blue Ribbon. All in all, Joe had things pretty swell.

But beneath the easy patter in his letters, there were undercurrents of envy. At one point Joe congratulated Jack on a medal Jack had won for valor in the South Pacific; among other deeds, JFK had saved the life of a badly burned sailor named Patrick McMahon. This had earned Jack fame as a war hero—as well as his brother's enmity. In a barbed compliment, Joe mentioned that he'd seen yet another magazine story about Jack, then added, "McMahon must be awful sick of talking about you." Born just two years apart, the brothers had grown up competing for everything—grades, girls, their father's affection. Joe almost always won, and it infuriated him to see his little brother beat him out for war glory, the most important competition of their young lives.

Joe had hopes of settling the score, however, and soon. Because in between Sunday Mass and Saturday boozing, he was training for a top-secret mission. Over the past year, Germany had erected several mysterious missile bunkers along the northern coast of France, just across the English Channel. If Hitler really did want to rain down atomic fury on London, these seemed like the perfect launch sites, and after the V-1 barrage started, Allied leaders were anxious to wipe the bunkers out.

The problem was, the bunkers were so large and so well reinforced that conventional bombs dropped from airplanes did no good. So officials had to get creative, and what they decided to do was turn the planes themselves into bombs. That is, they would fill them with explosives and fly them across the Channel as unmanned drones. Using crude remote control, they'd then ram the planes into the bunkers kamikaze style. The only hitch was that the planes couldn't take off on their own; someone had to rumble down the runway in these flying bombs to get them aloft, then arm them in midair before they exploded. Joe had volunteered to be one of those someones.

In the letters home to his brother, Joe of course couldn't reveal any details of the mission, but his excitement breaks through here and there. At one point, he brags that he's all but assured of winning a medal of his own. Still, knowing that his parents might read the letter, Joe hastened to reassure everyone that he was safe. "I am not intending to risk my fine neck...in any crazy venture," he said. It was a bald-faced lie. By the time he'd put pen to paper, several of Joe's fellow pilots had already suffered gruesome injuries: one had an arm ripped off while parachuting out, and another had plummeted to his death. Truth was, this was one of the craziest ventures of the war.

We all know how World War II ended, with two black mushroom clouds rising over the scorched remains of Hiroshima and Nagasaki. But most people don't realize how easily things could have gone the other way—how easily the war could have ended not with an American atomic bomb but a German one, obliterating not a Japanese city but London or Paris or even New York.

Many scientists on the Manhattan Project, in fact, were convinced that Germany had the inside track on the Bomb. After all, German chemists and physicists had discovered nuclear fission in the first place, and the Third Reich had founded its Manhattan Project (called the Uranium Club) in 1939, giving it a two-year head start. Germany had the world's best industrial firms as well, fully capable of processing the vast amount of raw material a nuclear bomb requires. No other country on earth could match its genius and industrial might—not to mention its diabolical urge to wage war.

This realization had two effects. First, it pushed American scientists to work maniacally hard on atomic bombs. Second, it convinced the Allies to sponsor a series of desperate missions to sabotage the Nazi bomb project. Spies, soldiers, physicists, politicians—all had roles to play. As one historian said, "Never, perhaps, have scientists and statesmen played for higher stakes, or has the sense of breathless urgency driven men to more extraordinary exertions."

The Bastard Brigade recounts these heroic, chaotic, and often deadly efforts—involving not only the likes of Boris Pash and Joe Kennedy, but courageous female scientists like Irène Joliot-Curie and Lise Meitner. Science had certainly contributed to warfare before 1939, but in World War II, the Allies gave scientists guns and helmets and dispatched them into combat zones for the first time. This shadow war paralleled the visible one in many ways, but the men and women involved more or less ignored the movements of troops, tanks, and airplanes, and instead stalked *ideas*—vast, world-changing scientific ideas.

Still, the Allies weren't above playing dirty when the mission called for it. The subject of the first chapter—the country's first atomic spy, an enigmatic baseball catcher named Moe Berg—stole his friends' mail, lied repeatedly to superiors, and went AWOL with alarming frequency. For him and others, no tactics were too extreme—air strikes, commando raids, Molotov cocktails, kidnappings—as long as they kept the Bomb out of Hitler's hands.

Unlike other histories of the Nazi atomic bomb, this story focuses on the Allies—putting us directly into the minds of the men and women confronted with, perhaps, the ultimate mission. Much of what follows comes from previously unpublished or overlooked sources, which provide new insight into some of the war's most fascinating yet unheralded characters. All the missions were top-secret, naturally, and those who volunteered for them often had dark motivations for doing so; in some cases they spent as much energy fighting each other as they did the enemy. But if they couldn't shake their personal demons, they never flinched when facing down the Nazi threat.

The Bastard Brigade starts in that "low dishonest decade" of the

1930s, with the birth of nuclear fission, and it continues through the epic manhunts of the very last days of the war. The Allies had sacrificed millions of lives conquering North Africa and Italy, not to mention gaining footholds in France and Germany. But with just a few pounds of uranium, they feared, Hitler could reverse the entire D-Day operation and drive the Allies off the continent forever.

So if the story that follows seems frantic, reckless, or even mad at times, there's good reason for that. Scientists and soldiers alike were convinced that a madman would soon acquire the superhuman power locked inside the atomic nucleus. And to prevent that, no price was too high to pay.

PART I

Prewar, to 1939

CHAPTER 1

Professor Berg

America's first atomic spy very nearly wasn't American at all. After fleeing pogroms in Ukraine in the 1890s, Moe Berg's father Bernard booked passage from London to the United States on a crowded, dirty steamer that reeked of bologna and unwashed bodies. And when he arrived in New York, the ghettoes and tenements there made steerage class seem luxurious. After hearing that foreigners who fought in the Boer War would get automatic British citizenship, he hopped the next boat back to London—only to find that the offer had expired. With great reluctance, he spent his last ten dollars returning to New York, resigned to becoming an American.

Bernard soon married a seamstress from Romania named Rose, with whom he had three children, and they opened a launderette on the Lower East Side. It was not a success. A chronic reader, Bernard often got so absorbed in his books while ironing that he burned holes in people's garments. Eventually he admitted his shortcomings and opened a pharmacy in Newark instead, installing his young family in the apartment above. (Because he worked so much—fifteen hours a day—he interacted with them by hollering through a tube that ran upstairs.) As the first Jewish family in their neighborhood, the Bergs suffered occasional discrimination (children would holler, "Hey, Christ-killer!"), but the pharmacy eventually became a social hub in the neighborhood. Bernard was especially renowned for his "Berg cocktails"—laxatives of castor oil and root beer. Before mixing one, he'd ask Mrs. So-and-So how far away she lived. Four blocks, she'd say. He'd then measure out a four-block cocktail and have her chug it. Go straight home, he'd warn her, and don't dally to talk. People learned the hard way that he wasn't joking.

Bernard and Rose's youngest child, twelve-pound Moe, arrived in 1902. With Bernard working all the time, the boy had complete freedom to pursue his passion, baseball. He'd toss around balls, apples, oranges, anything vaguely spheroid, at any hour, and even as a child, he was the best catcher in Newark. He'd squat behind manhole covers, holding a glove that looked as big as a pillow in his tiny hands, and let local cops fire heaters at him. "Harder!" Berg would cry. "Harder!" Finally one cop wound up and really smoked a pitch. Berg staggered back and almost toppled. But he held on—no adult could get one past him. Hearing of this prodigy, a local church all-star team scooped him up. They insisted he use a Christian pseudonym, Runt Wolfe, but Runt quickly became the squad's star.

The only person not impressed by Moe's baseball prowess was his father. A reluctant U.S. citizen, he never could embrace this most American of sports. He looked down on ballplayers as clods and contrasted them with his real heroes, scholars. But the thing was, Moe was pretty sharp in the classroom, too, graduating from high school at sixteen and winning admission to Princeton University. There, following one of his father's passions, he majored in Romance languages, taking six courses some semesters; he dabbled in Sanskrit and Greek to boot. When Berg later became famous, no quirk of his would attract more attention than his faculty for languages. Some admirers claimed he spoke six, all fluently; some said eight; others a dozen.

To his father's distress, Berg also played baseball for Princeton's Tigers. Ivy League games often drew huge crowds back then, up to twenty thousand people, and Berg blossomed into the team's star shortstop. It helped that he stood six foot one, and had huge mitts: "shaking hands with him was like shaking hands with a tree," an acquaintance remembered. In Berg's junior year the Tigers almost beat the world champion New York Giants in an exhibition game at the Polo Grounds, losing 3–2. He then led the Tigers to a 21–4 record his senior year—including an 18-game winning streak—and hit .337, including .611 against rivals Harvard and Yale. He and the team's second baseman that year, another linguaphile, would discuss on-field defensive strategies in Latin to prevent the other team from catching on.

Now, you might think that a tall, well-built, all-American short-stop at Princeton with a flair for Romance languages would be a pop-ular guy, and people did admire Berg. But mostly from afar; he had few real friends at school. In part, this was Princeton's fault. Most Princeton boys (it was an all-male university then) had attended fancy prep schools, and some showed up for classes in chauffeured cars. Berg, meanwhile, toiled to afford the $650 tuition, working as a camp counselor in New Hampshire each summer and delivering Christmas packages over winter break. The expensive habits he affected—smoking jackets, scented hair oil—didn't fool anyone. Being Jewish didn't help, either. His senior year the Princeton base-ball nine elected someone a little more suitable (read: WASPy) as team captain, which stung. And when it came time to join an eating club (the Princeton version of a fraternity), he got voted into one—on the condition that he not get pushy and advocate for other Jews. Humiliated, Berg refused to join.

But the isolation wasn't all Princeton's fault. Berg's essential trait, the one that defined the whole course of his life, was his furtiveness. He was handsome and witty. Men admired his erudition and athletic skill. Women cooed when he whispered in French and Italian. But he never attended parties, never asked anyone to dinner, never let any-one get close to him. He was an incorrigible loner, constantly push-ing people away, and he cultivated an air of inscrutability.

Two ball clubs, the New York Giants and the Brooklyn Robins (later the Dodgers), tried to sign Berg out of Princeton in 1923—in part because attendance was sagging, and they figured a Jewish star would provide a boost. But Berg hesitated; he'd set his heart on attending graduate school at the Sorbonne in Paris that year. He did finally sign, however, figuring that he could attend the Sorbonne during the off-season. (You know, like most ballplayers.) Of the two clubs, the Robins had the worse record, which meant that Berg could play immediately. So to his father's further shame, he signed a $5,000 contract ($71,000 today) that summer. A few days later in Philadelphia, in his first at-bat, he singled and drove in a run.

It was probably the highlight of his rookie season. Although a graceful fielder, with a blistering arm, he was young and skittish and made too many errors to play full-time. Worse, he struggled to adjust to Major League pitching. Although he rarely struck out, he didn't hit for power and was far too slow to leg out hits; a manager once cracked that Berg looked like he was running the bases in snowshoes. He batted just .186 in 49 games, and a scout that summer summed up Berg's prospects in four words: "Good field, no hit."

Instead of working on his hitting, Berg skipped town for the Sorbonne that winter. Tuition was cheap ($1.95 per course, $28 today), so he gorged himself on classes, sitting in on twenty-two. Topics included French, Italian, Latin of the Middle Ages, and "The Comic in Drama." Berg was particularly interested in tracing the "bastardization" of Latin as it spread through Europe. ("The farther Caesar's legions trekked from Rome," he later explained, "the more the pure Latin become diluted with the words and idioms of the people they were trying to subjugate.") He was a feisty student, too. Before one European history course—which covered the fraught decades leading up to the Great War—he declared, "If it becomes too one-sided, I'll tell the professor to stick the course up his—." But overall the classes more than fulfilled his expectations. In a letter home he declared that he would have paid five dollars to hear some of the individual *lectures*, they were that good: "For what I am getting out of it, I ought to endow a chair in the Sorbonne."

While in Paris, Berg also picked up a lifelong habit of reading several newspapers a day, often in different languages. Although he owned few possessions, newspapers were something he got territorial about. He'd haul them back to his room by the armful and read a few stories here, a few stories there. Then, according to some recondite filing system, he'd drape them over chairs, dressers, bathroom plumbing, even his bed, intending to pick them up again later. He called these half-read periodicals his "live" papers, and woe betide anyone who touched a live paper. Berg would explode in rage, flinging the pages and stomping off to buy a "fresh" one, no matter how late at night or how lousy the weather. Only when he'd finished a paper and pronounced it "dead" could people touch it. No one ever figured out why he got so upset over this—it was part of the mystery of Moe.

Unfortunately for Berg's baseball career, he gorged himself on more than newspapers in Paris, taking full advantage of the city's cuisine. A typical day started with chocolate and buttered croissants for breakfast, and for dinner he'd stuff himself at restaurants for fifty cents. Drink tempted him as well. In one letter home he declared, "[I] shall probably drink no more water. The wine is very strengthening." He made no attempt to exercise beyond walking, and gained at least ten pounds that winter. As a result, he showed up for spring training in March in dreadful shape and got demoted to AAA ball.

Thus began a long, frustrating stint in the minors, hopping from the Minneapolis Millers to the Toledo Mud Hens to the Reading Keystones. (The demotion must have vexed his father, too.) But during his second season of purgatory, Berg collected 200 hits and 124 RBIs, and in 1926 the Chicago White Sox snagged him for $50,000 ($700,000 today), a gigantic contract. Not wanting to miss another chance, Berg worked hard and rewarded the Sox by playing the best baseball of his life over the next few years.

Berg owed some of this improvement to his switching to a more natural position. He told different versions of the story over the years, but in August 1927, the Sox's starting catcher got injured in a collision at the plate. A few days later, the backup split a finger open during a doubleheader. Then the backup's backup, the last catcher on the roster, got knocked silly in another collision in Boston. The manager groaned: What the hell are we going to do now? Berg, sitting on the bench, apparently jerked a thumb toward a teammate, a chubby first baseman who'd caught in the minors. "You've got a catcher right here," he said. But the manager had his back turned and didn't see Berg's gesture; he only heard his voice and thought Berg was volunteering—or being a wiseacre. He turned and looked his budding shortstop up and down. "You ever catch?"

"In high school," Berg answered.

"Why did you quit?"

"Because of something a man said. He said I was lousy."

"Who was that man?"

"My coach."

"Well, get in there and let's see if he knew what he was talking about."

Berg said aye aye and began strapping himself into the catcher's upholstery. "If the worst happens," he announced to the bench, "kindly deliver the body to Newark."

The Sox lost the game, but Berg played well. That night, while teammates went carousing, he joined a mass protest against the Sacco and Vanzetti executions on the Boston Common, and when the Sox shipped out for New York the next day to play the dreaded Yankees, the manager penciled him in as the starting catcher. As Babe Ruth stepped into the batter's box in the bottom of the first, he smirked and said, "Moe, you're going to be the fourth wounded White Sox catcher by the fifth inning." Moe replied that he was going to call for a barrage of inside pitches on Ruth. That way, "We can keep each other company at the hospital." They both laughed. But the catcher laughed last: with Berg calling pitches, the mighty Babe struck out twice that day and never hit a ball out of the infield. Berg picked apart the rest of Murderer's Row in similar fashion, and added a single and an RBI in the decisive three-run sixth inning, helping the White Sox win 6–3.

Nevertheless, Berg's manager didn't trust his new catcher and kept scouring the East Coast for minor leaguers and semipro lads. History remains grateful that he didn't find anyone, because as Berg got more comfortable behind the plate, he developed into one of the top catchers in the American League. Teams quickly learned not to test his arm on the base paths, and with his experience as a shortstop, few pitches got by him; he once set an AL record by playing 117 straight errorless games. Aside from fielding, he excelled at the cerebral side of baseball. He catalogued every batter's weakness, and with his constant patter and diabolical pitch calls, he easily got inside their heads; pitchers rarely called off his signs. Catching proved an advantage at the plate as well. With a better understanding of how pitchers thought, he developed into a serviceable pull hitter, regularly stroking fastballs down the left-field line. Even his glaring weakness, a lack of foot speed, proved no real handicap now—catchers are *supposed* to plod. In 1929, his best season, Berg hit .287 in 107 games and collected 101 hits. He even garnered a few votes for MVP.

Incredibly, Berg did all this while attending law school at Columbia University in the off-season. When other players took trains back to Alabama and Texas to chop wood in October, Berg schlepped up to Manhattan and—having started classes three weeks late due to baseball commitments—busted his ass catching up on contracts and finance law. "I worked like a Trojan," he once said, "thinking always of February and the south [for spring training] once more." Teammates thought the whole arrangement queer, sportswriters amusing. The White Sox owners found it frustrating, since he often missed spring training in Shreveport while finishing up classes. But Berg insisted on attending, probably because of his father. Even when his son blossomed into a poor man's MVP candidate, Bernard refused to attend any games. Whenever someone mentioned the catcher around the pharmacy in Newark, he'd turn his head and spit. "A sport," he'd scoff. Law was vastly more respectable.

So after Berg took the New York bar exam one spring—a series of long essay questions—and reported to Chicago for the season, he kept checking the *New York Times* each day in the library to see if he'd passed. He finally saw his name listed—one of 600 to qualify, out of 1,600. "Think of the poor suckers who flunked," he gloated. "I was never happier in my life." He phoned Bernard with the news.

His father was terse: "You didn't have to call long distance. I read the papers." With that, he hung up on his son.

Six months after his best Major League season, Berg suffered a devastating injury. In April 1930, during an exhibition game in Little Rock, he dove back to first base during a pickoff attempt. His spikes got caught in the dirt, and he tore a ligament in his right knee and ended up needing surgery at the Mayo Clinic. He sat out a few months and tried coming back, but clearly wasn't healthy. A midseason bout of pneumonia weakened him further. All in all, the injury and the illness wiped out the next two years, limiting him to twenty games with Chicago and (after Chicago cut him) ten with Cleveland. With his future in baseball looking shaky, he began practicing law on Wall Street in the off-season to make money. He hated it.

In 1932, after two years of rehab, Berg had recovered enough to sign with the Washington Senators. But he simply didn't have the same spring in his legs anymore. His once adequate hitting deteriorated; he was slower than ever, becoming an outright liability on the base paths; and he simply couldn't squat for hours in the sun on that knee. So Washington demoted him to bullpen catcher. Berg would never again be mistaken for someone who collected MVP votes.

And yet, in a funny way, the knee injury was the best thing that ever happened to his career. It might sound strange to say that someone was born to be a bullpen catcher, but Moe Berg was. With his cerebral approach to the game, he proved a perfect mentor for young pitchers, and the lazy pace of bullpen life suited him perfectly. He didn't need to warm up or practice much, and could lounge around the clubhouse and leaf through "live" newspapers instead. (Fans would even bring foreign-language editions to the ballpark for him.) He also had plenty of time to gab with sportswriters, who found Berg irresistible—funny, chatty, highly quotable. The press fawned over him, and why not? Here was a big, lumbering, unibrowed catcher from Newark who'd attended Princeton and the Sorbonne and spoke seventeen languages. It made for scintillating copy.

Most columns about "Professor Berg" focused on his eccentricities: that he could read hieroglyphics and recite Edgar Allan Poe's entire poetic oeuvre; that he ordered applesauce instead of steaks or sandwiches for lunch; that he bought dictionaries "to see if they were complete"; that he traveled with eight identical black suits and never wore anything else; that he once polished off a book on non-Euclidean space-time in the bullpen during a doubleheader in Detroit, then called on Albert Einstein the next time he visited Princeton to discuss the matter further. (One writer thereby dubbed the catcher "Einstein in knickers.")

Altogether, Berg got more column inches than any benchwarmer in baseball history—something his more talented peers didn't always appreciate. In one of the all-time great putdowns in sports, a writer once asked a teammate about Berg's ability to speak so many languages. The teammate, having heard the question perhaps one time too many, scoffed, "Yeah, well, he can't hit in any of 'em."

Berg often played a curmudgeon for reporters, but he secretly adored media attention, in part because it won him several perks. For instance, he was one of just three big-leaguers selected to visit Japan in 1932 for a series of goodwill workshops on baseball. He taught the youngsters there the finer points of the sport: defending first-and-third situations, forcing ground balls with low pitches to set up double plays, even handling spitballs. For their part the Japanese players adored Berg and thought his dark complexion—and his unibrow—quite exotic. Berg later called Japan "heaven for umpires," because the players there were so polite to them.

The trip to Asia also gave Berg an excuse to travel more, and when his fellow ballplayers sailed back home, he headed west instead, touring Korea, China, Indochina, Cambodia, Siam, Burma, India, Iraq, Saudi Arabia, Syria, Palestine, Egypt, Crete, Greece, Yugoslavia, Hungary, Austria, Holland, France, and England. He no doubt returned to spring training out of shape again, but this time no one cared, since he had a fresh larder of tales to regale teammates and reporters with.

Privately, though, one leg of the trip disturbed him. On arriving in Berlin in late January 1933, he immediately picked up several newspapers. Every headline was the same: Germany had a new chancellor, a forty-three-year-old firebrand named Adolf Hitler. Berg then spent the day watching crowds of jubilant Nazis celebrate in the streets. Upon returning home, he told anyone who'd listen that Europe was headed for grief.

CHAPTER 2

Near Misses and Big Hits

Irène Curie wished that it would hurt less each time—that the pain and humiliation would fade. But every time she missed out on a major discovery, she felt the same sting.

Irène was the daughter of the pioneering physicists Marie and Pierre Curie. She was born in 1897, during one of their most productive periods, and often had to compete with their research for attention—something that didn't come naturally to a shy, retiring girl who sometimes hid behind doors rather than talk to houseguests. (One of the horrors of her childhood occurred when her parents won the Nobel Prize in 1903 for work on radioactivity and a mob of photographers stormed their house.) It didn't help that Marie, despite her many wonderful qualities, was a distant parent. Polish-born, she'd lost her own mother at age seven and felt uncomfortable with intimacy. Irène and her younger sister were largely raised by their paternal grandfather, and even when the girls clamored for Marie's affection—clinging to her skirt at night when she returned home late from the lab—she rarely hugged or touched them.

Marie grew even more remote after a family tragedy in 1906. In April of that year, while playing at a friend's house one afternoon, Irène received word that she'd have to stay there for a few days. No one explained why. Finally, late that night, Marie stopped by and mentioned something about Pierre hurting his head. "He will be away for a while," Marie said, which Irène didn't understand. Marie's siblings from Poland soon arrived, as did Pierre's brother, confusing the young girl further. It turned out that a carriage had struck and killed her father, which no one told her until after the funeral. The death might have knitted some families together, but Marie dealt with her grief by working even longer hours, and for years afterward she refused to say Pierre's name aloud.

Adolescence proved no easier for Irène. When she was twelve, Marie enrolled her in an alternative school where she taught math and science on Thursdays. The ten or so students there studied sculpture and Chinese as well, and participated in several sports. (No mere egghead, Marie believed strongly in physical education; the Curies swam and hiked and had a trapeze in their backyard.) The school sounded idyllic, a free-spirited alternative to the stuffy French education system, but Marie held her daughter to exacting standards. She once caught Irène daydreaming instead of working on a math problem, and when Irène admitted she didn't know the answer, Marie barked, "How can you be so stupid?" and flung Irène's notebook out the window. Irène had to trudge down two flights of stairs to retrieve it—and meanwhile solved the math problem in her head.

The years 1910–1911 were especially wretched in the Curie household. First, Irène's beloved grandfather died. Then a scandal involving Marie exploded in the French tabloids. She'd been carrying on with a married man, physicist Paul Langevin, and a newspaper printed excerpts of their love letters. ("When I know that you are with [your wife]," Marie wrote, "my nights are atrocious, I can't sleep.") One day the wife threatened to murder Marie in the street, and Langevin challenged the newspaper publisher to a duel. As things became increasingly sordid, both Marie and Langevin suffered humiliations, but Marie, as a woman, suffered more. Mobs threw rocks at her windows and screamed, "Go back to Poland!" Then, when Marie won a second, surprise Nobel Prize a few weeks later, the Swedish Academy asked her not to attend the award ceremony, to spare their king the embarrassment of shaking hands with an adulteress. Marie defied them and attended anyway, but grew so despondent over the scandal that she contemplated suicide. Unable to concentrate on research, much less raising children, she sent Irène and her sister to live with relatives.

It took the cataclysm of World War I to forge a real bond between mother and daughter. In August 1914, Irène and her sister were on holiday in L'Arcouest, a fishing village in northern France sometimes called "Port Science" for its popularity among researchers. Marie planned to join them in a few weeks. But as soon as the war broke out, she dropped those plans and turned all her attention to her precious gram of radium. She'd isolated this speck of radioactive element 88 after several years of backbreaking labor, boiling down eight tons of mineral ore in a cauldron in a shed. It was the basis of all her research, and frankly the most precious thing in the world to her. So instead of going to Port Science to get her daughters, Marie made a run to Bordeaux, in southwest France, to hide the radium from the invading Germans, hauling it in a special lead-lined case that weighed 130 pounds, roughly sixty thousand times more than the radium it was shielding.

Eventually France became stable enough for the Curie daughters to return to Paris. And here's where Irène finally managed to win her mother's respect. Drawing on her knowledge of science, Marie established a series of X-ray stations near the front lines to help surgeons locate shrapnel in soldiers' bodies; she also developed a fleet of vans with mobile X-ray units for the battlefield, which the army nicknamed "Little Curies." Irène insisted on volunteering for the work, and she proved so adept at it that, at age nineteen, she found herself running a field station in Belgium. She was close enough to the trenches to hear gunfire, and despite the risks to her health—the equipment was poorly shielded at best—she X-rayed thousands of soldiers and repaired the machines when they broke down. She also joined Marie on several harrowing trips to the front in the Little Curie vans. "We were often not sure of being able to press forward," Marie later recalled, "to say nothing of the uncertainties of finding lodging and food." But the hardship bonded them, and by war's end Marie could finally see her daughter as a real, independent woman.

Incredibly, in between trips to the front, Irène found time to earn a degree in physics from the Sorbonne. At war's end, she joined Marie's institute as a doctoral student and assistant researcher. (At the time, over half the scientists there were women, both because Marie made it a point to support women in science and because so many young men had died in the trenches.) Irène thrived in this atmosphere, and by the early 1920s had enough confidence to take on an assistant of her own—and, with him, to defy her mother for the first time in her life.

Frédéric Joliot couldn't believe his luck. When the war ended, he was just another junior scientist struggling to find a job, largely because he hadn't attended the "right" schools in snooty Paris. So when he applied to work at Marie Curie's institute, he hadn't gotten his hopes up. But as an outsider herself, Marie decided to take a flyer on this tall, thin youngster with a shark fin of a nose. (It helped that her former lover, Langevin, had recommended Joliot on the strongest terms.) The job offer stunned Joliot: as a child he used to clip out pictures of Curie from magazines, and he still revered her. He accepted in a heartbeat. Marie then introduced Joliot to his new boss, Irène.

The youngsters fell into a comfortable partnership, with Irène focusing on chemistry and Joliot on physics. Marie approved of this relationship, as it echoed the division of labor that had proved so successful for her and her late husband. What she didn't approve of—and was in fact stunned to learn—was that Frédéric also had his eye on a romantic relationship with the green-eyed Irène and had been courting her behind Marie's back.

Even more astounding, Irène reciprocated Joliot's feelings. It was a hopeless match, really, given their polar temperaments. He was impulsive, vain, outgoing, and well groomed, always wearing an impeccable white coat in the lab; she was reserved, stoic, and frumpy, sometimes taking naps right on the floor. But they bonded, deeply, over several things—losing fathers at a young age; a passion for social justice; and especially a love of nuclear science. You can see this most clearly in their lab notebooks, which at times read like scientific arias: one of them might start writing up an experiment and the other would pick up the thought midsentence, extending the duet in a different handwriting. After a few years of such intimacy Irène finally accepted Joliot's proposal of marriage, and on the morning of October 9, 1926, Joliot wedded and bedded his bride—or at least the former. Following the nuptials, they spent the afternoon in the lab.

Suspicious of the match, Marie Curie often introduced Joliot to others not as her son-in-law, but as "the man who married Irène." Among other things, she felt miffed that Irène and Joliot had changed their surnames to "Joliot-Curie" after marrying. On the one hand, the hyphenation seemed progressive and feminist, a declaration of equality. But cynics noted that Frédéric gained a whole lot more out of attaching "Curie" to his name than Irène did in attaching "Joliot" to hers. As a result, some colleagues began referring to Joliot as "Irène's gigolo." They did so both to put the upstart Joliot in his place and to insult Irène, who was in many ways the stronger, dominant partner. Nevertheless, the Joliot-Curies' marriage, and their research, thrived.

The couple endured their first setback as scientists in January 1932. A few years earlier, physicists in Germany had published some odd experimental results involving radioactive atoms. Radioactive atoms are unstable atoms: they break down and shoot out different

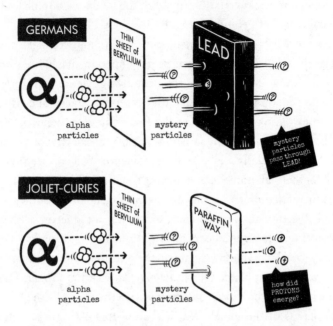

types of particles—a sort of subatomic shrapnel. Specifically, the Germans were working with so-called alpha particles. They directed a stream of these alpha particles at a thin sheet of beryllium metal. This in turn caused the beryllium to release a second type of particle. But the identity of this secondary shrapnel proved mysterious. For one thing, it was extremely energetic: it zipped along so fast that it could pass through four solid inches of lead. The most vigorous type of radioactive particle then known was called a gamma ray, so the Germans concluded that this must be a special type of gamma ray and wrote up a paper.

Two teams began doing follow-up work, including the Joliot-Curies in Paris, and thanks to Marie Curie's nepotism they had a huge advantage over their rivals. Curie had the best equipment in the world, as well as the most potent sources of alpha particles, including her two grams of radium. (In addition to the original gram she'd hidden during World War I, she'd received another gram as a gift from the women of the United States in 1921, to honor her role as a pioneering female scientist.) Marie in turn gave her daughter and the man who'd married her daughter exclusive access to these scientific riches. In fact, before marrying into the family, Joliot had had to sign a prenuptial agreement specifying that, if Marie died and he divorced Irène, the radium belonged to Irène alone. That's how valuable the stuff was: at least $100,000 per gram then, or $1.3 million today.

Radium decays into other substances over time, and by sifting through Marie's radium, among other sources, the Joliot-Curies isolated a sample of polonium, an element that releases an intense stream of alpha particles. They then re-created the German experiment and discovered something startling. Like the Germans, they let the alpha particles strike a sample of beryllium and knock loose "gamma rays." But they also extended the experiment by putting a block of paraffin near the beryllium and letting the gamma rays slam into it. To their amazement, the paraffin began coughing up protons, another subatomic particle. Protons are vastly heavier than gamma rays; so for gammas to knock protons loose, the gammas had to be moving at unthinkable speeds. It would be like shooting spitballs so hard they dislodged a boulder. Excited, the Joliot-Curies wrote up a paper about their work and mailed it off for publication. Irène was quite pregnant at the time (there were no safety standards about exposing fetuses to radioactivity), so after the paper appeared they took a well-earned vacation to the Curie family cottage near L'Arcouest. (And make no mistake, it was the *Curie* family cottage: Joliot's prenup also barred him from claiming any ownership of that.)

Meanwhile, the other person doing follow-up work, James Chadwick in England, was struggling. He worked in the skinflint Cavendish lab in Cambridge, with clunky apparatus and weak sources of alpha particles. He finally cadged a better source from a hospital in

Baltimore, which mailed him some nearly spent ampules of radioactive elements used to attack tumors. (There were no postal safety standards then, either.) By the time Chadwick received the ampules, the Joliot-Curies had published their work. But rather than resign himself to losing, he read their paper with a critical eye—and realized that their conclusions smelled fishy. He simply didn't believe that tiny gamma spitballs could dislodge huge proton boulders. He came to a different conclusion instead.

Scientists at the time believed that atoms were made of two particles: positive protons, which resided in the nucleus of an atom, and negative electrons, which swirled around the nucleus. But some theorists predicted the existence of a third particle, also residing in the nucleus—the neutral neutron. Chadwick wondered whether the strange beryllium "gamma rays" might actually be the first glimpse of neutrons. It would make sense: neutrons, being the same size as protons, could readily

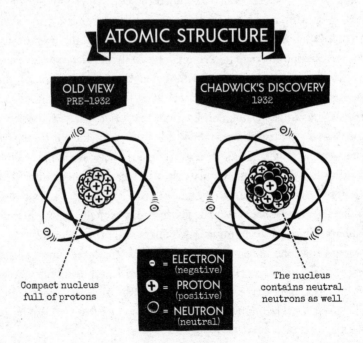

ATOMIC STRUCTURE

OLD VIEW
PRE-1932

CHADWICK'S DISCOVERY
1932

Compact nucleus
full of protons

⊖ = ELECTRON
(negative)

⊕ = PROTON
(positive)

○ = NEUTRON
(neutral)

The nucleus
contains neutral
neutrons as well

dislodge them. And because they were electrically neutral, neutrons could penetrate matter easily, even thick slabs of lead.

Chadwick spent the next thirty days running and rerunning experiments—sleeping just three hours many nights—and soon had solid proof of neutrons. Accordingly, he sent off a paper to *Nature* in February 1932. Upon returning from their vacation in Port Science, Irène and Joliot got hold of the paper and were mortified: they'd just missed out on discovering one of the three fundamental particles of the universe. It was the sharpest setback they could imagine—until things quickly got worse.

After fumbling the discovery of the neutron, the Joliot-Curies redoubled their efforts. Despite having given birth six weeks earlier, Irène dragged Joliot to a lab on an 11,000-foot peak in the Swiss Alps in April. This altitude made the lab an ideal place to study so-called cosmic rays, a stream of subatomic particles that streak down onto Earth from outer space. No one really knew what the rays were back then, and Irène and Joliot wanted to study them and see whether this neutron particle appeared within the shower.

Their work used a piece of equipment called a cloud chamber, a sealed basin with vaporous alcohol or water inside. When cosmic rays zipped through the chamber, they left behind a visible trail of droplets. By subjecting the basin to electric and magnetic fields, scientists could twist or bend the droplet trails, and from the shape of the twists and bends they could infer the size, speed, and electrical charge of the particles. An equipment geek, Joliot adored cloud chambers and would gaze at the trails for hours, fawning over the loops and whorls. Whenever a particularly lovely track appeared he'd gush, "Isn't this the most beautiful experience in the world?" To which Irène would reply, "Yes, my dear, it would be...if not for childbirth."

In the Alps, Irène and her husband saw some mildly interesting trails appear, including some odd spirals. The particle that created them apparently weighed the same as an electron, but the trail twisted in the opposite direction, like the trail of a positive particle. Regardless, neutral neutrons wouldn't leave such a trail, so after two fruitless months the couple dropped the project and returned to Paris with their child.

But that September, an announcement sent them racing back to their lab books. A physicist in California, also using cloud chambers, had found something called antimatter. Different combinations of the three fundamental particles—protons, neutrons, electrons—make up virtually everything around us, and we call this everyday stuff matter. But the universe also contains antimatter, which is basically matter's photographic negative. (If matter and antimatter touch, they obliterate each other in a burst of energy.) Like the Joliot-Curies, the California scientist had noticed an electron-sized particle tracing out unusual swirls in his chamber. Unlike them, he realized the significance of this: that he'd captured the first proof that antimatter exists. In particular, he'd found a particle called the positron.

When Irène and Joliot dug up their old lab notes, they could only groan. They'd seen the same tracks, the same evidence—and for the second time in a few months had missed a fundamental discovery. This time, their scientific aria was one of heartbreak.

If 1932 couldn't end fast enough for the Joliot-Curies, the next few years brought them some redemption. They resumed bombarding different sheets of metal with alpha particles, and they got a nice surprise when they tested aluminum in the autumn of 1933. Normally, a barrage of alphas produced only one type of secondary shrapnel, often neutrons. But bombarding aluminum foil produced both neutrons and positrons—a twofer. No one had ever seen double-barreled radioactivity like this, so the Joliot-Curies decided to prepare a report for a prestigious conference in Brussels in October. The attendees would include virtually every bigwig in nuclear physics—Bohr, Fermi, Dirac, Schrödinger, Rutherford, Pauli, Heisenberg.

The talk could have made their careers. Instead, it nearly ruined them. Because of their previous blunders the Joliot-Curies had gained a reputation as careless, and this new discovery—one that, conveniently, involved both new particles they'd missed before—seemed too good to be true. A brilliant Austrian physicist named Lise Meitner stood up after their talk and declared, with all the sternness of an Old Testament prophet, "That is not so." She'd run similar experiments in Berlin, she claimed, and had never seen such a thing. It was a damning assessment, and given Meitner's reputation, most scientists in attendance believed her.

Crushed, Irène and Joliot returned to Paris. Rather than hang their heads, however, they grew obsessed with proving their results valid. They thought about little else, discussing the experiments over

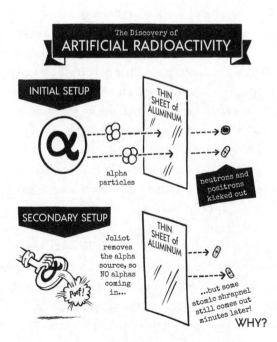

The Discovery of
ARTIFICIAL RADIOACTIVITY

INITIAL SETUP

α

alpha particles

THIN SHEET of ALUMINUM

neutrons and positrons kicked out

SECONDARY SETUP

Joliot removes the alpha source, so NO alphas coming in...

Poof!

THIN SHEET of ALUMINUM

...but some atomic shrapnel still comes out minutes later!

WHY?

every meal and late into the night. After weeks of tediously double-checking everything, Fortune's wheel finally spun their way. One morning in January 1934 Joliot rolled up the sleeves on his white lab coat and tried rearranging their experimental setup, just to see what happened. He started by pulling the alpha source farther back from the aluminum foil. Then, for no real reason, he removed the alpha source altogether. To his confusion, the radioactivity detector kept recording pings of shrapnel. And not just for a second or two, but for several minutes. This made no sense: the alpha particles were necessary to knock the shrapnel loose, and removing them should have halted everything. So why was the detector still registering hits minutes later? As he often did when perplexed, he called for Irène.

They set to work, and after a day of frenzied activity—which left their lab uncharacteristically messy—they realized what was happening. In all other known experiments of this type, when an alpha particle struck the metal foil, it immediately knocked something loose. In this case, however, the aluminum was absorbing the alpha particle and becoming radioactive only later, after a delay. That was intriguing, because alpha particles, in a technical sense, are simply a bundle of protons and neutrons. A little ball, really, with two of each. So if an aluminum atom absorbed an alpha particle, it gained two protons in the process. Atoms are defined by the number of protons they have, so if aluminum (element 13) absorbed an alpha with two protons, it must be changing into phosphorus (element 15); the phosphorus then released radioactive shrapnel and disintegrated. In other words, Irène and Joliot had seemingly discovered a way to convert one element into another element via artificial means. It was artificial radioactivity—scientific alchemy.

Poignantly, the very magnitude of this discovery made the Joliot-Curies hesitate. They couldn't quite trust themselves anymore, not after stumbling twice already. What if their detector was defective? What if they were misinterpreting their results again? What if, what if? Alas, they had an important dinner to attend that night and couldn't keep working. But they left instructions for a young German assistant in the lab—Joliot's cigarette buddy—to check every millimeter of their detector for shorts or other flaws.

The German worked all night running various tests, then left a note for Irène and Joliot. They rushed back to lab the next morning, as anxious as teenagers after a big exam. The counter, the German assured them, worked perfectly.

This convinced the flighty Joliot, who was ready to celebrate their discovery. Irène reserved judgment. Chemists are more tactile than physicists, and she needed to *see* that newly created phosphorus for herself, hold it in a vial. So she devised a plan. They pushed the clutter of the previous night aside and bombarded another sheet of aluminum foil for a few minutes. Instead of placing it in front of a detector, however, this time Irène plopped the foil into a beaker of acid, which began to bubble and hiss, releasing a gas.

If they really had created phosphorus, then that gas was phosphine (PH_3). Identifying phosphine was straightforward, but the nature of this setup complicated things, since the P in the PH_3 was itself radioactive and was disappearing at a rapid clip. So Irène had to work fast, collecting the gas and carrying out her entire analysis in just three minutes. A lesser chemist would have stumbled under the pressure. Irène didn't, and found definitive evidence of phosphorus. Alchemy was real.

At this point, watching his wife finish up, Joliot all but burst into song. He began running around the lab, leaping with joy. "With the neutron we were too late!" he shouted. "With the positron we were too late! Now we are in time!"

Still, within the Joliot-Curie family, no discovery counted until the big Curie, Marie, had weighed in. By early 1934, after years of exposing herself to radioactive substances, Curie was suffering from anemia and rarely visited the lab. That afternoon, however, upon hearing what her daughter and the man who'd married her daughter had discovered, the old lioness roused herself and barged into the lab. (She was accompanied, oddly enough, by her former lover, Paul Langevin, who'd since divorced his wife and remained a family friend.) Irène coolly reran the experiment for her mother, dissolving the foil in acid and collecting the gas. As Marie clutched the vial with the phosphorus inside, her daughter could see cracks and ulcers on her fingers from radiation damage. The old woman's eyes had clouded with cataracts as well, and she had to hold the Geiger counter close to hear the clicks of radioactivity. But when she did, she smiled a smile that could only be described as phosphorescent. Joliot later said, "It was without a doubt the last great satisfaction of her life."

Marie died a few months later. But in autumn 1935, the Joliot-Curies won the Nobel Prize in Chemistry for their work on artificial radioactivity. Remembering the media swarm that had engulfed her parents, Irène fled her home on the afternoon of the announcement and dragged her husband out to shop for a tablecloth. Still, she attended the ceremony in Stockholm that December and received her Nobel from the same king, Gustave V, who'd twice hung the medal around her mother's neck.

Fittingly, she and Joliot shared the Nobel stage that year with the man whose discovery of the neutron had so tormented them, newly minted physics laureate James Chadwick. But it was another winner that year — biologist Hans Spemann — whom most attendees would recall in later years, albeit with a shiver. Spemann was German, and at the end of his acceptance speech, he threw out a bizarre salute to the audience — his palm flat and arm extended at the shoulder. The world would soon know it as the *Sieg Heil.*

As with most relationship milestones, winning the Nobel Prize together changed things for the Joliot-Curies, especially for Frédéric. A colleague once dubbed him "the most ambitious man since Richard Wagner," and as soon as he returned from Stockholm he began sketching out plans to build what was then the most ambitious piece of scientific apparatus in the world, a cyclotron. These particle accelerators allowed scientists to study the subatomic world by smashing atoms together. Cyclotrons were also the best way to mass-produce radioactive isotopes.

There was just one problem. Cyclotrons were big, expensive machines, and Joliot and Irène's institute had no room to house one. As a result, Joliot had to transfer to a new lab in an abandoned power station a few miles away. And with this move, things changed between Irène and Frédéric. As one biographer wrote, they were "only a short walk from one another, but it wasn't the same as being in the same room, with their heads stuck together over a single experiment." The Joliot-Curies would be working apart for the first time in their professional lives, breaking up one of the world's most productive scientific teams.

Far from regretting this schism, however, Joliot pushed for it. As husband and wife, he and Irène were still on good terms, still much in love. Scientifically, though, he was tired of being Irène's gigolo. He wanted to break free from the Curie matriarchy, to become his own man. They could have their grams of radium and their family cottage—he'd have his cyclotron. He had no idea how badly this decision would burn him.

CHAPTER 3

Fast and Slow

During his Nobel Prize acceptance speech, Frédéric Joliot made a sobering prediction. Artificial radioactivity, he warned, could someday lead to "transmutations of an explosive character," using something called a "chain reaction." No one had ever applied that term to a nuclear process before, and Joliot no doubt assumed that the danger lay far in the future. But within a few years, those two words were on the lips of every nuclear scientist in the world, thanks largely to a group of high-spirited physicists in Rome.

Like the Joliot-Curies, the Italian team bombarded samples of various elements with radioactive shrapnel. The difference was that they used neutrons instead of alpha particles. The Italians were also more systematic, starting with samples of the lightest elements on the periodic table and working their way down.

It would have been an ingenious setup, except for one flaw: due to space constraints, all the equipment to irradiate the samples lay at one end of a long hallway, while all the detection equipment lay at the opposite end. Worse, many of their artificially radioactive samples decayed in mere seconds, far shorter than it took to walk down the hallway. So, making the best of a bad situation, the leader of the lab, Enrico Fermi, made each experiment a game, challenging his assistants to footraces to see who could get the samples to the detectors the fastest. (Colleagues wandering the hall quickly learned to yield the right-of-way.) The races kept morale high, and each scientist swore he was the fastest physicist in Italy.

One morning in October 1934, when the team was halfway through the periodic table, one of Fermi's assistants, Edoardo Amaldi, noticed something strange while bombarding a piece of silver. If he performed the experiment on a marble shelf, the silver sample he sprinted down the hallway produced only a few radioactive pings. But if he performed the experiment on a wooden table, the number of pings increased a hundredfold. This made no sense to him. Why would the table matter? He called in Fermi to show him. On a whim, Fermi placed the silver inside a block of paraffin and irradiated it again. When they took off sprinting this time, the detector down the hallway went mad, clicking almost too fast to count all the activity. It seemed, one of them remembered, like "black magic."

Baffled, the team broke for lunch. But while the other scientists concentrated on filling their bellies, Fermi kept chewing over the odd result. Foot speed aside, he was considered the fleetest thinker in science, and sure enough, by the time everyone reconvened in the lab, Fermi had solved the mystery. (Granted, a typical Italian lunch did last several hours.) The key, he announced, was neutron speed.

When neutrons slammed into a target, one of two things could happen. They could ricochet off, or they could be absorbed by the atoms of the target. And it was the speed of the neutrons, Fermi argued, that determined their fate. Neutrons normally move at incredible speeds (10,000 miles per second), and some elements were simply good catchers; like Moe Berg, they could snag anything thrown at them. But perhaps elements like silver were clumsier, and couldn't handle heaters. Perhaps they preferred neutrons traveling at more modest speeds (maybe 1 mile per second). Fermi used more sophisticated arguments, naturally, but the overall point was simple:

each version of an element preferred neutrons of a certain speed, and when it's bombarded with neutrons of that speed, it readily absorbs them and turns radioactive. Otherwise, it struggles.

But how, his assistants asked, did that explain the difference between the marble countertop and wooden table? Easy, Fermi responded. Imagine a neutron flying along. Perhaps it doesn't hit the target directly but rebounds off a nearby surface first. If the material that makes up the surface contains mostly heavy atoms, the neutron will bounce off without losing much momentum—just like a cue ball bounces off the cushions of a much heavier pool table without losing much speed. But if the material that makes up the surface contains lighter elements, then the neutron *will* lose momentum, just like a cue ball loses speed when it strikes another, similar-sized billiard ball. The key point is that wood and paraffin contain a much higher percentage of light elements, especially hydrogen, than marble does. So when silver was surrounded by those materials, the neutrons ricocheting around got slowed down quite nicely, allowing silver to catch them.

It was a virtuoso performance. Fermi had basically discovered a new law of physics over antipasto. But he wasn't done. According to his reasoning, a substance with an even higher percentage of hydrogen—say, H_2O—should slow down neutrons even more effectively. So the Italians decided to grab some water and test this theory. Why they didn't fill up a bucket in the nearest sink, no one knows. Instead, Fermi and Amaldi and the others dashed down the steps that afternoon like boys after the last school bell and made for the pond behind the institute. Normally they caught salamanders and raced toy boats there, but today they splashed right in, skimming off the pond scum and scooping up some water.

Back in the lab, Fermi proved right, as always: water slowed down neutrons brilliantly. And although they didn't yet realize it, this discovery would vastly expand the power of artificial radioactivity. The Joliot-Curies had showed how to turn a few elements radioactive. But with the discovery of fast and slow neutrons, Fermi could now make almost any element radioactive—a skill the world would soon curse.

CHAPTER 4

Crimea to Hollywood

The mob at the wharf was getting desperate. Hordes of murderous Bolsheviks were about to overrun the city of Theodosia, on the coast of Crimea. The ships moored there offered the only means of escape now, and thousands upon thousands of refugees were clamoring to get aboard. Only the barbed-wire barricades surrounding the wharf prevented a full-on riot.

A handful of Red Cross relief workers in Theodosia, including Boris Pashkovsky, spent the afternoon of November 12, 1920, loading supplies onto their ship and preparing to evacuate. Just as they were finishing up, a few rogue soldiers with sabers attacked them, desperate to steal the supplies for themselves. Pashkovsky's crew beat them back, but duty compelled him to stand guard and prevent looting. He left his post only once, with armed escorts, to bolt down a meal at the Red Cross compound in the city. While there, he kissed his new bride, Lydia, and warned her to stay put at their villa until the Red Cross trucks brought her down to the wharf for evacuation the next day.

Refugees, meanwhile, kept pouring into the city, many of them carrying their entire lives in filthy bundles in their arms. They could hear Bolshevik rifles firing in the distance; later, an ammunition dump exploded with a series of booms. As the mob at the dock grew and grew, Pashkovsky stood guard through it all that night and much of the next day. He never allowed himself to relax until the Red Cross trucks finally rolled through the dockyard gate, bringing dozens of relief workers to safety.

But when the truck doors opened, Lydia was nowhere to be seen. Not panicking quite yet, Pashkovsky ran up and asked where she was, yelling over the noise of the crowd. Was she hurt? Lost? His colleagues wouldn't meet his eyes. Dead?

Finally the captain of their evacuation ship, the SS *Faraby*, leveled with Pashkovsky. Lydia was missing. "That damn fool wanted to say goodbye to some friend," he said, and had run off to the woman's house. Several hours later she still hadn't returned, so they'd left her behind.

Pashkovsky immediately declared that he was going after her. Now the captain called *him* a damn fool: "You'll never find her. The people are crazy out there." He pointed to the barbed-wire barricade, to the snarling mob. Pashkovsky could see that there was little chance of fighting his way out, much less of getting back in time. He hesitated, and the captain warned him, "We cannot wait."

Pashkovsky looked again. *Why the hell hadn't she stayed put?*

Even before this point, the war had been a disaster for Pashkovsky. Although born in San Francisco in 1900—one of his earliest memories was of the 1906 earthquake—his family had roots in Russia and had returned to Moscow in 1913 when his father, a Russian Orthodox bishop, took a post there. Boris had just turned fourteen when World War I started, but this short, moon-faced, bespectacled lad joined a Russian artillery unit and fought in several battles against the Germans.

Confusingly, in the midst of the world war, a civil war broke out within Russia in 1917. On one side were the Bolsheviks, who commanded the Red Army. Given his religious roots, Pashkovsky despised these godless commies, and he rushed to join the opposition White Army, making him a two-war veteran by age eighteen. Over the next few years he served in various ground and naval campaigns, helping to capture the city of Odessa and suffering a five-inch bayonet wound on his left leg that left an impressive scar. He was also taken prisoner by the Reds and eventually won the Cross of St. George for his bravery.

But the war ended for Pashkovsky in 1920, when he married a young blue-eyed woman named Lydia and resigned from the army. By that point the White cause was looking grim anyway. While the Reds had ideological unity, the Whites were a loose confederation of tsarists, religious zealots, and Tolstoyan feudal lords—all of whom mistrusted each other almost as much as they did the Bolsheviks. Pashkovsky continued to support the Whites by joining the Red Cross and doing relief work in Crimea, but by the time he married Lydia, he said, the White Army "had degenerated into a rabble." Meanwhile the ranks of the Red Army continued to swell as they conquered city after city, leaving behind an appalling trail of dead and mutilated bodies.

In November 1920 the war came to a head. Two million Reds had pinned the remaining 60,000 Whites into Crimea, in the extreme southwest corner of Russia. Crimea connects to the mainland via an isthmus surrounded by swampy marshes, and despite the disparity in the size of the armies, the White generals judged the terrain favorable for a last stand, to halt the Reds' progress.

It was not to be. Like Moses parting the, well, Red Sea, an unlucky combination of strong winds and extremely low tides in early November all but drained the marshes; a cold snap then froze the ground, making it suitable for marching across. The battle was pitched and bloody, and the Whites absorbed several haymakers in a row, but their ranks eventually broke. More than 100,000 refugees and soldiers began retreating pell-mell into Crimea. They came on foot, on donkey carts, on camels. Many were ruined aristocrats, forced to steal boots and coats from not-quite-dead bodies. Cholera ran rampant. Because Crimea stood at the sea's edge, there was no more land to retreat to, and relief groups like the Red Cross tried to evacuate whomever it could—loading people on dinghies and coal barges, on yachts and rowboats and trawlers, anything that floated; passengers sometimes tore up their own clothes to make sails. At last, even the Red Cross could do no more, and its employees received orders to pack up and flee. Pashkovsky was ready to, until he found Lydia missing.

When he declared that he was going after her, the captain warned, "We'll not wait for you. You just better forget about her."

Pashkovsky knew this was prudent advice. Assuming he even found her, the screaming, spitting mob would block his path back to the ship. He'd be trapped in Theodosia, and when the Reds finally caught up with him, they would hang him as a White Army veteran—if he was lucky. Thousands of others had endured worse deaths.

"I'll take my chances," he told the captain.

He had an idea which friend she'd run off to, and after fighting through the mob, he arrived at the woman's house. There Lydia was. When she saw her husband—saw that he'd returned for her—she was no doubt overwhelmed with emotion. But there was no time for tenderness. Riots had already broken out within the city, killing several, and other people had committed suicide rather than face the Reds. The young couple ran back to the wharf to find that the crowds had swelled even more, now packed forty yards deep around the barbed wire. The din was such that Pashkovsky didn't even bother shouting for help. Being rather short—he stood five-foot-five—he had no hope of someone spotting him amid the throng either. He finally climbed on top of an abandoned wagon and began waving his arms, hoping the sailors preparing to launch could see him.

After several minutes the captain spotted him—or so Pashkovsky thought. He couldn't tell in the confusion. But he saw the captain pull an officer aside and gesture in his direction. Sure enough, the captain was growling, "There's that damn fool." Despite his disgust he gathered his troops and ordered them into an assault column.

Then they flung open a gate and went plunging into the crowd, rifles and bayonets at the ready.

The scene had all the makings of a massacre, and a few people did challenge the sailors. But upon seeing a group of bluejackets in formation, most of the crowd fell aside. And after several tense minutes of shoving and shouting, the sailors managed to corral the Pashkovskys and bull their way back inside the barbed-wire perimeter.

The young couple no doubt got an earful from the captain at this point. Eerily, though, the mob had fallen quiet. Two people outside the barricade had been rescued, and the rest of them were left to fend for themselves. Thousands upon thousands of people now knew for certain that no one would save them, and they fell silent in despair.

In the weeks after the Pashkovskys sailed, the Red Army shot fifty thousand people in and around Crimea, and although skirmishes continued in Siberia and elsewhere, the losses there effectively eradicated the White cause. Pashkovsky later described his last days in Russia as every bit as traumatic as those after the San Francisco earthquake—filling everyone present "with a lifetime of dread that it ever should happen again." From that day forward, Boris Pashkovsky was a dedicated foe of communism.

After a few years in Berlin, Pashkovsky and Lydia, with a newborn son, struck out for the United States in 1924. As part of the immigration process, he shortened his name to Boris Pash. Although he first settled in Massachusetts, where he picked up a teaching degree, Pash longed to return to California, and in 1926 he landed a job at the world-famous Hollywood High School.

The contrast with his hardscrabble war days could not have been more stark. Hollywood High was one of the most affluent schools in America and definitely the most glamorous. California had strict child-labor laws, including for movie stars, so teenage actors and actresses had to attend classes between shoots, and Hollywood High was the most convenient location. An arrogant Mickey Rooney used to park his blue convertible on the school lawn and exit amid throngs of adoring girls. Classmates remember Lana Turner as so bewitching that even teachers stared. Judy Garland was crushed when a publicity tour prevented her from walking the stage at graduation. (Maggie Hamilton—the Wicked Witch of the West herself—appealed to MGM on Garland's behalf, but to no avail; the teenager got her diploma in the mail.) Other Hollywood High alums starred in *King Kong, It's a Wonderful Life, Psycho, Gentlemen Prefer Blondes,* and *Scarface.* It wasn't just starlets, either. Olympic gold medalists, a Nobel Prize winner, Judge Wapner—no school in America could boast of more famous alumni.

Boris Pash, however—known to students as Doc—kept a lower profile. He taught physical education and science (mostly physiology), and tried his hand at coaching every sport the school offered: football, basketball, soccer, swimming, volleyball, track. He seemed especially smitten with baseball, that all-American game. Hollywood High teams were known as the Sheikhs (after a Rudolph Valentino flick), and while his Sheikh squads often had lackluster records, Doc regularly appeared alongside them in yearbook photos in ball caps or letter sweaters, grinning like a fool. (Comically, many players in the photos towered over him, even those on sophomore teams.) Little did Pash suspect that coaching baseball would again entangle him in war.

Pash led something of a double life in Hollywood. He joined the U.S. Army Reserve as a second lieutenant over summer break in 1930—his third professional army—and started taking classes in military intelligence, including scouting, patrolling, sabotage, and document analysis. These skills became unexpectedly relevant in the late 1930s. By then, Pash's hair had grown sparse and his belly rounder. He'd become chair of the physical education department, and in 1938 his Sheikhs put together one of the best baseball nines in school history and won the conference title. Riding the momentum, he decided to organize an extracurricular trip for some of the students. Unusually for the time, Hollywood High's student body included a mix of ethnicities, and over the years Pash had coached several Japanese students on the baseball diamond. A recent tour of Japan by Major League baseball players (including Moe Berg) had proved quite popular there, so Pash decided to gather up a dozen Nisei (children of Japanese immigrants) from Southern California and tour the island as well.

While Doc was planning the trip, a student named Hashimoto pulled him aside to warn him off. The Japanese military was expanding rapidly, Hashimoto explained, and the government there still considered young men of Japanese descent to be citizens, no matter where they'd grown up. Any boy over the age of eighteen on the Hollywood High team would probably be seized and impressed when he set foot in Japan.

Pash, dumbstruck, didn't believe Hashimoto. Japanese people weren't belligerent—they were some of the sweetest people he knew. Nevertheless, he contacted an FBI agent, who interviewed Hashimoto at the Hollywood High gym under the guise of being a sponsor for the traveling team. Then, putting his military intelligence training to work, Pash began his own investigation. The local Japanese consulate revealed that it kept lists of all young Nisei men and, creepily, tracked their movements. Pash also discovered that many of his former Nisei students had picked up Japanese passports after graduation and joined the Japanese military, considering it their duty. Even Hashimoto succumbed to the pressure: not long after graduation, Pash got a postcard from him that read, "Banzai. I am in the Japanese navy."

These defections troubled Pash. He loved the Nisei boys: they were fine ballplayers, sharp and easy to coach, and fine citizens to boot. Still, he couldn't help thinking that Japan and the United States were quickly becoming adversaries.

CHAPTER 5

Division

Otto Hahn was convinced that the Joliot-Curies, or whatever the hell they called themselves, had bungled things once again. Which meant that he would now have to waste several months proving them wrong.

The trouble started in early 1938, when Irène Curie (now working without her husband) began a series of experiments on two obscure metals, uranium and thorium. In the first decades of the twentieth century, if you thought about radioactivity at all, you thought first of radium. Scientists coveted radium for its ability to produce a steady stream of alpha particles for experiments. Element 88 won fame beyond the lab as well: because it glowed alluringly in the dark, it found its way into watch dials, pajama buttons, roulette wheels, and fishing lures. Ingesting it even became a health fad, a supposed cure for everything from bad breath to depression, and drugstores carried radium-infused hair tonics, bath salts, face creams, condoms, and suppositories. (It got you coming and going.) At peak demand its price reached $180,000 per gram. In contrast, a metal like uranium was junk—the mineral cruft you sifted through to get at the precious radium.

Still, as junk, uranium and thorium were cheap, and Irène began bombarding them with neutrons to transmute them into other elements. There was just one problem: she had a devil of a time determining what elements they were changing into. Some tests indicated actinium, ele-

48

ment 89. Others pointed to lanthanum, element 57. Then her results transmuted yet again, and she claimed to have found so-called transuranic elements, artificial elements heavier than uranium.

All this back-and-forth looked dubious to Hahn, the premier nuclear chemist in Germany, if not the world. Hahn had a wicked tongue, and upon seeing all the elements Irène claimed to have found, he suggested that maybe she'd actually discovered a magical new element, "curiosum." Things got even testier when Hahn bumped into Frédéric Joliot at a conference in Rome in early 1938. Hahn had previously torn Irène's experiments apart in a private letter; especially galling was his accusation that she relied on outdated methods to detect radioactive species—methods pioneered by Marie. The Joliot-Curies didn't bother responding, which irked Hahn, and at the conference he aired his frustrations. If Irène doesn't retract her work, he warned, I'm going to expose her. With his wife's scientific honor at stake, Joliot couldn't back down, and he dared Hahn to try. As a Nobel Prize winner Joliot had more cachet, and Hahn realized he'd walked into a trap. As he grumbled to a colleague, "This damned woman. Now I have to go home and waste six months proving that she was wrong."

As it turned out, it took longer than six months. Hahn worked in Berlin with a brilliant physicist named Lise Meitner—the woman who'd stood up at the conference in 1933 and slammed the Joliot-Curies. It was an odd collaboration to say the least. On the one hand, they were quite devoted to each other. When Hahn's institute had denied Meitner lab space simply because she was a woman, he'd defiantly nailed a shack to the side of the building and worked with her in there; he later asked her to be godmother to his son. On the other hand, the two had stiff, even stilted personal relations. In several decades as colleagues, Hahn remembered, they'd never once shared a meal or even taken a walk together. In the scientific sphere, though, they were every bit as intimate as the Joliot-Curies: Hahn the chemist specialized in finding and isolating radioactive elements and Meitner the physicist in interpreting what it all meant. In this capacity, everyone regarded her as the "intellectual leader" of the team. After he'd walked into that trap in Rome, Hahn knew he'd need Meitner to unravel the mystery of curiosum.

But before they could get started, Adolf Hitler ruined everything. Meitner was part Jewish, and when Germany annexed her homeland of Austria in March 1938, she found herself subject to Hitler's racial laws. She considered fleeing, but her passport was now worthless, and Jews like her couldn't obtain new ones. Things finally came to a crisis in July 1938, when a Nazi flunky denounced her at a scientific meeting—pointing a finger at the sixty-year-old Meitner and sneering, "This Jewess endangers the institute." He demanded her arrest.

Hahn was also at the meeting, and by any standard of decency he should have loosed his wicked tongue upon this toad. Alas, though, Hahn himself was in trouble with the regime. Although impeccably Aryan, he'd been berating the Nazis for years and had gone out of his way to aid Jews. As revenge—and a warning—the Nazis had included a caricature of Hahn in a crude, hateful art exhibit in Munich titled "The Wandering Jew." Hahn got the message: lie low or else. So when the flunky denounced Meitner, his longtime partner, he stayed silent, and even met with institute leaders about her behind her back. As Meitner wrote in her diary, "[Hahn] was, in essence, throwing me out."

Luckily, Meitner had friends with more backbone. Within days of her condemnation a scientific journal editor named Paul Rosbaud and a Dutch physicist named Dirk Coster had converged on her Berlin apartment to help her escape. It was a ticklish situation, since Meitner actually lived next door to the Nazi who'd denounced her at the meeting. What's more, he'd warned the Gestapo (the secret police) that she might flee. On the night of her escape, she stayed at her institute until 8 p.m. correcting proofs of a paper, as if this were a normal night. Then she slipped back home, where Rosbaud helped her pack two small suitcases. You can imagine the tension they endured that night, peeking out the windows and flinching at every noise. A little sheepish, Hahn showed up as well and, trying to make amends, gave her a diamond ring he'd inherited from his mother, which she could sell or use to bribe someone in an emergency. She slipped it on her finger, as if they were engaged. They then drove to Hahn's house to hide out.

As they passed through the dark streets, Meitner began seeing Gestapo agents in every shadow. "One dare not look back," she later said, "one cannot look forward." The next morning they drove to the train station, which was swarming with guards. Meitner had always been tough—female scientists had to be tough in that era—but after several days of unrelenting stress she broke down on the platform. Rosbaud and Coster had to drag her aboard the car, and when the train pulled out, a Nazi porter began harassing her. To ease her mind, Coster had her remove Hahn's diamond ring and hid it in his pocket.

After several hours of torment they finally reached an obscure border crossing in northeast Holland. Coster had made arrangements with Dutch officials there to let her in without a passport, but she probably got past the German guards only because they assumed that "Frau Professor Meitner" was a professor's *wife,* and therefore no one important. From Holland, Meitner fled to neutral Sweden, where she knew hardly a soul. Over the next few months Hahn tried to ship her clothes, books, and linens to Stockholm, but German authorities— furious she'd escaped—blocked the items from leaving the country. Not even a toothbrush holder would reach the Jewess, they declared.

Understandably, Hahn didn't get much research done during the middle of 1938. He'd lost his scientific better half and intellectual leader, and the ongoing political crisis in Germany banished all thoughts of chemistry anyway. What finally spurred his return to the lab was, once again, Irène Curie. On October 20 she published yet another iffy paper on uranium transmutation. Hahn was exasperated, and he and an assistant decided to redo her experiments and put an end to this curiosum nonsense.

Instead, the experiments left Hahn more muddled than ever. After bombarding a sample of uranium with neutrons, he began searching for new radioactive substances inside. He found something that resembled radium, and dashed off a paper to let everyone know. After spending more time on the problem, though, he disavowed the paper, claiming instead to have found elements that behaved like lanthanum and barium. But that discovery only raised more questions than it answered. He knew these elements weren't really lanthanum and barium, because that was scientifically impossible. In every other case of transmutation so far, the original element had turned into another element nearby on the periodic table. Uranium (element 92) might spit out a bit of atomic shrapnel, for instance, and become thorium (element 90). In contrast, lanthanum and uranium sat thirty-five boxes apart on the periodic table—an impossible leap, since no known piece of atomic shrapnel was that big. Barium was thirty-six boxes distant.

Elementary logic, then, ruled out lanthanum or barium. But if uranium wasn't turning into those elements, what the heck was it turning into? Hahn was starting to appreciate Irène's confusion. So as he always did, he reached out to Meitner and wrote her a letter in mid-December 1938 outlining his odd results.

No one would have blamed Meitner if she'd told Hahn to do something anatomically impossible. But she couldn't resist this scientific puzzle—nor resist sticking it to the Joliot-Curies again. Moreover, she was cold and lonely in Sweden and desperate to reconnect with the scientific world. She was actually leaving for a Christmas holiday outside Stockholm with her nephew when the letter arrived, and she took it along. The nephew, a fellow physicist, found her frowning over it one morning at breakfast. She'd been stuck for days, she admitted, and they decided to take a walk in the snow and discuss the matter further.

No doubt the truth had been tickling Meitner's subconscious for some time—and she realized it suddenly on the walk. Hahn had said the transmuted elements acted like lanthanum and barium. So maybe they *were* lanthanum and barium. But for that to happen, the uranium nucleus had to have broken apart—not just spit out a little atomic shrapnel, but cracked in half. It was an unthinkable idea, one that virtually every other scientist alive would have rejected. Even

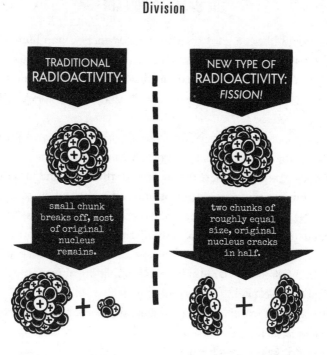

TRADITIONAL RADIOACTIVITY:

NEW TYPE OF RADIOACTIVITY: *FISSION!*

small chunk breaks off, most of original nucleus remains.

two chunks of roughly equal size, original nucleus cracks in half.

Irène Curie, a Nobel Prize winner, couldn't make that leap. Meitner could, and she concluded that, however unlikely it sounded, Hahn had split the atom.

Perhaps to punish him, Meitner didn't immediately tell Hahn her conclusion. When she did, the news staggered him. How could this be true? Yet he trusted Meitner—she didn't bungle things—and her conclusions reinforced a decision he'd made a week earlier. He'd recently called up Paul Rosbaud, the editor who'd helped Meitner escape Germany, to ask if he could rush a short paper about his work into print. These experiments thrilled Rosbaud, and he agreed to pull an already typeset paper from the next issue of *Naturwissenschaften* (*The Natural Sciences*) and swap in Hahn's. The date was December 22, 1938.

However excited most physicists were at this news, the more politically astute ones felt something else: a chill of foreboding. Uranium is fairly common in the Earth's crust, so it's easy to mine in large quantities. Physicists also knew (per $E = mc^2$) that splitting uranium atoms would liberate gobs of energy, more energy than humankind could safely handle. Until late 1938, this had only been a theoretical worry. No more. From that instant forward, uranium was no longer a second-rate radioactive element. From that instant forward, only uranium mattered.

A few scientifically savvy Nazis later accused Hahn and Rosbaud of treason for their actions that day. The duo hurried the paper into print, they charged, to alert the enemies of the Reich and prevent the Nazis from hoarding the secret of uranium. That might sound like typical Nazi paranoia, but for once the Nazi paranoia was correct: Rosbaud did indeed want to warn France, Great Britain, and the United States about this monumental discovery. He despised the Nazis and feared that war might break out at any moment; delaying publication by even one issue might cost the world dearly.

As it was, the world would suffer enough. December twenty-second happened to be the solstice in 1938, and as one historian commented, "The world's winter had begun."

Despite the treason, the world at large would remain ignorant of Hahn's paper until the new issue of *Naturwissenschaften* appeared several weeks later. But rumors about Hahn's discovery—soon called uranium fission, after the process by which bacteria divide—went rippling through the world of physics in the meantime.

It all started in early January 1939, when Lise Meitner's nephew ran into Danish physicist Niels Bohr in Copenhagen and confided what Tante Lise had discovered. Upon hearing this, Bohr slapped his forehead like a cartoon character. "What idiots we've been!" he cried. The nephew made Bohr promise to keep the discovery secret, since Hahn's paper had merely laid out the chemical evidence for fission. Meitner wanted to write her own paper about the physics involved, and Bohr's blabbing could undermine her. Bohr vowed to remain silent.

Then promptly broke his vow. A few days later he sailed off for a sabbatical in the United States, and practically before the ship pulled anchor, he spilled the beans to a colleague onboard, who was equally stunned. Bohr and the colleague then scrounged up a blackboard from somewhere on the ship, and despite the choppy waters in the Atlantic—Bohr was seasick the whole voyage—they spent the entire nine days working out the implications of fission. In his excitement, Bohr forgot to mention the need to keep mum. Not bound by any promises, the colleague started yammering the moment he landed in New York on January 16, and pretty soon the entire U.S. physics community knew.

Realizing he'd screwed up, Bohr made an official announcement at a formal dinner in Washington, D.C., on the twenty-sixth. He'd been scheduled to talk about low-temperature physics that evening, but to hell with that—this was much hotter. And while he tried to ensure that Meitner got proper credit, everyone stopped listening after he revealed the basic discovery. Even before he'd finished speaking, in fact, several local physicists pushed their way out of the lecture room—some still wearing tuxedoes from dinner—and ran off to fire up their own experiments. One physicist in Northern California, reading an account of Bohr's talk in the newspaper the next day, actually jumped up in the middle of a haircut and raced to his lab. Within the nuclear physics community, the news of fission would forever divide the world into Before and After.

Enrico Fermi was one of the politically savvy scientists who shivered at the news of fission. After several years of slow neutrons and fast footraces in Rome, Fermi had watched Italy sink into fascism under Benito Mussolini. By 1938 he could no longer imagine a future there for him and his wife, Laura, who was Jewish, and they began making plans to leave. They finally got their chance in the fall of that year, when Niels Bohr pulled Fermi aside at a scientific meeting.

As a Scandinavian, Bohr had close contact with the Nobel Prize committee in Stockholm, and he told Fermi that he might expect a little award in a few weeks. Normally, this would have been a serious breach of etiquette: the Nobel committee works in secret and guards the identity of winners. But Bohr had a good reason for this peccadillo. The Italian lira was fluctuating wildly in value then, and Italian laws would probably prevent Fermi from converting the money to a more stable currency. Bohr wanted to know if Fermi wished to delay his prize a year, when the economics might be more favorable.

Fermi then let Bohr in on a secret of his own. He was fleeing Rome for Columbia University in New York, so getting the prize money immediately would be a huge help. Bohr said not to worry. A few weeks later Fermi did indeed win the Nobel in physics, and he and Laura left Rome for Stockholm with little more than a suitcase apiece, never to return.

Sadly, with Fermi's departure, his rambunctious little physics team—most of which was Jewish—underwent its own fission. One scientist fled to California, another to Paris. Top lieutenant Edoardo Amaldi (who'd first noticed the difference between samples irradiated on a marble shelf and a wooden tabletop) also wanted to flee, but as a good Catholic he was in no personal danger in Italy. Colleagues therefore begged him to stick things out; there were few jobs available for foreigners abroad, and others needed those slots more than he did. So Amaldi pasted a smile on his face and accompanied Fermi to the train station to see his mentor off. He returned home miserable, and was soon drafted into the Italian army, destined for the front in North Africa.

Coincidentally, Bohr spilled the beans about fission not long after Fermi had arrived in New York, and with his agile, leaping mind, Fermi saw exactly where the discovery would lead. A colleague remembers him standing in his office one gloomy January day, chatting about the brave new world of fission. Fermi then turned toward the Ozymandias of the Manhattan skyline and cupped his hands into a ball. "A little bomb like that," he murmured, "and it would all disappear."

Division

The mood in Paris was equally grim, albeit for different reasons. Upon seeing that bastard Hahn's article on fission in *Naturwissenschaften,* Frédéric Joliot had locked himself away for two days to study it. He finally emerged dark-eyed and haggard, and broke the news to everyone at his and Irène's labs: they'd been scooped again. They'd seen the same evidence as Hahn and Meitner, the same fission by-products, and simply hadn't understood.

Echoing Bohr, Irène wailed, "What assholes we've been!" She then turned on her husband in fury. If you hadn't run away to your own lab and abandoned me, she yelled, we would have made this discovery ourselves. Several people who worked in the lab silently agreed. There had been nothing wrong with Irène's chemistry, but she lacked the physics knowledge that Joliot had, knowledge crucial for interpreting her results. He'd simply been too distracted with his cyclotron, his midlife scientific crisis, to pay his wife's work any attention. Joliot was mortified.

Nothing if not ambitious, however, he decided that his Paris team would take the next crucial step in fission research. Despite Fermi's prophecy, Meitner and Hahn hadn't actually proved that you could make a bomb with uranium. All they'd proved was that bombarding a uranium atom with a neutron would split the atom and release energy. But what then? The big question was whether, aside from energy, the fissioning atom released anything more. In particular, did it release more neutrons? If so, then other uranium atoms in the vicinity might absorb these neutrons and become unstable, too. Those atoms would then undergo fission themselves and—here's the key point—release still more neutrons. These secondary neutrons would destabilize more uranium atoms, which would release tertiary neutrons, and so on. As Joliot had predicted in his Nobel speech, it would be a nuclear chain reaction.

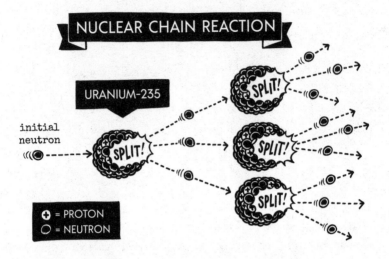

It all depended on one thing: the number of neutrons released. If a uranium fission liberated just one neutron, there was no reason to get excited. Each step would cause just one additional atom to fission, and the chain reaction would proceed slowly or even fizzle. But if uranium atoms released two or more neutrons every round, watch out. One fission would lead to two, two to four, four to eight, eight to sixteen, and so on—an uncontrollable cascade of energy. Joliot's course was clear, then: bombard samples of uranium, and measure the neutron multiplication.

Having already been scooped on several discoveries, Joliot put a gag order on his assistants: no one could discuss their new experiments with outsiders. Ironically, though, these precautions ended up exposing Joliot's secret. A physicist at Columbia University named George Placzek had recently visited the lab in Paris, and to make extra-double-sure that Placzek didn't blab about anything, Joliot's team sent him a cable in early 1939 reading JOLIOT'S EXPERIMENT SECRET. Unfortunately for them, in one of those minor blunders that rewrite history, someone got one Columbia physicist with a funny Eastern European name confused with another Columbia physicist with a funny Eastern

European name, and the message ended up down the hall on the desk of Leo Szilard—who read it and froze in horror.

Szilard had actually dreamed up the concept of nuclear chain reactions way back in 1933. He therefore knew better than anyone that chain reactions could lead to bombs. Moreover, as a Jewish refugee from Hungary, the idea held a special horror for him. Above all else in life, Szilard feared the Third Reich, and although Joliot's telegram never mentioned uranium or fission, Szilard guessed instantly what the "secret experiment" must be.

Never one to hold his tongue, Szilard eventually wrote a letter to Paris outlining his fears: Germany had the top scientists in the world, along with the best industrial plants. If anyone could weaponize nuclear physics, it was the Nazis. He therefore asked Joliot to be prudent. Do all the research you want, he said; you can even submit papers to journals to establish priority for discoveries. But please, *please,* don't publish anything related to chain reactions. Don't tip off Hitler.

Joliot wasn't the only one Szilard was keeping an eye on. Szilard's fellow refugee at Columbia, Enrico Fermi, had also plunged forward with fission research, and Szilard made the same plea for secrecy to him. At first the Italian blew him off. (Fermi's exact response was one word: "Nuts!" Apparently his fondness for American slang had outpaced his command of it; he probably meant something like "Nuts to that.") But Szilard kept pressing Fermi, and Fermi eventually relented. He even withdrew an already accepted paper from a research journal to keep the information secret.

Joliot proved less accommodating. If we censor our scientific work out of fear, he told Szilard, then Hitler will have destroyed yet another basic freedom. It was a powerful argument, but it's hard not to suspect Joliot of having selfish motives here: given his reputation for blowing discoveries, he desperately hoped to redeem himself.

To Szilard's dismay, Joliot published results in April 1939 showing that uranium atoms release an average of 3.5 neutrons every fission, well above the minimum for a chain reaction. This was actually a mistake; the currently accepted figure is 2.5. Joliot never achieved a self-sustaining chain reaction, either; his efforts petered out quickly. But the overall point stood: nuclear chain reactions, and therefore nuclear bombs, were possible. And far from keeping all this a secret, Joliot began discussing wild plans to build and detonate a nuclear weapon in the Sahara desert—the world's first, if short-lived, atomic bomb project.

Things moved at a gallop after that. No one had heard of uranium fission before January 1939; by December, more than a hundred papers on the topic had appeared worldwide. This explosion of research caused nothing but grief for the chemist at the center of things, Otto Hahn. The Nazis despised him even more now for leaking the secret of fission. He also despaired over the Pandora's box he'd opened: when he first saw Joliot's work on neutron multiplication— and realized that his own discovery might lead to the most destructive bombs in history—he decided to kill himself.

Ultimately Hahn changed his mind, but he couldn't shake off his anguish. He enlisted a few colleagues in a scheme to seize all the stocks of uranium in Germany and dump them into the sea, a plan he abandoned only when someone pointed out its futility: in seizing part of Czechoslovakia that spring, Hitler had inadvertently acquired the richest uranium mines in Europe. The realization that he could not stop nuclear fission put Hahn back on suicide watch. He'd split more than the atom—he'd divided the world.

CHAPTER 6

Spinning out of Control

Even as scientists like Fermi were fleeing Europe in the late 1930s, other physicists were wrestling with the morality, and wisdom, of returning there. Two old friends in particular, Samuel Goudsmit and Werner Heisenberg, faced tough choices on this front, and while both men debated long and hard over what to do, they knew their decisions would end in pain either way.

Of the two, Goudsmit was the less likely scientist. "In my high school days," he once said, "I wanted to solve mysteries, and there were three professions where one could solve mysteries: the police, archaeology, or science." Science eventually won out, but it was a close thing. After growing up in a typical Jewish household in The Hague he decided to attend college in Leiden and, to his parents' disappointment, abandon the family businesses for scholarly pursuits. (His father sold bathroom fixtures, his mother designed frilly hats.) At Leiden his rich black hair earned him the nickname *luizebos,* or mop-top, and he proved a sharp if erratic student, regularly failing exams in subjects he didn't care about. He was also prone to enthusiasms: he suddenly began studying hieroglyphics one year, eventually becoming fluent in them, and took an eight-month course on scientific detective work that included units on fingerprints, forgeries, and bloodwork. It took him some time to settle on physics, and when he did so, he made the great discovery of his life almost too easily.

That discovery was quantum spin. The technical details needn't concern us, but spin describes the intrinsic angular momentum of particles such as electrons and neutrons; along with mass and charge, it's one of their fundamental properties. When Goudsmit started off, however, he had no idea he was on the trail of something that important. He'd simply heard about some odd new experimental results and wanted to solve the mystery of what produced them. So he and a fellow student, George Uhlenbeck, finally sat down one summer day in 1925 to hash things out. They groped about with no real plan, trying this and that, and the idea of spin fell out almost accidentally. But by the end of the session, they knew they really had something. Invigorated at solving the mystery, Goudsmit finally looked up from his papers after hours of labor—and noticed an ominous black cloud sweeping across the sky. He and Uhlenbeck later learned that a tornado had touched down in Holland, but they'd been so wrapped up in the equations they hadn't noticed.

Spin would soon send whirlwinds of its own through quantum physics. The two students dashed off a short paper and showed it to their advisor, who scratched his head and said, *Well, it's either brilliant or nonsense. Let's publish it and find out which.* Sure enough, several eminent physicists hated the idea—it seemed too strange. But others took a shine to spin, including Werner Heisenberg, who sent Goudsmit a letter congratulating him on the "courageous" work. An ecstatic Goudsmit ran to Uhlenbeck to show him the message. Uhlenbeck just blinked. Who's Heisenberg?

This ignorance stunned Goudsmit. Heisenberg was *Heisenberg,* the top young physicist in the world. Goudsmit already revered the man, and the letter cemented his affection. The two eventually met and became friends, and when Heisenberg later visited Holland, he stayed with Goudsmit's parents, eating dinner at their house and accompanying Goudsmit to the seaside to watch fireworks. Thanks in part to Heisenberg's support, the world physics community embraced spin, and mop-top's reputation soared.

Still, even Heisenberg's endorsement couldn't get Goudsmit what he really wanted—a job. Professorships in Europe were scarce in the 1920s, and young scientists basically had to wait around until some old goat croaked, then scramble like hyenas for the spot. In particular, the job Goudsmit most coveted, at a university in Holland, virtually required a Nobel Prize as a prerequisite. Really, only American universities, especially the University of Michigan, seemed serious about hiring him—which put him in a difficult position. For Dutch people at the time, he said, the only reason you'd move to America was "to evade the police or the draft," and Ann Arbor in particular seemed like the other side of the world. But the university paid well, and his partner Uhlenbeck had already accepted an offer there. So even though, as he later said, "I always felt like a deserter for having left," Goudsmit scooped up his young wife Jaantje and sailed for America. Robert Oppenheimer himself—future head of the Manhattan Project—met the Goudsmits and Uhlenbecks in New York and introduced them to their first American delicacy, corn on the cob.

Goudsmit's first home in Ann Arbor consisted of one room in a rooming house, with no bathroom or kitchen. He and Jaantje had two windows, one overlooking a sick ward, one a cemetery. And sadly, his general view of Ann Arbor didn't improve much over the years. He was a dandyish Jew with a dopey smile and a cynical sense of humor that clashed with midwestern earnestness. People couldn't quite warm up to him as a foreigner, either: they asked him repeatedly about tulips and wooden shoes, and his nickname of "Uncle Sam" no doubt just reinforced how un-American he felt. Other prejudices he encountered were more pernicious. Goudsmit cringed to learn that the university had dorm rooms set aside "for gentiles only." Then a history professor announced one semester that he planned to weed out all the Jews in his class by midterms. To make matters worse, Goudsmit didn't fit in with other Jews in the area, either.

Most were Mitteleuropean types who looked askance at him because he didn't speak Yiddish. As a result of all this, Goudsmit and Jaantje found Ann Arbor lonely. The highlight of their social week was Friday night, when he'd invite his graduate students over for pancakes. At one point, Goudsmit saw an advertisement for a professorship in Egypt and applied immediately. Even the Sahara, he figured, couldn't be worse than Michigan.

Worst of all, his scientific career was languishing. In contrast to Heisenberg, Goudsmit struggled to live up to his early promise, making no new discoveries in the late 1920s. Outwardly he blamed this on working in the intellectual desert of Ann Arbor, light-years away from the scientific hubs of Berlin and Amsterdam and Paris. Deep down, though, he knew he simply didn't have the intellectual chops to contribute much to modern physics. "It was outside the capacity of my brain," he once admitted. He began to fear he'd gotten lucky in discovering spin, and by 1931 he was already referring to himself as a has-been in speeches. When Heisenberg won the Nobel Prize in 1932, Goudsmit was no doubt happy for his friend and idol, but he felt a stab of envy as well. If he could only win a Nobel of his own, he could return to Holland to work. But year after year passed with no call from Stockholm.

His misery culminated over the next few years. A beloved mentor back in the Netherlands committed suicide. As the Great Depression deepened, paychecks became sporadic, and he fretted about being laid off. He was forced to give up a beloved dog (it barked too much), and someone broke into his lab and stole some equipment. He also missed out on an important discovery in 1937, perhaps the worst year of all. Hoping to revitalize his career, he'd recently switched from theoretical to experimental work, and because the University of Michigan owned a cyclotron, he decided to focus on the hot field of neutron research. (Specifically, he wanted to investigate the magnetic properties of neutrons.) The only problem was, he needed large blocks of time on the cyclotron, and a visiting professor was hogging it week after week. Worse still, an assistant of Goudsmit's, who should have been on his side, began working with the other scientist behind his back, a humiliating betrayal. This situation led to several confrontations between the scientists—confrontations Goudsmit usually lost. After a while he could barely stomach going in to work: "Every day I had to start with a pep talk like a football coach," he recalled. "Unfortunately I did not use enough swear words." The dispute cost Goudsmit dearly. A team in Denmark was doing similar research, and although Goudsmit had started a year earlier, the Danish team had regular access to a cyclotron and ended up finishing first and scooping him.

Frustrated with Midwest life, Goudsmit took off for a sabbatical in Europe in 1938. He visited several top institutes and dozens of colleagues across the continent. He also spent time in Paris, a city he adored, and threw a coin in the Trevi Fountain in Rome to ensure a swift return. Best of all, he saw scads of relatives in Holland. (He complained in his cynical way about the endless rounds of birthday parties and dinners, but his affection nevertheless shines through.) In taking the sabbatical Goudsmit suffered a huge pay cut (from $5,000 to $1,000; or $85,000 to $17,000 today), but the trip was worth it for his mental health. To his shock, he even received two job offers while abroad—including the very post he'd been coveting in Holland. They'd apparently decided to overlook his lack of a Nobel Prize, and if he accepted, he could return home at last.

But should he? Even in remote Michigan he'd heard rumblings that Europe was no longer safe for Jews: he'd in fact upbraided the university library for subscribing to pro-Nazi periodicals, and had resigned his membership in a German physics society for not denouncing attacks against Jewish members. And what he saw firsthand on his sabbatical only deepened his anxiety. As he entered Germany and Italy, border guards confiscated his copies of *The New Yorker* and *The Saturday Evening Post,* declaring them decadent propaganda, and private papers were stolen from his hotel room in Rome. Moreover, he was present in Holland when his friend Dirk Coster smuggled Lise Meitner across the border—a blatant reminder of the threat he faced. Most frightening of all, he saw Nazi mobs marching in the streets of various cities. "The main German export" nowadays, he declared in one letter, is "propaganda of hatred." He began calling the people there Germaniacs.

Still, he couldn't quite let go of his old dream of winning a professorship in Holland, and after returning to Ann Arbor in October 1938, he spent the next month talking himself into accepting the offer. His main concern was this: if he said no, the university would feel snubbed, and he'd never work in Holland again. Besides, were things really *that* bad abroad? Germany was awful, yes, but he was going to Holland—"an oasis in the European desert," as he called it. He also suspected that Germany's military might was overrated. France had always had a formidable army, he told friends, and would stand up to the Germaniacs in any war. Hitler was probably bluffing anyway. "As a Hollander with an objective outlook," he wrote to one friend, "I still bet 6 to 1 there will be no war in 1939."

However badly he wanted to believe that, fear of Germany finally won out. Hitler had already annexed Austria and much of Czechoslovakia, and if he hadn't respected those borders, why would he respect the border of meek, defenseless Holland? More importantly, Goudsmit had a daughter now, Esther, and no matter how much he coveted a professorship in his homeland, he couldn't stomach the thought of exposing her to Hitler. With a heavy heart he wrote to the university and turned down the job. His parents were crushed by his decision.

Having closed the door on Europe professionally, he now made plans to extricate his parents from the Continent to shield them from danger. He knew it wouldn't be easy. In getting married, Isaac and Marianne had overcome serious social barriers—she came from a wealthy family, while he was the nephew of their maid—and both now ran successful businesses in The Hague, businesses they were reluctant to abandon. His father carved furniture and sold bathroom fixtures; Goudsmit always joked about the "beautiful, shiny, hand-polished mahogany toilet seats." His mother ran a shop that designed women's hats. (Apparently the young Samuel had had a gift for spotting fashion trends, and

Marianne often took his advice on new designs. "That one ought to have a flower instead of a feather," he'd say, and he was invariably right.) Life in Holland was all Isaac and Marianne knew, and they couldn't fathom moving to America. But Goudsmit pressed them to reconsider and began applying for immigration visas.

In the meantime, perhaps to mask his disappointment over the lost job, Goudsmit threw his energy into another task: organizing Michigan's annual summer physics camp.

A decade old at that point, the camp drew top scientists to Michigan from across the world for seminars and lectures — Robert Oppenheimer, Paul Dirac, Niels Bohr, Wolfgang Pauli. Everyone stayed in a frat house near campus and took field trips together on their days off, visiting lions and bears at the zoo and heading to the local beach to drink malted milks and zip down the waterslide in black Tarzan-esque swimsuits with shoulder straps. (Less successful were the baseball games: the scientists proved pretty clumsy.) Surrounding himself with Europeans eased Goudsmit's isolation, and after returning from Europe he put together his best lineup yet for the 1939 camp, securing both Enrico Fermi and Werner Heisenberg, two Nobel Prize winners. Goudsmit was especially happy to land Heisenberg. He insisted that his old friend stay at his home, and he sweetened the deal by offering Heisenberg — a passionate musician who sometimes lectured on Mozart's operas — a piano to play during his stay. Other scientists were no less enthused. One told Goudsmit that he was considering flying to Ann Arbor in an airplane, an unheard-of extravagance then, just to meet the man.

By the time the camp rolled around, in the last week of July, people could barely contain their excitement. Seminar topics included cosmic rays and uranium fission, and Goudsmit had scheduled plenty of time for drinks and socializing. All in all, it was shaping up as the best session ever. And it might have been, if not for one thing. Just like in Europe, the specter of Adolf Hitler loomed over Ann Arbor that summer. Despite all the exciting science, people were distracted and edgy, and however gamely they tried to confine themselves to physics, the chatter over drinks inevitably turned to war, war, war. This was especially true whenever they bumped into Heisenberg, who had to field a number of pointed questions about Germany's conduct.

To be sure, Heisenberg was no politician, and he privately loathed the Nazis. But he was also arrogant, and like many intellectuals he assumed that his genius in one field qualified him to hold forth on other subjects. His patriotism got the better of him, too, and he kept moaning about how bad the whole political mess was for the good name of Deutschland. As if the German *Volk* were the real victims here. Not surprisingly, Heisenberg found himself in sparring matches night after night. How can you live there anymore? his friends demanded. Doesn't your conscience revolt? Normally a confident man, Heisenberg found himself wrong-footed again and again. Little did people know, these were not simple questions for him to answer.

Heisenberg had arrived in Michigan shaken, having recently endured several run-ins with Nazi officials over alleged thought crimes. Specifically, they'd accused him of promoting "Jew physics" at the expense of "Aryan physics."

He'd first heard of this nonsense while attending a lecture by Albert Einstein in 1922. When Heisenberg arrived at the auditorium, a protester shoved a red leaflet into his hand; it denounced Einstein as a fraud perpetrated by the Jew-run media. Heisenberg didn't pay much attention—there's always some nut at meetings, he figured—but then Einstein canceled the talk, citing threats of violence. Heisenberg later discovered that Philipp Lenard, a Nobel Prize–winning physicist from Germany, had organized the protest. And, emboldened by his success in stopping the lecture, Lenard soon expanded the attack on Einstein in articles and speeches, denouncing his theory of relativity as a "plot" hatched by Jews and Bolsheviks to undermine good, solid, Aryan physics. Above all Lenard and his ilk objected to the abstract and highly mathematical nature of relativity, which they contrasted with the tangible, intuitive physics of their youth. The whole uproar left Heisenberg baffled.

Unfortunately for him, Heisenberg proceeded to undermine good, solid, Aryan physics even more over the next few years—especially with the publication of his Uncertainty Principle in 1927. More than any other idea, Uncertainty marked a break between the classical physics that Lenard adored and the quantum mechanics that was quickly replacing it. As a result, even though Heisenberg was as German as mustard, Lenard and other Aryan physicists narrowed their eyes and grumbled whenever they saw him.

Characteristically—he was equal parts jolly and naïve—Heisenberg remained oblivious to their hostility. He therefore felt blindsided in January 1936 when another Nobel laureate, Johannes Stark, took a swipe at him in a newspaper polemic against "Jew physics." Stark slammed Heisenberg again in a speech in February, calling him spiritual kin to that Jew Einstein. Shortly afterward, a state-run scientific institute reneged on an agreement to hire Heisenberg for a key post. Shocked, Heisenberg wrote a rebuttal to the article, but he was somehow able to convince himself that all was still well within German science.

That fond hope ended, abruptly, in July 1937—a year as painful to Heisenberg as it was to Goudsmit. Upon returning home to Munich after a trip, a friend alerted him to another article by Stark in *Das Schwarze Korps (The Black Corps)*, the official SS newspaper. Its title was "White Jews in Science." This full-page screed, rife with grammatical errors, rehashed several old arguments about decadent Jewish physics, then added something new and ugly. As if Jews weren't bad enough, Stark fumed, we now have Aryan folks acting like Jews. "Common slang," he wrote, "has coined a phrase for such bacteria: the 'white Jew.'" Stark then singled out Heisenberg and bashed him for sheltering Jews and foreigners in his institute, as well as for refusing to join other Nobel Prize winners in declaring support for Der Führer. Reading all this, Heisenberg sank down in his chair, slack with disbelief. He then shot right back up upon reaching the conclusion: "These representatives of Judaism in German spiritual life...must all be eliminated just as the Jews themselves."

Finally roused to action—he would not stand for an attack on his scientific honor—Heisenberg brought the article to the attention of Heinrich Himmler, head of the SS. Himmler's father and Heisenberg's grandfather had known each other as fellow teachers in Munich. They also belonged to the same hiking club, and Himmler and Heisenberg's mothers eventually became friendly. So in late July 1937, Mrs. Heisenberg paid a call on Mrs. Himmler, and dropped off a letter that her son had penned in his defense. Mrs. Himmler didn't want to bother her little Heinrich at work, but Mrs. Heisenberg talked her around. "Oh, we mothers know nothing about politics," she said with a laugh. "But we know that we have to care for our boys. That's why I've come to you." This persuaded Mrs. Himmler, who agreed to pass the letter on.

Months passed before Himmler responded. Having no integrity himself, he assumed that Heisenberg was simply agitating for a better job, and tried to buy him off with a professorship in Vienna. To Himmler's bafflement, Heisenberg refused. There's something rotten in German physics, he insisted, and someone needs to stand up for what's right. Himmler shrugged—scientists were such kooks—and agreed to open an investigation.

Heisenberg was thrilled. But his wife Elisabeth, who was less naïve, blanched when she found out. *Why in God's name would you invite such scrutiny?* she asked. *They're going to pry into every nook of our lives.* She was right. The man in charge of the investigation— Reinhard Heydrich, one of the slimiest Nazis of all, a major architect of the Holocaust—immediately planted spies in Heisenberg's classes and bugged his telephone. He also dragged the physicist in for long interrogations in dimly lit rooms, digging for dirt far beyond the scope of any scientific dispute. These sessions terrified Elisabeth, who knew that one misstep, one misunderstood word, could land their whole family in a concentration camp. Still, with his scientific honor at stake, Heisenberg willingly endured the hardship.

Hearing that Himmler had opened an investigation into Heisenberg of course emboldened others to keep attacking him. Some accused him of being a sexual predator or, the ultimate slur, a closet homosexual. The scandal soon became front-page news, and Heisenberg became what the Germans called "hot cobblestones"—someone so controversial that no one wanted to touch him. He began having nightmares in which Nazi troops barged into his bedroom to arrest him.

At last, a full year after the initial appeal, Himmler cleared Heisenberg in a letter, signing it with his signature green pencil. In short, the SS determined that Heisenberg was an apolitical savant and therefore not worth attacking. Heisenberg was ecstatic with the ruling, and the public broadsides ceased.

But it proved a Pyrrhic victory. At its core, the dispute with Lenard and Stark was a dispute over the future of physics, and by appealing to the Nazis to settle it, Heisenberg had legitimized their authority—as if Himmler alone had the wisdom to resolve scientific matters. Worse, although the SS gave Heisenberg its blessing to continue teaching relativity and quantum mechanics, they forbade him from mentioning the names of any Jews who'd helped develop the theories. (Imagine teaching relativity without Einstein!) Shamefully, Heisenberg agreed to this restriction, telling himself that the ideas, not the names, were the important thing. Having compromised once, though, it became easier to do so again later.

Even after this crisis passed, things remained tense for Heisenberg. Like most young men in Germany, he belonged to a paramilitary group, which he actually found invigorating. (He loved shooting machine guns, and once quipped that army service was merely "mountaineering, complicated by the presence of sergeants." The man was essentially a boy scout with a hypertrophied brain.) But in

September 1938 war almost broke out in Europe after Germany seized part of Czechoslovakia. Relishing a fight, Germany began mobilizing its forces, and Heisenberg very nearly had to march off to the front. Only the compromises of Neville "Peace for Our Time" Chamberlain, the British prime minister, saved him from almost-certain death.

After that, Heisenberg knew that war would erupt sooner or later, and in late spring 1939 he bought a shabby mountain chalet in southern Germany (grandly named the Eagle's Nest) where his wife and children could flee if necessary. He then left for a speaking tour of the United States. Ostensibly he was promoting a new theory about cosmic rays there; in reality he was saying goodbye to old friends, people he knew he wouldn't see once the war began. This included Samuel Goudsmit in Michigan, where Heisenberg arrived exactly one year and one day after the SS had given him its official blessing. Heisenberg was no ideologue, but given his ties to Himmler, you can't really blame the physicists in Ann Arbor for having a few questions.

In some ways these questions were unfair. Heisenberg had no way of knowing how strong Germany's military was, nor when the war would likely start. Yet he tried to answer anyway, and the only thing he succeeded in illuminating was his own muddled thinking. He told some people that Germany would crush the rest of Europe, while at other times he moaned about Germany's inevitable defeat. He also fielded questions about science, of course, but with the discovery of uranium fission, it became increasingly hard to separate politics from physics. At one point, while watching Fermi and Heisenberg go at it, an onlooker whispered to a colleague, "Everyone in this room expects a big war, and the two of them to lead fission work on opposite sides, but nobody says so."

All the tension of that sweltering summer week in Ann Arbor

finally culminated in a heated exchange at a party. A graduate student hired to mix drinks there remembers the discussion ranging widely, "but the crucial part of their argument," he later wrote, "was whether a decent, honest scientist could function and maintain his scientific integrity and personal self-respect in a country where all standards of decency and humanity had been suspended."

Eventually Goudsmit, who could be blunt, cut to the chase. Given all the abuse you've suffered in the newspapers, he asked Heisenberg, given your disdain for Nazi leaders, given Kristallnacht and the political prisons and everything else—why not flee Germany?

Heisenberg hesitated. He'd been hearing that question over and over during his American tour—in California, Indiana, New York, Chicago. He'd even received surreptitious job offers from Columbia and Princeton and the University of Chicago recently, meaning he could easily emigrate and leave Germany behind. But things weren't so simple for him. Someone needs to stay behind and defend German science and German values, he told Goudsmit. And as a Nobel Prize winner and internationally renowned physicist, he felt he had enough prestige to influence German politicians. Sooner or later they would approach him for advice on technical matters, he argued, and when they did so, he'd set them straight.

Fermi all but laughed in his face: "These people have no principles. They will kill anybody who might be a threat—and they won't think twice about it. You have only the influence they grant you." Heisenberg refused to believe that. Besides, he still felt loyal to his students back in Munich. "If I abandoned them now, I would feel like a traitor," he said. Then he added, "People must learn to prevent catastrophes, not run away from them."

On and on they went, and the longer they argued, the more agitated Heisenberg got. Back in Germany, the Nazis had accused him of disloyalty for defending Jew physics. He was now being squeezed from the other side: accused of *undue* loyalty to the Reich for not leaving Germany. Secretly, too, Heisenberg feared that if he emigrated to the United States, the government would force him to work on atomic bombs for use against Germany, against his own people—and he simply couldn't face that.

Finally, somebody asked him again why he didn't just leave Germany. His answer this time was brief: "Germany needs me." He seemed to fancy himself a messiah.

People eventually gave up on Heisenberg that night, and the science summer camp ended not long afterward. But before it did, Heisenberg and his old friend Goudsmit put aside their differences for a few minutes and—knowing they might not see each other ever again—posed for a picture in front of Goudsmit's home. Each man put on a good face and smiled.

On the voyage home, Heisenberg realized at some point that he was the sole person returning to Germania, the single one willing to go back. It felt incredibly lonely. So after arriving home, he put a copy of the photograph with Goudsmit on his desk—a reminder of the last, good, late-summer days before the inevitable fall.

CHAPTER 7

Banzai Berg

The Washington Senators made the World Series in 1933, and while "Professor" Moe Berg never once stepped onto the field, he was perfectly content to watch from the bullpen and gab with teammates and reporters.

Much more memorable than the World Series for Berg was the chance to return to Japan in November 1934 with a team of Major League all-stars. No one knows why a schlub like Berg got tapped for the tour—the team included legends like Babe Ruth, Lou Gehrig, Lefty Gomez, and Jimmy Foxx, with Connie Mack managing—but his teammates loved having him around. So did the Japanese, who were mad for baseball and fondly remembered his workshops from two years earlier. A hundred thousand people reportedly greeted the players as they arrived in Tokyo, and Berg got as many cheers as anyone. "Banzai Babe Ruth!" the crowds roared. "Banzai the great Gehrig. Banzai catcher Berg, the linguist!" At one point Berg greeted a few people in Japanese, which drew an incredulous stare from Ruth. At the start of their sea voyage two weeks earlier, he'd asked Berg if the catcher spoke any Japanese; Berg claimed not to. So what gives? Ruth now asked. Berg shrugged. "That was two weeks ago."

The American all-stars played seventeen games in a dozen cities on the tour, including one at the base of Mount Fuji, drawing crowds of up to fifty thousand. One fan walked eighty miles just to see a game, lugging a samurai sword the whole way; he presented it to a Cleveland center fielder who homered that day. The Americans won all seventeen contests (most in lopsided fashion: they were playing Japanese college and semipro teams), and Ruth emerged as the tour's undisputed star. Although thirty-nine years old and in the December of his career, Ruth clobbered thirteen home runs, and the trip made him as famous in Japan as he was in the United States. After one ovation he turned to Berg, choked up, and said, "Moe, this has to be one of the greatest days of my life."

Berg and the Babe actually became quite close on the trip — partly because they shared an eye for Japanese women and liked visiting geisha houses. One night, drunk and feeling his oats, Ruth began harassing one of the geisha, who clearly wasn't enjoying the attention. Berg considered himself more chivalrous, so he scribbled out something for her to say, in phonetic Japanese characters, in case Ruth got handsy again. He did, and she smiled and let him have it: "Fuck you, Babe Ruth."

For his part, Berg didn't play much on the tour. He got beaned in the hip his first at-bat ("Vicious clout, Moe," his teammates razzed him), and after that he mostly lazed around the dugout, finishing with two hits total in the seventeen games, fewer than some pitchers on the team collected. Berg made a far greater impact off the field as a goodwill ambassador. In between games he donned kimonos (one had catcher's mitts on it, another red baseballs) and attended several cultural events, chatting with Japanese movie stars and giving talks at universities. He also shot footage for a documentary about the trip, using movie cameras he'd borrowed from a company in New York. The one Berg incident that no one could forget occurred on a tour of the Japanese emperor Hirohito's palace. Naturally, the tour called for strict decorum; some players remember being told that the emperor was a veritable god, and that they should avoid even looking at him. You can imagine their shock, then, during a reception afterward, when they glanced up to see old Moe Berg standing next to the man-god, shooting the breeze with him. Although amused, Hirohito seemed no less surprised.

Despite the adoring crowds, there was an uneasy political undercurrent to the tour. The U.S. government had promoted the games as a diplomatic tool to smooth Japanese-American relations. Japan had recently seized Manchuria in China, and its military clearly had ambitions to take more territory, a policy the United States opposed. The Japanese government in turn opposed the upstart Americans' meddling in Asian affairs, and a militant faction within the government began kicking scientists and other westerners off the island, denouncing them as spies.

No wonder, then, that Japanese officials looked askance at the all-stars and monitored their activities; manager Connie Mack swore that his hotel telephone was bugged. It wasn't just government officials who hated the Americans, either. After the tour finished, three young men representing the ultranationalist War God Society accosted a Japanese newspaper mogul who'd arranged for one game to take place inside a stadium dedicated to the emperor. The young men felt that the presence of filthy foreigners had defiled the shrine, and they ambushed the mogul at the gate of the newspaper building one morning, stabbing him in the neck with a sword and leaving him for dead. Only the quick action of a nearby witness saved his life.

In truth, though, the Japanese government was right to be suspicious of the all-stars. One of them was indeed spying on the island. Moe Berg.

While devouring newspapers in Japan, Berg came across an interesting item one morning near the end of the tour. The daughter of the American ambassador had recently given birth to a child at St. Luke's hospital in Tokyo. The all-stars had a game seventeen miles north of Tokyo that day, but Berg decided to skip it, feigning illness. None of his teammates believed him (he rarely got sick), but they were also familiar with Berg's inscrutable moods, so they didn't inquire too closely. As soon as the other players departed, Berg parted his hair Japanese-style and donned a kimono. Sneaking down to the street, he grabbed a bouquet for the ambassador's daughter from a florist, then hopped a cab to St. Luke's.

In the hospital lobby he slipped past the security guards and caught the elevator for the fifth floor, where the ambassador's daughter was staying. But instead of visiting her, he dumped the flowers when the coast was clear and ducked back into the elevator, now riding it to the seventh floor. There he found a spiral staircase and proceeded to the roof, where he scaled a bell tower. By tradition, no one in Japan could look down upon the emperor's palace, so most buildings topped out at a few stories. St. Luke's, at seven stories, was among the tallest. And as soon as he'd gained this vantage point, Berg reached into his baggy kimono and removed the lunchbox-sized movie camera strapped to his chest. He trained it on the cityscape around him, taking special care to linger over industrial sites: ammunition plants, railway lines, oil refineries, warships in the harbor. In all, he got twenty-three seconds of footage before packing up and scurrying down to the lobby. He never did meet the ambassador's daughter.

During the game that afternoon, the Americans briefly fell behind 5–4 before rallying and winning in a 23–5 laugher. They told Moe all about it back at the hotel. But he never reciprocated and told them what he'd been up to. Even today, no one knows why Berg engaged in this freelance espionage. A few people close to him swore that U.S. officials, including possibly the secretary of state, put him up to it. Or perhaps Berg simply fancied himself a spy and enjoyed the danger of going undercover. (Mastering new languages, visiting exotic cities, rendezvousing with mysterious strangers—this was already Berg's life.) Regardless, when the United States and Japan went to war seven years later, Berg dug out that twenty-three seconds of film and sent it to U.S. intelligence agencies. It proved quite valuable, some of the only extant footage of the Japanese capital. American officials were so impressed, in fact, that they soon dreamed up a much more dangerous assignment for the catcher.

Berg was washed up by the time he returned from Japan, slower than ever and running to fat. Yet he managed to hang around the majors for several more seasons, and an appearance on a national radio quiz show in early 1938, NBC's *Information, Please,* made him a national sensation. In a virtuoso half-hour performance he answered questions about Halley's comet, chop suey, Nero's wives, poi, the Dreyfus Affair, and Kaiser Wilhelm, among other topics; he even got the trick question right. (What's the brightest star in the sky? The sun.) NBC got twenty-four thousand letters in response, and baseball commissioner Kennesaw Mountain Landis later told Berg, "In thirty minutes you did more for baseball than I've done the entire time I've been commissioner." The added attention did chafe Berg some. During his rare appearances on the field, fans in opposing stadiums would heckle him with questions. ("Hey, Berg, is it walruses or walri?") But after *Information, Please,* Berg was a bona fide celebrity, palling around with the Marx brothers and Nelson Rockefeller and Will Rogers. Whenever FDR attended a Senators' game, he waved to the catcher from the stands, and Berg's teammates remember him keeping a tuxedo in his locker for postgame soirées at foreign embassies. (Given his linguistic skills, Berg proved quite popular at these events. According to one reporter, he "kissed the hands of more women than Valentino.") It was the best life Berg could imagine: good pay, short hours, leisure time to lounge around hotels. As he once told an interviewer, "I'd rather be a ballplayer than a president of a bank or justice on the United States Supreme Court."

Paradoxically, fame only deepened the aura of mystery surrounding Moe. Not even friends knew where he lived during the off-season, and only one teammate ever got to see inside his home. (He reported that Berg had more newspapers lying around than the *New York Times* building did.) During this era Berg also developed a reputation, perhaps justly, for seducing married women—society types who wouldn't entangle him in a real relationship. No one ever knew for sure, though, since he never spoke a word about his love affairs—or about anything private. Teammates used to rib him for being so secretive. Are you a spook, Moe? Is that it? He just smiled his enigmatic smile and slipped away.

The last team Berg signed with was the Boston Red Sox, in part because he adored the city's restaurants, theaters, and bookstores. Always savvy, he restricted himself to playing only on days when his knees didn't ache and he was feeling spry. That way, he could make a decent showing and preserve the legend of Moe Berg. He'd developed a few parlor tricks over the years as well. He always tried to play on Ladies' Day at the ballpark, and if someone hit a pop-up behind home plate, he'd hang onto his mask until the last second. Then he'd fling the mask into the air, catch the ball in his glove, and snag the mask with his other hand. The dolls swooned.

Despite his popularity in Boston, Berg was long past his expiration date by 1939, his fifteenth season. He caught just fourteen games that year, and every time he got in, he'd feign confusion as he stepped onto the field: "It's been a while, fellas," he'd say to teammates. "Do they still get three strikes out here?" Boston happened to have a hot-shot rookie named Ted Williams that season, and one time Berg got lucky and knocked the horsehide off the ball, bouncing it off the fence. After a teammate knocked the lumbering catcher home, Berg caught his breath, turned to Williams in the dugout, and said, "That's the way you're supposed to hit them, Ted. I hope you were watching."

On August 30 that year, the most popular journeyman in baseball history went out in style. The Sox faced pitcher Fred Hutchinson of the Detroit Tigers that day, and for his final Major League hit, Berg socked a fastball into the left-field bleachers—just the sixth home run of his career. "As Mighty Moe rounded the bases," one reporter wrote, "some sarcastic observers in the press box remarked that poor Hutch had suffered the ultimate in humiliation. A loud foul by Prof. Berg, they said, was a batting streak for him, and now that he had made a home run—his first in several years—it was plain that Hutchinson should be hurried to safety."

Everyone enjoyed watching Moe chug around the bases one last time. But Berg couldn't enjoy the moment much. Nazi Germany and Soviet Russia had signed their notorious nonaggression pact a week earlier, all but condemning Europe to war. Less than forty-eight hours after Berg's unlikely homer, in fact, Germany invaded Poland and kicked off World War II. "Europe is in flames, withering in a fire set by Hitler," Berg moaned that summer. "And what am I doing? Sitting in the bullpen, telling jokes to relief pitchers." At long last he'd decided that his father had been right: there was more to life than playing baseball.

CHAPTER 8

On the Brink

Because the United States didn't enter World War II until late 1941, concerned Americans like Moe Berg could do little more than watch events unfold from afar. A few, though, got to see the panic of the war's early days up close, including the son of the U.S. ambassador in London, Joseph Kennedy Jr.

This was actually Joe's second European war. As a student at Harvard, Kennedy was exactly the kind of Ivy League hotshot that Moe Berg had despised in college: Joe had a full-time valet and private tutors in every subject, and received up to $150 in "pocket money" each month ($2,800 today). Still, he was a serious student, and after writing his senior thesis on the Spanish Civil War in 1938, he decided to pursue international politics. Attempts to land a job on his father's staff in London ran afoul of U.S. nepotism laws, so as the next best thing, Kennedy Senior sent Joe on a tour of the Continent to gather intelligence—Warsaw, Leningrad, Copenhagen, Prague. Berlin made an especially strong impression: Joe found the strength of the German army both stirring and menacing, and he reported that the Nazis "have the most powerful propaganda I have seen anywhere." At one point, while cruising through northern France in a Chrysler convertible with his sister Kick (Kathleen), Joe passed within a few miles of Mimoyecques, an obscure village of fifty people. Joe probably didn't even notice the place, but Mimoyecques would soon become a cursed name for the Kennedy clan.

During a family holiday after this tour, Joe took up the new sport of bobsledding, and almost set a world speed record on a precarious course where several others had died. He then broke his arm skiing down a ridiculously steep trail and went back to London to recuperate. He soon grew restless there, however, and the more he thought about it, the more he wanted to go to Spain, then in its third year of civil war. Despite speaking no Spanish, Joe desperately wanted to augment his thesis and actually see the conflict, which pitted a popularly elected but volatile group of communists and anarchists against a right-wing junta of royalists and military honchos led by Francisco Franco. Joe's father vetoed the idea, in part because Joe held a diplomatic passport: this meant that if he got into trouble, the U.S. State Department would get tangled up in the affair. Moreover, the war was messy and dangerous, with massacres aplenty.

Undeterred, Joe secretly secured a regular passport. And a few weeks later, when Kennedy Senior returned to Boston for a spell, Joe snuck off to Spain. When his father received a telegram from Valencia, he feigned anger, but in truth his son's bravery pleased him. He even convinced a few newspapers to cover the boy's adventures; one dubbed him "Don José."

By the time Joe arrived in Spain, Franco's forces had bombed parts of Valencia flat. Packs of dogs ran wild, and although he supported Franco overall, Joe was troubled to find elderly women cowering in pathetic shelters, little more than dirt tunnels beneath fallen buildings. Other people had grown fatalistic. During one air raid Joe watched in disbelief as a bootblack sat unmoving in the town square, placidly polishing shoes while craters opened up around him.

After scouting Valencia, Joe persuaded a military bus to take him to Madrid, the white-hot center of the conflict; the trip took almost twelve hours, and the bus caught fire at one point when a soldier carelessly tossed a cigarette onto the floor. Upon arriving, Joe camped out in the American embassy, which the U.S. staff had abandoned but which enterprising Spaniards had taken over, barricading themselves inside and raising chickens, pigs, and cows on the lawn.

Madrid was in desperate straits. Franco had been laying siege to the city for some time, and with no food coming in, many people were subsisting on a few ounces of lentils or rice per day. One former American diplomat, Joe learned, had starved to death in the streets. His body was so infested with lice and other vermin that no one had the stomach to go through his pockets.

Given this suffering, Joe was surprised to find the theaters packed every evening, partly because people needed diversion and partly because there was nothing else to spend money on. He particularly admired a performance by a young "Spanish Shirley Temple," who hoofed right through an aerial bombardment without missing a step. As he always did, Joe attended church each Sunday, where the priest celebrated Mass in civilian clothes; otherwise he risked being shot by communists, who hated the Catholic Church. (Leftists there used churches as ammunition depots, and had recently vandalized a nativity scene by dressing Baby Jesus in fatigues and jamming a pistol into his hand, a sight that outraged many.) Joe recorded all these impressions in lively dispatches to his father. "To contradict the customary postcard message," he joked in one, "I don't wish any of you were here." The ambassador was so proud that he took the letters to British prime minister Neville Chamberlain and read them aloud at 10 Downing Street.

Indeed, Don José was having a hell of a good time in Spain, right up until he almost got executed. In early 1939 the fragile coalition between the communists and anarchists finally crumbled, and a sort of meta–civil war broke out between them in Madrid, making life there even more dangerous. "It was regular trench warfare," Joe noted, "with hand grenades, tanks, and machine guns in the streets. This went on for five days." Despite the gunfire (and dead bodies) in the streets, Joe joined a team of right-wing resistance fighters carrying out clandestine missions (e.g., springing prisoners who'd likely be butchered if Franco took the city). They secured an American car and began motoring around with several sets of fake credentials. At each roadblock they had to guess the sympathies of the soldiers guarding it in order to show the right ones.

One afternoon Joe and a companion guessed wrong. They'd been driving down a walled road with no outlet, and at the checkpoint they handed over fake Red Cross papers. The soldiers rejected them and forced the duo out of the car. "Against the wall! Come on, quick!" They lined them up as if to shoot them.

Joe's companion, convinced that "everything ends here," thought of his wife and daughter and began whispering his final prayers. Not Joe. When the militants began taunting them, asking, "Where are you from, sweeties?," Joe took the opportunity to explain, in broken Spanish, that he was an American and that he was very important. The soldiers, confused, lowered their guns. *Un yanqui?* They demanded Joe's passport, looking it over from every angle. They couldn't believe it. But after a few tense minutes, the militants waved the duo on. "*Está bien.* Scram!"

Without a word, the pair got into the car and sped off. As soon as they turned the corner, Joe glanced at his companion and burst out laughing. *That was a close one, eh?* All in all, Joe treated it like some prank gone awry in Harvard Yard. But life was just like that for Joe. He was young and invincible, and nothing could touch him. When Franco finally marched into Madrid, effectively ending the war, Joe said adiós and returned to London for his next adventure.

Life in London had changed, however, and not for the better as far as the Kennedys were concerned. The problem lay with Ambassador Kennedy, who found himself increasingly marginalized in government circles.

After making a pile of money in the 1920s, partly through (then legal) insider trading, the hardworking and conniving Kennedy Senior had set his ambitions on politics. When Franklin Roosevelt got elected president in 1932, he became the first chairman of the Securities and Exchange Commission. Hoping to parlay that post into something better, he made a play for secretary of the treasury a few years later, but was denied. As a consolation prize, he asked Roosevelt for the ambassadorship to the United Kingdom in 1938. In a way, the appointment made sense, but Roosevelt hesitated. The United Kingdom was culturally British, mostly Protestant, and quite reserved. Kennedy was proudly Irish, fiercely Catholic, and both argumentative and foul-mouthed; no one ever described him as diplomatic. Sensing Roosevelt's reluctance, Kennedy forced his hand by leaking word of his imminent appointment to a newspaper. The gambit worked—Kennedy got the post—but it made Roosevelt suspicious. FDR finally concluded that sending Kennedy abroad might be the best move after all: the man, he said, was "too dangerous to have around."

To be frank, Kennedy was a terrible ambassador—stubborn, opinionated, prone to spouting off. He proved especially disastrous on the topic of the Third Reich. Although he didn't endorse Nazism per se, he thought that fascism had a lot going for it: he once offered to introduce Joe Junior to Mussolini, and regularly said, in public, that America needed to adopt fascism to survive in the modern world. Worse, perhaps influenced by Joe's reports on German military prowess, he argued that Germany would rout France and England in any war, a decidedly defeatist attitude. In private, Kennedy opposed a war for more selfish reasons. He considered war bad for business and worried about the family fortune taking a hit. More important, he had two sons of military age who would surely have to fight in any conflict. Joe Junior's hijinks in Spain were one thing. Facing the mighty Wehrmacht was another.

Given this tumult of emotions, Ambassador Kennedy began preaching two ideas. First, appeasement: that European countries should give Hitler whatever he wanted, to avoid a conflict at all costs. Second, isolation: that the United States had no business getting dragged into a European war. Isolationism was actually popular back home, but Roosevelt loathed the idea and told his ambassador to zip it. Kennedy refused, and at times even blamed Jewish groups—who wanted to fight the Nazis—for antagonizing Hitler and putting his sons' lives in danger. It didn't take long for the ambassador to earn a reputation in the press as a defeatist and a coward, and Roosevelt began cutting him out of dealings with British officials, a humiliating development. FDR felt he had no choice, though: Kennedy Senior was volatile to the point of being dangerous. In the words of one historian, "he moved from neutrality, to appeasement, to defeatism, to surrender, to the exchange of democracy for fascism—and all before a single shot had been fired."

Those shots were not long in coming, however. On September 1, 1939—not long after Kennedy Senior had brought all nine of his children to London to live with him—Nazi Germany invaded Poland and ignited World War II. By the third, Great Britain had resolved to fight Germany, and that morning, Joe, Jack, Kick, and their mother, Rose, met the ambassador outside the American embassy to accompany him to the House of Commons; at noon, Prime Minister Chamberlain would ask for a formal declaration of war. Joe was wearing a baggy black pinstriped suit with a striped tie and a handkerchief in his pocket. Jack looked svelte in a subtler, lighter-colored suit. Kick was wearing a dress and hat and white gloves; she looked ready for Easter.

Suddenly an air-raid siren sounded. In the street people stared at each other, then turned to run. The dreaded Nazi air force, the Luftwaffe, was coming for London already! A year earlier, Kennedy Senior had quietly ordered a thousand gas masks for the embassy in preparation for just such an attack. But with his children's lives at stake he abandoned the embassy entirely, since it lacked a sturdy basement and might collapse if struck by bombs. Instead he herded

Joe, Jack, Kick, and Rose into the basement of a nearby building, where they huddled in a dressmaker's shop. Half an hour later the all-clear sounded and they hustled off to Parliament, arriving just before noon.

Only to hear another siren at 12:05, which sent them scrambling into the building's basement. Both "raids" turned out to be false alarms; the Luftwaffe would not bomb London for another year. But the experience, as well as Chamberlain's speech — "Everything that I have worked for," he said, "everything that I have hoped for, everything that I have believed in during my public life has crashed in ruins" — left Kennedy Senior shaken. Over the next week he began shipping his children back to America, booking passage on different ocean liners and airplanes to mitigate the risk of losing them all in a single attack. Whatever his faults, Kennedy Senior adored his children.

He was right to worry about their lives, but wrong about the threat. Whereas he feared Panzers and Messerschmitts, the real danger to the Kennedys turned out to be Werner Heisenberg and an obscure element called uranium.

PART II

1940–1941

CHAPTER 9

The Uranium Club

Two weeks after World War II started, the German military summoned eight physicists and chemists, including Otto Hahn, to a secret conference in Berlin. Most of them arrived nervous, fearful that the Wehrmacht would dispatch them to the front for some ghastly project. They'd packed their bags and said goodbye to their families that morning with heavy hearts.

You can imagine their relief, then, when physicist Kurt Diebner was waiting to greet them. To be sure, most of them despised Diebner personally. The man was not a properly credentialed professor, and he worked in the lowbrow field of military ordnance—two strikes against him in the snobbish world of German science. Plus, there was just something pathetic about him. His clothes fit poorly, his glasses were always slipping off his nose, and he sucked up shamelessly to elites: he'd once joined a fencing club simply to make social contacts, and persisted in the sport despite picking up several scars on his face—visible reminders of his unseemly striving. Yet however much they sneered at Diebner, the scientists shook his hand warmly that day. The man was no soldier, no fanatic, and whatever this conference entailed, they wouldn't be headed to the front. They had no idea he'd be proposing something far more lethal.

In keeping with their distrust of intellectuals, the Nazis provided few draft deferments for scientists in 1939. (Of the roughly thousand people exempted from military service then, most were actors, painters, dancers, and singers whom Hitler admired.) This handful of chemists and physicists were an exception. Why? Because Diebner had convinced his bosses to gamble on an ambitious project: building a nuclear fission bomb. They would call themselves the Uranverein, the Uranium Club.

Eyebrows shot up around the room at the mention of fission bombs, and several scientists voiced objections. The project seemed a boondoggle, likely to divert money and resources away from more pressing needs. There was no proof nuclear bombs would work, either. But Diebner didn't share this pessimism, and over the course of the meeting, this awkward, striving man changed the course of history by persuading his colleagues to at least try. Physicists in the United States and Great Britain would be doing their own fission research, he reminded them, and no good German could stand to let the Yanks or Limeys beat them. If nothing else, their draft deferments depended on their cooperation. Faced with this choice, the scientists signed on.

Although he'd achieved his main objective, the meeting ended sourly for Diebner. In filling out the club's roster, he'd left off one obvious name: Werner Heisenberg. When asked about this, he explained that Heisenberg worked in theoretical physics, whereas building an atomic bomb would require engineering skills and applied science. In truth, Diebner simply loathed Heisenberg and had omitted him as a snub. Heisenberg's clique of theoretical smartypants looked down on military scientists like Diebner as second-rate, and Heisenberg had once dismissed his work as "amateur." Diebner also worried that the charming, brilliant Heisenberg would outshine him and take over the project. So when another scientist at the meeting suggested inviting Heisenberg to join the Uranium Club, Diebner refused. The majority supported Heisenberg's inclusion, however, and rather than jeopardize the whole project, Diebner gave in. He would soon come to rue this decision.

Heisenberg received orders to join the Uranium Club on September 25, and traveled from Leipzig to Berlin the next day. However reluctant to work with Diebner, he was mighty glad to secure a deferment. And there was one member of the club he was dying to see, his estranged friend Carl Weizsäcker.

Michelangelo himself, if asked to sculpt an allegory of haughtiness, could not have improved upon the visage of Carl Friedrich Freiherr von Weizsäcker—the arched nostrils, the raised chin, the tightly drawn mouth. Heisenberg had first met him in Copenhagen in 1926, when he was twenty-five and working with Niels Bohr there. Weizsäcker was fourteen, the descendent of a long line of lords and the son of a top German diplomat in Denmark. But he had a scientific bent as well, and when his mother came home one evening from a gala and began raving about some nice young physicist playing the piano there, Weizsäcker was floored: he'd read about Heisenberg in magazines, and he quickly secured an introduction.

Despite their age difference they got on well, perhaps because Heisenberg was laddish and Weizsäcker overly mature. They began to see more of each other, and an encounter a few months later cemented their friendship. Heisenberg was swinging through Berlin one day, where the Weizsäcker family had moved. He had to switch train stations, and he asked his young pal to join him on the taxi ride across town. There, crammed into the backseat, Heisenberg revealed a revolutionary idea he'd just developed, the Uncertainty Principle. Like many teenagers, Weizsäcker had a romantic streak, and this sneak peek at one of the greatest and most confounding discoveries of twentieth-century science left him intoxicated. He soon vowed to dedicate his life to theoretical physics.

After that, the two became inseparable. They both loved the outdoors and began taking long hikes and ski trips together, discussing the latest developments in physics as they tramped around Europe. Frankly, Weizsäcker wasn't in the same class as Heisenberg intellectually. Few were. But the young man was sharp and insightful, and proved a valuable junior partner. Moreover, he offered his mentor something important—political savvy. Heisenberg could be obtuse, even idiotic, on matters beyond physics. The diplomat's son was more worldly and could help his friend navigate the ruthless world of German science. Weizsäcker sometimes referred to politics as an "unclean medium," but he willingly played the game, and played it well.

The friends' estrangement began in 1932. After spending more and more time with the Weizsäcker family, the thirty-year-old Heisenberg developed an unseemly crush on Weizsäcker's sixteen-year-old sister, Adelheid. One day he declared his love for her, shocking the entire Weizsäcker clan. Frau Weizsäcker visited Heisenberg's apartment afterward to chew him out and forbade him from visiting her household. Not even the Nobel Prize he won a few months later improved their opinion of him.

Although disturbed by this, Carl Weizsäcker continued to see Heisenberg and take trips with him. (Like Moe Berg, they happened to visit Berlin together in late January 1933, where hordes of Nazis were celebrating the appointment of Hitler as chancellor of Germany.) But Heisenberg kept stupidly, stubbornly pursuing Adelheid, even traveling to other countries to call on her when the family moved. The elder Weizsäcker finally ordered Heisenberg to desist, then married off his daughter to a German military officer. Heisenberg was devastated. The flip side of laddishness is childishness, and he all but cut the Weizsäckers, including Carl, out of his life.

Over the next few years, both men suffered personal crises. After quickly marrying a young woman named Elisabeth (whom he met while playing piano at a dinner party), Heisenberg was dragged into the Aryan physics debacle, enduring several rounds of interrogation. Weizsäcker's political advice would have been especially welcome during this period. Weizsäcker, meanwhile, watched his father move up the ranks of the Nazi Party; he eventually became the number two man in its foreign service department and helped lay the groundwork for Hitler's war. The family paid the price for this when Weizsäcker's younger brother Heinrich was killed in Poland on the second day of the war, one of the very first of the fifty million casualties to come.

Despite his grief, Weizsäcker attended the second meeting of the Uranium Club in late September. In truth, he found topics like cosmic rays more stimulating than nuclear fission. Always calculating, however, he realized the political importance of the topic, and he argued sotto voce among his colleagues that conducting bomb research would give German scientists prestige within the military and access to more funding. If they played things right, he added, they might even be able manipulate the Nazis for their own ends—a highly dangerous game.

Weizsäcker was also instrumental in persuading Otto Hahn— who had moral qualms about fission research—to join the club. He was perhaps less sure about his old friend Heisenberg: the two men had barely seen each other during the previous few years. They nevertheless had a warm reunion when Heisenberg reported for duty, and both would become key members of this new and potentially lethal clique.

In one of its first acts, the Uranium Club assigned each scientist to one of two projects. The first involved enriching uranium.

Two main types of uranium atoms exist in nature, uranium-235 and uranium-238. Each type—called an isotope—has the same number of protons (ninety-two). They differ in the number of neutrons, with uranium-238 having three extra. In most situations this difference makes no difference at all, since uranium-235 and -238 behave the same way in chemical reactions. But those extra neutrons make a huge difference in *chain* reactions. When a neutron strikes uranium-235, it undergoes fission and releases more neutrons, kicking chain reactions into

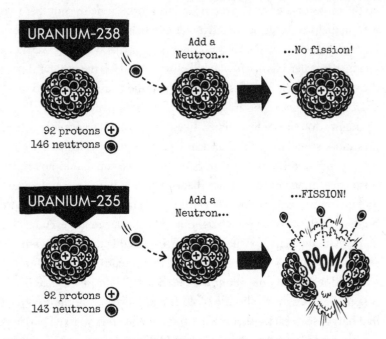

high gear. In contrast, when a neutron strikes uranium-238, the atom simply absorbs it. The nucleus doesn't fission and no extra neutrons get released. So if you want to make a fission bomb, you need as much uranium-235 and as little uranium-238 as possible. Otherwise, the 238 will absorb too many neutrons and the chain reaction will peter out.

The problem is, the fission-happy 235 isotope is rare in nature: for every 140 uranium atoms out there, just one is uranium-235. The obvious solution, then, is to separate the 235 and concentrate it, producing what's called enriched or "supernatural" uranium.

But enriching uranium is easier said than done. Most of the tricks chemists use to separate substances rely on chemical differences between them: you bathe them in acid or something, and one substance dissolves away while the other remains behind. But those tricks fail with uranium-235 and -238 because, again, they're the same element and behave the same way in chemical reactions. Scientists therefore had to invent new ways to separate uranium, all of which were cumbersome and required ridiculous amounts of energy; in some cases, they were separating isotopes literally atom by atom. No less a scientist than Niels Bohr declared atomic weapons impossible for this very reason. Indeed, the enrichment of uranium seemed the single biggest obstacle to making nuclear weapons, which is why the Uranium Club dedicated a whole team to the problem.

The second Uranium Club project involved building a nuclear reactor, which the Germans called a Uranium Machine. Reactors are basically small-scale chain reactions in the lab. Because the uranium in them isn't enriched, reactors won't explode in a mushroom cloud the way bombs do. But you can use them to study how chain reactions work, which is vital knowledge for making bombs.

Heisenberg joined the Uranium Machine team, and he absolutely threw himself into the task. Still stinging over the "Jew physics" debacle, he wanted to prove to German officials that he wasn't a mere theoretical wizard; he could do solid, practical work and build useful things, too. Heisenberg was also ruthlessly competitive. He loved to crush people at Ping-Pong and other games, and saw physics as no less of a bloodsport. Producing the world's first self-sustaining chain reaction would be quite an honor for Germany.

By early December 1939 Heisenberg had produced two secret reports on chain reactions, one focused on energy production and one on bombs. To begin experimental work, he opened a small lab in a wooden shed attached to a scientific institute in Berlin, just off a lane with cherry trees. Inside the shed was a six-foot-deep pit lined with brick—the world's first test chamber for nuclear reactors. Because they were working with radioactive materials, scientists entering the shed had to wear goggles, overalls, and face masks, an ominous sight. And if this wasn't enough to deter folks from snooping around, they code-named the place the Virus House.

Two years before the start of the Manhattan Project, then, the Uranium Club had scientists working on two key aspects of nuclear weapons: enriching uranium and producing a self-sustaining chain reaction. The German atomic bomb project was off to a rip-roaring start.

Carl Weizsäcker, the diplomat's son, soon provided a third key ingredient. In the early days of the war, he often read journals like *Physical Review* while riding the subways in Berlin. This earned him dirty looks from other passengers, since the articles were written in English, the language of the enemy. (Later in the war, one German scientist was sentenced to death for listening to the BBC.) But Weizsäcker cared little what hoi polloi thought. He did some of his best thinking on the subway, in fact, never more so than one fateful morning in July 1940.

As noted, uranium-235 undergoes fission when a neutron strikes it, while uranium-238 does not. So it makes sense that most nuclear scientists focused on studying 235. But in doing so, they overlooked something. When 238 absorbs a neutron, it becomes uranium-239, a slightly fatter version of itself. It then undergoes what's called beta decay. During this process a neutron in the nucleus transforms into a proton and an electron; the atom spits the electron out, while the proton stays behind.

Intriguing stuff. But while most scientists stopped at this point, Weizsäcker saw further. After the atom of uranium-239 underwent beta decay and gained a proton, it would become a new element, element 93. And Weizsäcker realized that this new element 93 would also be unstable and would also undergo explosive fission. Moreover, you could make this element cheaply in a reactor, since you wouldn't need to enrich anything beforehand; you could just bombard the common 238 isotope with neutrons. Best of all, because 93 was a different element than uranium, chemists could easily dissolve it out and isolate it. It had all the advantages of uranium-235 and none of the drawbacks.

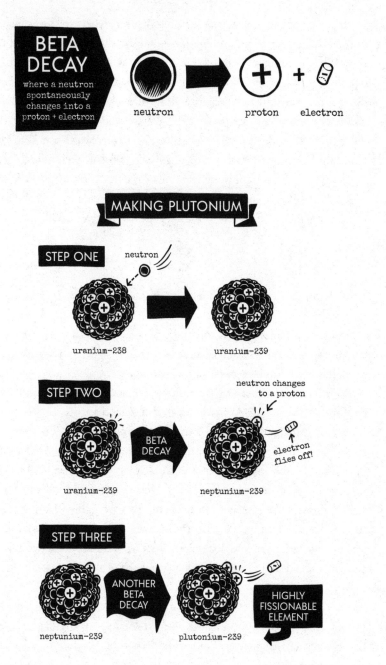

BETA DECAY
where a neutron spontaneously changes into a proton + electron

neutron

proton   electron

MAKING PLUTONIUM

STEP ONE

neutron

uranium-238   uranium-239

STEP TWO

neutron changes to a proton

BETA DECAY

electron flies off!

uranium-239   neptunium-239

STEP THREE

ANOTHER BETA DECAY

HIGHLY FISSIONABLE ELEMENT

neptunium-239   plutonium-239

Neutrons of thought were soon pinging everywhere in Weizsäcker's brain—a full mental chain reaction—and by the time he reached his train stop, he had it all worked out. In truth, Weizsäcker bungled some details here. Element 93, now called neptunium, also undergoes beta decay and transmutes one more time into element 94, plutonium, which is the really dangerous stuff. Weizsäcker nevertheless got the most important thing right—that the Germans could use 238 in a Uranium Machine to mass-produce a fissile element, then quickly separate that element out. The son of one of the most important officials in the Third Reich, in other words, had just dreamed up a cheap and much more practical means of making fuel for nuclear bombs.

Aside from the striving Diebner, the diplomat's son Weizsäcker, and the brilliant but obtuse Heisenberg, one other member of the Uranium Club played a crucial role in its history. Unfortunately for Walther Bothe, his contribution involved no great insight or breakthrough, but a monumental boner.

Bothe occupied a strange position in German science—half legend and half loser. He'd made himself a legend for his bravery and sangfroid during World War I. After earning his Ph.D. in 1914, he joined a German machine gun unit and was taken prisoner in Russia; he spent the next five years in a desolate camp in Siberia. Unbowed, he continued to do independent research in mathematics and theoretical physics there, even constructing his own table of logarithms to aid his calculations. He also found time to learn the Russian language and to court and marry a Russian woman, with whom he returned to Germany in 1920 as a war hero—a stirring example of dignity in trying circumstances.

But the 1930s were rough on Bothe. Exactly like the Joliot-Curies—in fact, before the Joliot-Curies—he'd turned up strong evidence for the existence of the neutron in several experiments, only to misinterpret his results. Unlike the Joliot-Curies, he had no other major discoveries to offset that embarrassment. To compound his troubles, the luster of his marriage had tarnished and he was growing increasingly restless with domestic life. Worst of all, as a mild opponent of the Nazis, he found himself marginalized within German science, pushed aside for colleagues who had more ardor for the Aryan agenda. This triple weight of professional, personal, and political pressure finally proved too much: his health collapsed and he suffered a nervous breakdown, eventually checking himself into a sanatorium.

Needing a break from Germany, the balding, mustached Bothe sailed to the United States in the summer of 1939 for a speaking tour. He planned on spending the trip alone, focused on science, but on the voyage over he met Ingeborg Moerschner, a vivacious and flirty blonde thirteen years his junior. To his astonishment, Moerschner reciprocated his affection, and long-dormant feelings of love (and lust) soon quickened inside him. He spent every waking hour on the ship mooning over her—and many nonwaking hours as well, as the middle-aged physicist and young Nazi civil servant became lovers.

In a delightful coincidence, both were headed to New York first, so Bothe accompanied her to the World's Fair in Queens, where they strolled the grounds arm in arm. By the sheerest luck, their schedules overlapped again a few weeks later in the Bay Area, where she was taking a job at the German consulate in San Francisco and Bothe planned to study cyclotron design at the University of California at Berkeley. So they rendezvoused there and went sightseeing. Bothe still had a wife and two children back home, but he pushed all that out of his mind and happily succumbed to folly.

The overlap in their schedules was no coincidence, of course. She'd been spying on him from the moment they'd met onboard, reporting his comings and goings to Berlin. (Bothe's foreign wife and eagerness to travel to America no doubt put him under suspicion.) Bothe remained oblivious to this, however, and when he returned to Germany late that summer, he was a changed man. He even showed some zeal for Moerschner's Nazi Party. During the first meeting of the Uranium Club, when other members hemmed and hawed about developing a nuclear bomb, Bothe rallied them and urged their support. "Gentlemen," he said, "it must be done."

The club assigned him a crucial task within the reactor project. As Enrico Fermi had proved a few years earlier, the speed of neutrons plays an important role in fission. In particular, scientists knew that uranium-238 prefers fast neutrons, while the fission-happy uranium-235 can work with slow ones. The trick to building a reactor, then, is reducing the speed of neutrons, to decrease the chances of capture by 238 and boost the chances of capture by 235. For technical reasons, scientists referred to reducing the speed of neutrons as "moderating" them.

Fermi had slowed neutrons down with pond water and paraffin, but scientists suspected that other substances might work equally well as moderators, especially graphite. (Most people today think of graphite as the "lead" in pencils, but this is actually pure carbon.) Bothe set out to answer two questions about graphite. First, how well did it slow neutrons down? As a small atom, carbon seemed promising in this regard, but he had to verify that experimentally. Second, did graphite absorb neutrons? Some elements have a bad habit of gobbling up stray neutrons and snuffing out chain reactions. Bothe needed to determine whether carbon was one of them.

He worked with massive hunks of graphite, spheres three feet wide. These had a "chimney" bored into them, and the experiments started when he dropped a neutron source down the chimney into the center of the sphere. These neutrons would fire outward, colliding with carbon atoms as they pinballed their way toward the surface. Eventually they'd either get absorbed on the inside or emerge from the sphere and ping a detector. Based on the number and the speed of the neutrons that emerged, Bothe could determine how effectively the graphite was slowing them down and how many neutrons it absorbed. Pretty straightforward.

Unfortunately, Bothe couldn't keep his mind on the work. He'd never quite broken off the affair with Moerschner, who had moved to Lisbon after the U.S. government shut down the Nazi consulate in San Francisco. (It was riddled with spies.) And Bothe's wife Barbara soon found out about the dalliance, reportedly after Moerschner called his house one night and Barbara picked up. Despite Barbara's wrath—or perhaps because of it—Bothe clung to the affair, and by his own account he found himself constantly distracted. On the first anniversary of their meeting, he wrote to Moerschner, "I have been speaking of physics the entire day, while thinking only of you." He also referred to himself as a "drunken teenager," and recalled playing Beethoven's "Moonlight Sonata" over and over while pining for her.

Given his mindset, it's no wonder his experiments didn't go well. According to theory, the average neutron pinballing its way through a graphite sphere should travel twenty-eight inches before being absorbed; Bothe's neutrons averaged just twenty-three. He'd tried to buy the purest graphite he could, but he suspected that contaminants— such as boron and cadmium, both of which gobble up neutrons by the bushel—might still be lingering inside. So he tried to develop purer sources of carbon. His top idea involved burning simple syrup and other sugars normally used in sweets, scorching them into a black, smoldering mass. At one point, in other words, a key ingredient in the Nazi nuclear bomb was candy.

These experiments failed, too, and Bothe finally approached an industrial firm to obtain one more graphite sphere for a definitive test. He asked for the purest stuff they had, a ball nearly four feet wide. Not quite trusting their assurances, he ran some purity assays of his own, charring a small chunk of the sphere and testing the ashes. It seemed sufficiently clean. You can imagine his shock, then, when the experiments failed yet again. In this ultrapure graphite, neutrons traveled only fourteen inches on average before being gobbled up, half of what theory predicted. Against all expectations, graphite seemed to be a terrible moderator, and Bothe wrote up a report to this effect.

It was one of the most consequential blunders in science history. Another German scientist, working outside the club, read the report and realized that the lovesick, loopy Bothe had committed a mistake during the purity assay. The ashes he'd examined were no doubt pure carbon; but the process of heating and burning the graphite would have driven off all the impurities that gobble up neutrons, leaving no trace behind in the ashes. This scientist wrote a report to counter Bothe's, but for some reason it never gained traction, perhaps because of bureaucratic rivalries between different groups. As a result, Bothe and the Uranium Club probably never knew.

Given the high cost of graphite and Bothe's seemingly definitive experiments, the Uranium Club ruled out carbon. Instead they focused on a different moderator, so-called heavy water. Scientifically, this represented a modest shift. Politically, it was a radical change, and would soon leave dead bodies scattered across the wilds of northern Europe.

CHAPTER 10

Heavy Water

In September 1939, in a speech in Danzig, Adolf Hitler promised to unleash upon the Allies "a weapon which there was no way to defend against." Lord knows what fantasy he was indulging, but a few Allied officials jumped to an immediate conclusion: atomic bombs. In support of this theory, they cited the fact that Germany had already banned the export of uranium from the Reich. Equally disturbing, French intelligence agents reported that the Nazis had taken a keen interest in heavy water.

Just as uranium comes in different varieties, 235 and 238, so does hydrogen. Most hydrogen atoms are simple, with a single proton for a nucleus. One of every 6,400 hydrogens, however, has a proton and a neutron; this is called heavy hydrogen (or deuterium, D). Fuse two heavy hydrogens onto an oxygen atom and you have heavy water (D_2O). Heavy water looks just like regular water, though it does have different properties. It's denser, and if consumed in large quantities (several gallons), it can damage DNA and interfere with basic metabolism. Compared to regular water, it also intensifies nuclear chain reactions. That's because regular water, when mixed with uranium, tends to absorb some of the neutrons flying around, which hampers chain reactions. Heavy water doesn't absorb neutrons, merely slows them down, so it's perfect for priming uranium-235.

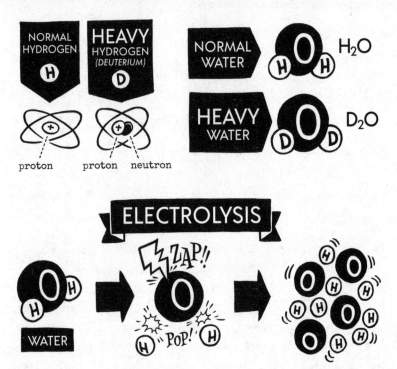

Given the scarcity of heavy water—it makes up just one of every 41 million water molecules—only one firm in the world bothered producing it in 1940. This firm, Norsk Hydro, managed a series of power stations on a bleak plateau a hundred miles west of Oslo, an area of Norway with an ugly reputation: one sixteenth-century chronicle said of the people there that their "chief delight is to kill bishops, priests, bailiffs, and superiors." One of those power stations, called Vemork, was the biggest hydroelectric plant in the world. It sat on a forbidding ledge near a massive waterfall, and most of the electricity it produced got channeled into making fertilizers and explosives. The remainder was diverted into a nearby building for electrolysis. This process involved zapping huge tanks of water with electric currents to separate the H_2 from the O. And as a by-product of electrolysis, they collected heavy water in a series of specialized fuel cells in the basement (more on these cells later). Unfortunately, the market for heavy water proved anemic. Between 1934 and 1938, Vemork sold just eighty-eight pounds total, for around $4,000 per pound. In modern terms, they made craft batches of the stuff, rarely selling more than a third of an ounce per order.

You can imagine the flabbergastation of Vemork officials, then, when they received an order for several hundred pounds of D_2O from Germany in January 1940. Kurt Diebner, the pathetic, striving head of the Uranium Club, sent an agent to Norway to buy up the entire stock of heavy water there—408 pounds—plus sign a contract for 220 more pounds per month. Given that heavy water had no real use beyond nuclear research, the Vemork officials asked the agent about the Nazis' plans for it. The agent was evasive, never giving a straight answer, and the officials could only assume that something nefarious was in the works—probably a radioactive poison gas. They managed to put the agent off, and when the firm's general counsel visited Paris a month later for a meeting, he alerted an official there named Jacques Allier.

Although nominally a banker, Allier worked as a spy for French military intelligence during the war. He also understood nuclear science: Frédéric Joliot had approached him earlier about buying up stocks of Norwegian heavy water for his and Irène's experiments in Paris. (Allier's bank owned two-thirds of Norsk Hydro, so he was in good position to carry out negotiations.) Allier thanked his Norwegian colleague for the intel on Germany and dashed off to tell Joliot. The banker and physicist met in secret in a hotel a few hundred yards from the Arc de Triomphe, and decided they needed to take drastic action.

So in one of the most daring heists of World War II, the tall, balding, bespectacled Allier shipped out for Oslo in early March 1940. He traveled under the name of Freiss, his mother's maiden name, and the bank gave him permission to spend up to 36 million francs

($25 million today) to secure every ounce of heavy water at Vemork. If he couldn't secure it—or if he did secure it and then got double-crossed or captured—Joliot had given him a vial of water laced with cadmium. Cadmium gobbles up neutrons at a fantastic rate, and even a few drops would irredeemably poison a tank of heavy water.

To heighten the drama, the Nazis caught wind of Allier's mission and sent a cable to their agents in Norway: "At any price intercept a suspect Frenchman traveling under the name of Freiss." These were pretty unambiguous instructions. Arrest him, kidnap him, murder him in the street—nothing was verboten. Allier knew what he was facing, too: friendly Norwegians tipped him off to the German dragnet. He plunged ahead anyway, trusting his wiles to elude capture.

After arriving in Oslo on March 4, he called Vemork officials from a pay phone and arranged a rendezvous on a street corner. Negotiations began the next day, with Allier prepared to spend all 36 million francs—fifteen times the product's market value. But the payment proved unnecessary. To keep the heavy water out of Nazi hands, the Vemork officials offered to turn over every ounce they had to France, as well as all future stocks, for free. Moved by their generosity, Allier disclosed the reason for the sudden interest in heavy water—not poison gas, but explosives of unimaginable strength. The Norwegians nodded soberly. One then told Allier: "If later, by bad luck, France should lose the war, I shall be shot for what I have done today. But it is with pride that I run that risk."

A few days later Allier and the officials crossed the desolate, frost-bitten plateau west of Oslo and arrived at Vemork. At the suitably clandestine hour of midnight, they began decanting the heavy water into stainless-steel canisters for transport. A local welder had fabricated them secretly inside his home over the previous few days; they were custom-built to fit inside suitcases.

On the drive back to Oslo, the smugglers took evasive maneuvers to lose any tails. Allier no doubt had the vial of cadmium ready, too, to poison the stocks if the Nazis pounced. But the coast looked clear, so they hurried to a French safe house and rushed the canisters inside—a risky move, since the house stood next to a Nazi military institute. For the next two days they hid indoors and plotted how to smuggle the heavy water out of Norway. They considered ordering a submarine to the Oslo harbor, but that would violate Norwegian neutrality in the war. Their only other plan seemed daft, even desperate, but with no other options, they made arrangements for a caper at the Oslo airport on March 12.

Allier and a fellow spy arrived at the airport early that morning with tickets to Amsterdam. They also had several suspiciously heavy suitcases, which they made a huge fuss about at the ticket counter. Nazi agents watching the airport took notice and alerted the German air force. Unbeknownst to the Nazis, however, Allier and his colleague had also purchased tickets to Scotland using aliases. Because the Scotland and Amsterdam flights left at roughly the same time, Allier hoped that the planes would be parked next to each other on the tarmac. Sure enough, when he and his companion finally finished up at the counter and walked outside, they saw the propellers spinning on both planes, ready to take off. They began loading their bags onto the Amsterdam one.

As they did so, a taxi came roaring up to the airport gate. A man who'd purchased tickets to Scotland demanded to be let onto the runway or he'd miss his flight. Airport security being a little more casual then, the guards on duty agreed. The taxi drove up and parked between the planes, out of view of everyone in the terminal. In a great rush, the man flung open the taxi's trunk and began tossing his bags onto the Scotland-bound plane. During the uproar, Allier and his partner managed to duck onto the Scotland flight themselves with their fake tickets.

Both planes took off a few minutes later. The Amsterdam passengers had a memorable flight. As soon as their plane left Norwegian airspace, German Luftwaffe fighters intercepted it and diverted it to Hamburg. Upon landing, Nazi officials flung open the cargo hold and seized Allier's luggage. It was filled with gravel.

Meanwhile, the heavy water—concealed inside the luggage of the man from the taxi—was soaring toward Scotland. Before it reached British airspace, however, Allier noticed a tail following them, an unidentified plane. He talked his way into the cockpit (again, lax security back then) and leveled with the pilot, explaining that he was a French agent fighting for the Allies. The pilot thanked him and began taking evasive action, flying inside clouds as much as he could. No one ever figured out what the second plane was—possibly Luftwaffe—but the pilot eluded it and landed safely.

The precious heavy water was hustled through customs and rerouted to France. By March 18 every last canister was tucked away in the bombproof cellar of Joliot's institute, safe and sound and secure. Provided the Nazis never reached Paris.

CHAPTER 11

Phoney to Real

After Joe Kennedy Jr. left London in September 1939, his father insisted that he re-enroll at Harvard and pursue a law degree. Kennedy Senior even hired a Massachusetts Supreme Court justice to tutor him. But Joe found the work dull, and after the war in Europe took a dark turn in the spring of 1940, he began to think seriously about enlisting in the military.

Before the spring of 1940, World War II had been something of a dud. Hitler had invaded Poland, yes, and Great Britain and France had declared war on Germany, making them official belligerents. But no actual battles or skirmishes had taken place, and the cheeky British press began referring to this state of affairs as the Phoney War. They would soon rue their cleverness. In April 1940, the Reich suddenly invaded Denmark and Norway, with Holland and Belgium and much of France falling in May. Along the way the Wehrmacht routed the British army and nearly annihilated it at Dunkirk. Most frightening of all, the air-raid sirens in London—last heard crying *wulf* in September 1939—began wailing in earnest at teatime on September 7, 1940, when a thousand German planes swept in to bomb the city, kicking off the famous Blitz—nine months of unrelenting air attacks. The phony war had become all too real.

Joseph Kennedy Sr. was still the American ambassador in London at the time. His clashes with President Roosevelt had continued, but Roosevelt refrained from firing him for a simple reason: If Kennedy was going to spout off about appeasement and isolation, better he do so abroad than at home. But when the Blitz started, Kennedy gave in completely to despair, telling one member of his staff, "I'll bet you five to one, any sum, that Hitler will be in Buckingham Palace in two weeks." A furious FDR kept Kennedy in London long enough to eliminate any chance that he'd challenge him for the 1940 presidential nomination, then curtly recalled him in October.

Far from feeling chastened by the dismissal, Kennedy began preaching appeasement and isolation even more widely. And in doing so, he managed the neat trick of infuriating both liberals and conservatives in Washington. Kennedy Senior had long harbored hopes of sitting in the White House himself someday, but his performance in 1940 dashed those dreams. Both Joe and Jack watched helplessly as their old man became a pariah in government circles, effectively committing political suicide.

Joe took this especially hard, since he harbored political ambitions of his own. Mind you, these ambitions didn't arise out of any deep-seated convictions. As a matter of fact, Joe was pretty muddled politically. One night in 1934 he'd come home to dinner and declared himself a communist, which enraged his father. (*When you sell your boat and horse,* he'd roared, then *you can call yourself a communist.*) Swinging to the opposite pole a few years later, Joe began praising Hitler's eugenic sterilization laws as forward-thinking. Groaning, his father reminded Joe that his sister Rose, who suffered from behavioral and emotional problems, would probably run afoul of such laws, and that maybe he should rethink his stance.

Despite these, shall we say, evolving beliefs, Joe Junior knew a few things for sure: that he as a Kennedy had a God-given right to hold office, and that war heroes had a leg up in getting elected. So with his father's political fortunes waning, Joe decided that becoming a war hero was the surest way to land a Kennedy in the White House.

CHAPTER 12

Mad Jack

Several members of the German atomic bomb project considered heavy water every bit as vital to their ambitions as uranium. The French heavy-water heist in March 1940 therefore left them fuming. But the invasion of Norway the very next month quickly bucked the Germans up. Oslo fell immediately, and although the feisty men and women of central Norway continued to fight for weeks, when the resistance there collapsed in May, the Third Reich seized the Vemork power plant. Adolf Hitler now commanded the only heavy-water production facility in the world.

Meanwhile, German Panzers were steamrolling France as well, roaring past the Maginot Line and crushing every pocket of resistance they encountered. Although still at some remove from the front, the people of Paris panicked. Thousands fled, and once sober government officials began tossing whole cabinets of documents out their windows to the lawns below, to burn in gigantic bonfires. Anarchy reigned.

Although calmer than most, Frédéric and Irène Joliot-Curie received orders in mid-May to evacuate the heavy water in their possession. Under no circumstances could the Nazis get their talons on it. So one night at 10 p.m., two of Joliot's assistants grabbed a pistol, loaded the canisters into a Peugeot truck, and began rumbling south. Jacques Allier, the banker-spy who'd pulled the airport heist, had made some calls and arranged for the assistants to deposit the canisters in a bank vault 250 miles away. Refusing to say what the material was, the assistants registered the heavy water as "Product Z." It stayed there five days before the bank manager got itchy and demanded its removal. The assistants located a temporary home in a nearby women's prison, then transferred it again to a maximum-security jail. Some of the inmates helped haul the canisters inside, placing them in a reinforced cell on death row normally reserved for violent offenders.

Joliot and Irène lingered in Paris as long as they could. (And meanwhile evacuated their children, Pierre and Hélène, to the family cottage in the fishing village of L'Arcouest, where young Irène had spent the beginning of World War I.) With the Germans just fifty miles away, the Joliot-Curies finally retreated on June 12. As they drove south, the smoke from burning oil refineries filled the sky. They carried with them some equipment from their lab, as well as Irène's birthright—the 130-pound lead-lined case containing the gram of radium that the women of America had given Marie Curie.

Early on the sixteenth, the Joliot-Curies reached the prison town, where they hoped to set up a new lab with the heavy water. But while Joliot was taking a stroll that afternoon, scoping out the town, a car pulled up and Allier jumped out. The French army had crumbled even more quickly than expected, and they had to evacuate again. Allier ordered Joliot to send the heavy water to Bordeaux, in western France, and from there ship it to England for safekeeping. Although reluctant, Joliot agreed. Allier then went to the jail to retrieve the canisters—not without difficulty, since the jailor refused to relinquish them at first. He came around to Allier's point of view when the banker shoved a pistol into his ribs. The doughty prisoners once again helped out by loading Product Z onto a truck.

Two other Joliot assistants escorted the D_2O this time, leaving the prison at dawn on May 17 and winding their way through the hilly countryside between central France and Bordeaux. Exhausted, they arrived at 11 p.m. and checked with military officials about an evacuation ship. They were assigned to the *Broompark,* a Scottish coal steamer under the command of Charles Henry George Howard, the 20th Earl of Suffolk.

The next morning, the assistants wandered down to the dockyard to find the *Broompark.* What they found instead was bedlam. Half a million refugees had swamped Bordeaux, every last one of them desperate to flee. To make matters worse, the diabolical Germans had mined the harbor, and every so often a squadron of planes swept through to strafe. The day before, an aerial bomb had bull's-eyed an ocean liner and killed three thousand passengers.

The situation seemed even more hopeless when the assistants found the *Broompark* and met the captain. Although a peer of the realm — his family line predated the House of Windsor — the Earl of Suffolk was known to most people as Mad Jack. That day the assistants found him stripped naked to the waist on deck, whacking his thigh with a riding crop and showing off his myriad tattoos (a bizarre affectation then). He had a woman hanging on each arm, one blond and one brunette, and was cracking dirty jokes in an excruciating French accent. With his noticeable limp and full beard, he looked like an "unkempt pirate," one witness recalled. Swallowing their misgivings, the assistants asked him if the *Broompark* planned to sail soon — they needed time to load the canisters. Never fear, said Mad Jack. The *Broompark* should have sailed already, but he'd taken the crew out the night before and gotten them stinking drunk, buying round after round until they'd collapsed. They were currently nursing the worst hangovers of their lives, and it would be at least a day before their stomachs were shipshape. So there's plenty of time to load everything, he assured them. This was what passed for a plan in Mad Jack's world.

That same day, May 18, the new prime minister of Great Britain, Winston Churchill, gave one of his most famous speeches, the "finest hour" talk. ("Let us therefore brace ourselves to our duties, and so bear ourselves that, if the British Empire and its Commonwealth last for a thousand years, men will still say, 'This was their finest hour.'")

Few people remember that he also warned against "a new dark age made more sinister...by the lights of perverted science." Those unlucky few who understood the threat of atomic weapons must have swallowed hard.

The *Broompark* finally sailed at 6 a.m. on June 19 with 101 souls aboard, each one clutching the inner tube of a car tire as a life preserver. In addition to the heavy-water canisters, Mad Jack had taken on two crates of diamonds from Amsterdam and Antwerp worth $15 million ($250 million today); they represented the bulk of the European diamond market. To protect this precious cargo, the earl had also scared up two 75mm guns and three machine guns. He hadn't found any ammo yet, but he wasn't worried. Bordeaux sat at the mouth of an inlet, seventy miles from open ocean, and he planned on stopping at a city along the coast for bullets and shells.

When they arrived at the coast, around noon, the tide was changing, and while Mad Jack limped off to ask about ammo, disaster nearly struck: a ship anchored next to the *Broompark* drifted into a mine and blew sky high. The scientists were actively trembling by this point, but when Mad Jack returned, he slapped them on the back and told them not to worry: he figured they had at least a fifty-fifty shot of reaching England alive. In fact, the explosion gave him an idea for a project. They would build an "ark" out of wood scraps to save the heavy water and diamonds in case the worst should happen. Work would take their minds off bombs and torpedoes anyway.

Leading by example, Mad Jack stripped to the waist again and lit up two cigarettes, which he smoked simultaneously using a special filter. He then grabbed a hammer and began pounding nails for the ark. Others joined in. A born raconteur, perhaps he took advantage of the crowd to regale everyone with the story of his life. Craving more adventure than the family estate near Oxford could offer, he'd sailed for Australia in a merchant ship while still a teenager, and had begun acquiring tattoos on far-flung isles as souvenirs. After a few years bouncing back and forth between Australia and England (a period that included his dismissal from the Royal Navy for insubordination), he'd decided to study chemistry and pharmacology in Edinburgh. Lest anyone should think he'd become conventional, he married a notorious nightclub dancer from Chicago. A few years later, in June 1935, he fell ill with rheumatoid arthritis, which hobbled his legs and left him with a limp. When World War II broke out, the army wouldn't take him, so he volunteered to work as a scientific spy in Paris. He quickly proved the most flamboyant secret agent in Europe, throwing champagne-soaked parties at the Ritz and showing off the twin .45 automatics he kept in shoulder holsters beneath his tuxedo jacket. He'd named them Oscar and Genevieve. (One of Joliot's assistants recalled thinking, "All this was completely in keeping with the ideas of British aristocracy I had gathered from the works of P. G. Wodehouse.")

But when his country needed him, Mad Jack proved himself worthy. The *Broompark* finally reached open ocean on the night of June 19. By that point most of the passengers—many of whom had to sleep on piles of coal belowdeck—were covered in black soot. They looked like chimneysweeps, albeit ones wearing suits and dresses. Even worse, several ships near the *Broompark* were bombed by German planes on the trip north. Yet despite the terror and discomfort, many passengers had the time of their lives. By sheer force of personality, Mad Jack made the whole voyage seem, as one passenger had it, like a "schoolboy adventure," limping up and down the deck, cracking jokes, handing out mugs of champagne, which he insisted was the best cure for seasickness. Joliot's assistant later remembered their situation in almost comic terms: "You know, the story of two scientists fleeing for their lives from an implacable enemy and carrying the world's supply of rare material which will enable them to master a new force of nature? It was preposterous, it was dime-novel stuff." And as in any novel worth its dime, the heroes survived their harrowing journey, landing unharmed in England on June 21.

The canisters of heavy water soon ended up in another jail, one with the Dickensian name of Wormwood Scrubs prison. After a few weeks in solitary there, they got a magnificent upgrade when the mad earl personally delivered them to Windsor Castle for safekeeping. Over the next few years these well-traveled cans of water would play a key role in experiments for the Allied atomic bomb project.

Sadly, Mad Jack blew himself up later in the war while engaged in a hobby of his—defusing unexploded German bombs that landed in England, an activity he liked to do while smoking. And despite his heroics on the *Broompark*, he did fail in one important aspect of his mission. In addition to the heavy water, he was supposed to seize another vital scientific asset in France. But Frédéric Joliot had eluded his grasp.

Joliot had arrived in Bordeaux a few days after his assistants. No matter how much the mad earl begged, though—at one point he seized Joliot's arm and tried to drag him up the gangplank—Joliot refused to board the *Broompark* and sail for England. He had several reasons for this. He didn't speak much English, and he knew he'd probably have to work under James Chadwick there, the man who'd beaten the Joliot-Curies to the discovery of the neutron. Even more important, Joliot was worried about his wife.

Irène had always been sickly, suffering from whooping cough and other ailments as a girl; her parents also had a bad habit of leaving lab coats covered in radioactive dust lying about, which further weakened her immune system. She'd grown even sicker after the birth of her daughter, developing both tuberculosis and anemia. (In an echo of her wedding day, Irène had worked all morning on her due date, taking only the afternoon off to give birth.) Ultimately the stress of the war broke her, and rather than accompany Joliot to Bordeaux, she'd checked herself into a sanatorium in western France to recover.

Even if he'd adored England, Joliot couldn't leave his sick wife behind—especially not after he'd already abandoned her once to open his cyclotron lab. Moreover, he found her contempt for the Nazis inspiring. "She was convinced," said her biographer, "that they would not dare lay a finger on a Curie," no matter how bad things got. Joliot needed such strength, and was determined to stick by her.

Moreover, Joliot still felt a sense of duty to France. Echoing Werner Heisenberg, he told Mad Jack that France needed him and that he wanted to salvage what he could of French science during the German occupation. In fact, he planned to return to Paris as soon as he could. Little did he suspect that the head of the Uranium Club would be waiting there for him.

CHAPTER 13

Compromise

After hiding out for several weeks to let the chaos in Paris subside, Joliot returned to his lab there to find two unwelcome visitors. One was Kurt Diebner, the striving, bespectacled founder of the Uranium Club. Diebner coveted the cyclotron in the basement of Joliot's institute and had come to assume control of it. These particle accelerators were necessary to study nuclear reactions, and the Nazis didn't have access to one—until now.

Joliot certainly didn't like Diebner, but he considered the other guest, Diebner's boss, outright loathsome. Erich Schumann was the grandson of composer Robert Schumann, and he'd never quite escaped the penumbra of his ancestor. Schumann did research on sound waves and the acoustics of musical instruments; he also penned military ditties for the Wehrmacht, which earned him healthy royalties. As an ardent Nazi, he'd been appointed the administrative head of the Uranium Club, even though Werner Heisenberg and others thought him a buffoon and mocked his work as "piano-string physics." (They despised him personally, too. "He might serve schnapps to visitors," one colleague recalled, "but it was always the cheapest.") Schumann considered nuclear bombs ridiculous—he called the idea "atomic poppycock." But he was cynical enough to pretend there was something to them, just to cover his ass, and he'd occupied Joliot's lab in Paris shortly after the city fell.

When Joliot turned up in early August, Schumann and Diebner called a meeting of his staff. Speaking through a translator, they worked hard to make a good first impression, praising them for their contributions to atomic physics and wishing for a fruitful Franco-German collaboration. Schumann then asked Joliot to join him, Diebner, and the translator in a nearby office for a tête-à-tête.

The moment the door closed, Schumann wheeled, glaring. He ripped into Joliot for not reporting to Paris sooner, then upbraided him for the theft (as he saw it) of the heavy water. We saw the bills of lading in Norway and know you stole it, he told Joliot. Now, where is it? Thinking fast, Joliot told Schumann that the canisters had been destroyed at Bordeaux, sinking aboard one of the ships that the Germans had bombed. All the heavy water had gurgled out into the harbor, he said, lost forever. Schumann didn't believe this, but had no proof otherwise. He also asked Joliot about several tons of uranium ore that had gone missing from Paris; in an obvious whopper, Joliot told them it was probably in North Africa. More invective followed.

As uncomfortable as things were for Joliot, the man who felt the most awkward here was the translator relaying all the insults, Wolfgang Gentner. Gentner was actually Joliot's old cigarette buddy—the young German assistant who'd spent a night in his and Irène's lab checking the detector after the discovery of artificial radioactivity. A few years later Gentner had been drafted into the German army and now worked for Schumann, whose conduct that day left him mortified. So as soon as the interrogation ended, Gentner tried to make things right. He lingered just long enough to catch Joliot's eye, then whispered that his old boss should meet him at the café on Boulevard Saint-Michel at 6 p.m.

That night they rendezvoused in a back room of the café. Gentner no doubt apologized to Joliot first thing—he deplored the circumstances of their reunion. Then he got to the point. The Germans had originally planned to dismantle Joliot's cyclotron and take it to Berlin as war booty. But Gentner had persuaded Schumann to keep the delicate machine in Paris and send the German scientists there instead. Gentner asked Joliot if he would accept this compromise—an intact lab, but one under Nazi control. He added that he could probably get himself named head of the lab, to shield Joliot from Schumann.

Perhaps Joliot lit a cigarette as he considered the offer. He certainly couldn't let the Reich steal the only cyclotron in France. But the thought of Nazi vermin infesting his lab disgusted him. And collaborating with the Germans, even decent ones like Gentner, could get him branded a traitor in France. Then again, he could probably fight the Germans most effectively through his research, and in that case he'd need his lab intact. Joliot finally agreed to the occupation on two conditions: Gentner had to run the lab, and the Germans could do no military research there, only pure science. Gentner said he could arrange that. And without lingering a moment longer—they'd already run a huge risk in meeting—the two men parted. It was Joliot's first undercover act of the war. It would not be his last.

As Joliot feared, when word of his compromise leaked out, people declared him a quisling. And not just in France—scientists in Great Britain and the United States wrung their hands over the news as well. Germany had already occupied Vemork, giving it access to heavy water. In conquering Belgium, it had stumbled into thousands of tons of uranium ore, which Belgium had imported from its colonies in Africa. Now, with the capitulation of Joliot, the Nazis gained a cyclotron, too. Germany had always had the intellectual firepower for a crash nuclear bomb program. It now had the equipment and the raw ingredients as well. And every nuclear physicist in the Allied world knew it.

CHAPTER 14

Harvard Highs and Lows

Samuel Goudsmit wished he could enjoy Harvard more. He'd joined the faculty in January 1941, and on paper the appointment certainly looked prestigious. What scientist wouldn't enjoy saying he worked at Harvard? Still, he couldn't quite feel content there, because he knew his hiring was a fluke: Harvard had brought him on to fill in for a professor doing war work; Goudsmit was basically a substitute teacher—"pinch-hitting," as he put it. Besides, everyone knew that the most prestigious posts then weren't at Ivy League universities, they were in defense labs. The United States hadn't yet entered World War II, but its best and brightest were already laboring away on rockets and radar and other vital technologies. Meanwhile, here he was, teaching Newton's laws to the bored sons of business magnates. On the whole he still enjoyed Harvard, but his old fear of being a has-been had stalked him from Michigan to Massachusetts.

Characteristically—he was a magnificent griper—Goudsmit unburdened himself in letters to friends. He complained about the damp climate of Cambridge. He complained about the stuffiness of the dress code at Harvard, all suits and ties. He complained about serving on too many committees, about having too many blue books to correct, about being stuck teaching Saturday morning classes. Even the best moments at Harvard were tinged with melancholy. One evening in May 1941 he chaperoned a formal dance for a fraternity and couldn't help but note the differences between the Harvard affair and similar events at Michigan. A senator's daughter showed up that night, for one thing. Harvard boys drank better liquor, too. Not that Goudsmit was complaining about the latter—he indulged freely. But he couldn't quite let go and enjoy himself, no matter how much he wished to. "I could not get it out of my mind that it was just a year ago that the Huns invaded Holland," he recalled. "No amount of liquor and dancing was able to cheer me up." In describing one partygoer there—"a sad-looking, brooding type of man who got sadder after each drink"—he might as well have been describing himself.

Beyond the damage to Holland, the anniversary of the invasion haunted Goudsmit on a personal level, too. In late 1939, a full year after returning from his sabbatical in Europe, he'd finally started looking into visas for his parents to immigrate to the United States. Isaac was almost seventy years old, and Marianne was going blind; he needed to get them out soon. But he'd gotten distracted with other things—research, his growing daughter, planning the next physics summer camp—and had put off finishing the applications for a few weeks here, a few weeks there. As a result, his parents didn't receive their visas until May 6, 1940. Four days later the Germaniacs invaded Holland, trapping his parents behind enemy lines.

Suddenly frantic, Goudsmit began writing letters to The Hague, then to immigration officials in the United States. Was there any news of them? Would the Reich still honor the visas? He even offered to pay for their departure, since "their money is probably valueless now." But no one knew anything for sure. Phones and telegraph wires were dead, mail service was suspended. Hitler had drawn a cloak around the entire country, making everyone inside vanish. Goudsmit was in agony. How had he ever, ever believed that Holland would be an oasis?

A few letters eventually reached Isaac and Marianne, and he learned that they were safe for the time being. But no Dutch citizen could leave the Reich now; the visas were useless scraps of paper. Worse than useless—every time he thought of them, they tormented him. *What if I hadn't waited a year to start applying? What if I hadn't put off dropping by the immigration office that week? What if I'd just gone over and gotten them myself, visas be damned?* What if, what if, what if...

CHAPTER 15

Maud Ray Kent

While Samuel Goudsmit agonized over his parents in Holland, the rest of the scientific world was agonizing over the fate of another disappeared soul, Denmark's Niels Bohr. After the Wehrmacht overran Copenhagen in April 1940, all contact with Bohr had ceased, and many of his friends and colleagues feared the beloved physicist dead.

Bohr was okay, it turned out—shaken but unharmed. To reassure everyone, he arranged through various underground channels to have Lise Meitner—still in exile in Sweden—send a telegram to England. It read, "Met Niels and Margrethe [Bohr's wife] recently. Both well but unhappy about events. Please inform Cockcroft and Maud Ray Kent."

The telegram caused three successive reactions. First, relief: Bohr was alive. But relief soon gave way to confusion. The "Cockcroft" mentioned was John Cockcroft, a physicist at Cambridge and future Nobel Prize winner. It made sense for Bohr to reach out to him. But who the Dickens was Maud Ray Kent? Someone finally asked Cockcroft if he knew—at which point confusion gave way to panic. Cockcroft loved doing British-style cryptic crosswords and other word puzzles, and after staring at the name for a moment, he realized that the letters were an anagram: Maud Ray Kent stood for "radyum taken." That was to say, the Nazis had seized the radium in Bohr's institute, no doubt in service of their atomic bomb project.

126

The news dealt a serious blow to British scientific morale. Intelligence agencies were already awash in rumors about the German interest in nuclear weapons, and Bohr's anagram ratcheted up the tension even higher. The British government responded by throwing together an advisory board on nuclear weapons that became known as the MAUD committee. (The name was inspired by Bohr's message, though someone later made a backronym of it—military application of uranium detonation.) The committee made its most important contribution to the war in July 1941, when it released a secret report showing just how scarily plausible nuclear bombs were. A mere 25 pounds of enriched uranium, it estimated, could explode with a deadly force equal to 3.6 million pounds of TNT. (If anything, this underestimated the true power of atomics: if 25 pounds of uranium-235 fissioned completely, it would unleash the equivalent of 385 million pounds of TNT.) What's more, the report implied that such a weapon might reasonably be available in two years—which, if the Germans opened up the throttle, would hand Hitler the Bomb in mid-1943.

The report shook the scientific world. Before this, many prominent physicists, while acknowledging the grave potential of nuclear bombs, had always expressed doubt that they were really possible. In a classic case of hope steering logic, they argued that some subtle factor that everyone had overlooked so far, or some new law of nature, would surely render such weapons unfeasible. The MAUD report murdered that hope. Committee member James Chadwick, the physicist who'd discovered the neutron in 1932, said that when he finally grasped the true threat of atomic weapons, he had to start gobbling pills in order to sleep at night, and never again in his life fell asleep without them.

The MAUD committee recommended that Great Britain start a crash program to build a nuclear bomb. At the same time, given the massive industrial effort required, the committee acknowledged that the strapped British economy couldn't support the project alone.

Great Britain should therefore partner with the United States, which had more natural resources and was safely distant from the threat of German bombers. To this end, MAUD dispatched copies of the report to Lyman Briggs, a top nuclear scientist in the United States who'd been working with them semiofficially. He was told to distribute it to American officials and report back.

A full week passed without any word from Briggs. Then a second week. Soon a month and a half had slipped away. Finally, one MAUD member who was visiting the States in late August decided to look into the delay. To his disgust and amazement, he remembered, "I called on Briggs in Washington, only to find out that this inarticulate and unimpressive man had put the reports in his safe and had not shown them to members of his committee." A terrified Briggs, in other words, had read the report and, like a five-year-old covering his ears during a thunderstorm, decided to lock the scary document away and pretend it didn't exist. Even years before it was built, the Bomb was reducing otherwise brilliant people to lunacy.

Despite Briggs, British and American scientists eventually convened and agreed to collaborate on an A-bomb. And not a moment too soon: by then Germany had been working on nuclear weapons for two years.

PART III

1942

CHAPTER 16

Resistance

Walther Bothe couldn't catch a break. The lovelorn physicist had already made a hash of his graphite research earlier in the war. Now he was bungling his way through a series of cyclotron experiments in Paris. Cyclotrons were intricate but temperamental machines, and no matter how many times Bothe and his assistants checked the valves and cooling lines and wires before each run, something always overheated or shorted out at the worst possible moment—flushing weeks of work straight down the bidet. This wasn't all Bothe's fault, but as head scientist he had to take responsibility. The failure tasted especially galling because he'd already tried to build a cyclotron back in Germany and had failed. It seemed he had some sort of cyclotron curse on him.

He didn't, of course. In turning his lab over to the Nazis, Frédéric Joliot had insisted they do no military research there. Bothe had nevertheless barged in and started running fission experiments, all but biting his thumb at the gentlemen's agreement. Equally galling, the Germans had stamped the machine with wax seals featuring Nazi eagles and swastikas, an insult Joliot's crew would not stand for. So whenever Bothe started an experiment, a French technician would sneak away and, say, turn off the water for the cooling lines, allowing the machine to overheat. Or he might discharge its huge electromagnetic coils all at once, producing a pulse of current strong enough to melt the copper wires. In abusing the cyclotron this way, the French scientists derailed their own research, too, but the choice between a busted cyclotron and an atomic Hitler was no choice at all.

However satisfying it was to watch Bothe flail, Joliot—and Irène, who'd rejoined him in Paris—were having a miserable war. Paris was growing gloomier every week, with widespread shortages of everything. In this greatest culinary city in the world, people resorted to eating stray cats to fill their bellies, despite the risk of catching plague from the rodents the cats ate. The Nazis also seized all available fuel for their munitions plants. As a result Irène and Joliot couldn't afford to heat their home, where temperatures sometimes hovered in the low thirties; a glass of water left on the nightstand would nearly have frozen. The cold in turn exacerbated the family's poor health. Hélène had rubella; little Pierre had both rubella and mumps; and Irène, already suffering from consumption, was knocked flat with bronchitis and anemia. Despite this misery, she gamely tried to continue her research, even if it meant lying down while working, and she spent months at a stretch in sanatoriums. Just like her mother, she was losing some of her most productive years to ill health.

But ill health didn't stop Irène from fighting the Germans in her own way. As mentioned earlier, upon fleeing Paris in 1940 she'd carried a lead-lined case containing the $100,000 gram of radium she'd inherited from her mother. Knowing the Germans coveted radium, she'd decided to stash it somewhere, but didn't want to use a vault or a safe—the first places the Nazis would look for loot. So she risked storing it in an ordinary cellar in a town near Bordeaux—hoping that, like Poe's purloined letter, the Germans would overlook it.

However clever this seemed at the time, Irène became increasingly uneasy as the war progressed. If the Germans caught wind of the radium's location, they would seize it for their own nuclear research. On top of that, this was her scientific inheritance, and as a Curie woman she wouldn't stand for the Germans stealing it. So in June 1941 she set out to recover her birthright.

Under normal circumstances Joliot would have accompanied her, but the Nazis were watching him too closely. A woman, on the other hand, wasn't worth their notice, and when Irène requested permission to visit Bordeaux and salvage some "lab equipment," the Nazis waved her and two assistants through. Irène still had to be sly: to justify her cover story, she grabbed some scientific instruments and other supplies while there. But amid these errands, she managed to slip into the cellar and recover the 130-pound case without attracting notice. She then had to smuggle everything back home on the train, enduring every German checkpoint. Arriving back in Paris, she found no cars available at the station, so despite her exhaustion she had to load the radium case and other equipment onto a horse cart and start pushing. "It was thus," she recalled laconically, "that the radium was returned to the laboratory." The Germans never suspected a thing.

Irène showed her mettle again later that month. On the morning of June 29 the Gestapo arrested Joliot, taking him by surprise over breakfast and hauling him down to their headquarters, an imposing stone building. His interrogators accused him of several crimes, including being a communist and stirring up student unrest. Thinking quickly, Irène ran off to find Wolfgang Gentner, the former assistant now running Joliot's lab. She begged him to intercede. Rather bravely—he easily could have landed himself in jail—Gentner marched downtown and demanded Joliot's release. He claimed that Joliot was doing vital military research, and that officials in Berlin would be furious. This was complete garbage: Joliot was in fact sabotaging Nazi research, as Gentner well knew. But the Gestapo guards, trembling at the mention of Berlin, quickly released him.

Despite this attempt to intimidate him, Joliot did join an underground communist cell shortly afterward to aid the French resistance. At first he merely passed on bits of scientific intel. As part of the wartime blackout, the windows at his institute had been painted over, so at night he would sneak into the labs of Bothe and others and read their secret reports. Over time this snooping gave way to riskier jobs, as he became a courier between different cells. He and other agents would rendezvous in theater lobbies, in dentists' waiting rooms, even amid the fossils of the National Museum of Natural History, and whisper messages back and forth.

It would take the deaths of several colleagues to push Joliot into active resistance. One was a sixty-one-year-old physicist and long-time friend of Marie Curie. Like Joliot, he worked for an underground communist cell and had been helping to sneak downed British pilots into neutral Spain. (If shot down in France, British pilots were instructed to seek out either communists or Catholic priests, the most trustworthy foes of the Nazis.) The Gestapo eventually infiltrated the group, and the physicist was arrested and beaten to death the day before Christmas. When his widow claimed his body, his clothes were torn and covered with feces, and he had broken bones and scald marks on his skin—signs he'd been tortured. Joliot and Irène were horrified.

Then, in May 1942, four more scientist-communists, one of whom was publishing an anti-Nazi newspaper, were swept up and shot without trial. Jolted again, Joliot officially joined the French communist party and committed himself to resisting Nazi rule. Given that he had a family, this was not an easy decision. And although they'd always shared everything with each other before, he and Irène decided that he would tell her nothing of his resistance work, to protect her when the inevitable crackdown came.

In the meantime Joliot set about the work of an underground agent. For example, he helped stage a fake car accident involving Marie Curie's former lover, Paul Langevin, who'd been in and out of Nazi prisons despite his advanced age. Joliot then swaddled Langevin in bandages to conceal his identity and got him across the border into Switzerland. Joliot also secured fake identity cards for Jews in hiding, passed along tips about impending raids, and, perhaps most importantly, smuggled weapons. A former assistant now worked in a police unit that investigated French sabotage against the Reich. His job involved impounding radios and guns and bomb parts seized in raids, for the purpose of destroying them. In reality, he packed everything into suitcases and smuggled it to Joliot, who redistributed the goods to the freedom fighters. Thanks to this recycling program the underground in Paris never lacked for arms or equipment.

More gravely, Joliot helped settle the score for his and Irène's murdered colleagues. When the French underground caught traitors — double agents secretly working for the Nazis — Joliot reportedly made the final decision about whether to execute them. Just eight years earlier, he'd danced around his laboratory after discovering a new secret of nature. Now he was killing men after uncovering theirs. War had made a hard man of this once cheerful physicist.

CHAPTER 17

The Fire Heard 'Round the World

While nuclear science limped along in Allied nations, things in Germany were hopping by June 1942, when Werner Heisenberg ran the most sophisticated nuclear fission experiment the world had ever seen. Over the previous few months, Heisenberg had built three different Uranium Machines in Leipzig, each more powerful than the last. Most of them used a "sandwich" design with alternating layers of heavy water and uranium ore (which the Germans code-named "Preparation 38," based on its chemical formula, U_3O_8). But for his newest reactor, L-IV, Heisenberg changed things up. Instead of flat layers, L-IV used four spherical shells, each one nested inside another. In addition, instead of using uranium ore, he swapped in pure, powdered uranium metal. (This powdered metal was kept separate from the heavy water by thin shells of aluminum, for reasons that will soon be clear.) Although just thirty-two inches across, the apparatus weighed three-quarters of a ton, and it generated so much heat that technicians kept it immersed in a tank of (normal) water at all times.

The L-IV experiment began when someone dropped a ball of radium and beryllium through a chimney into the innermost shell of heavy water. Radium emits alpha particles, which struck the beryllium atoms

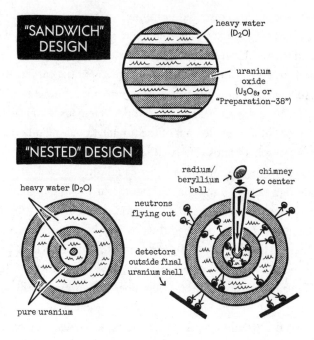

and dislodged neutrons. These neutrons plowed into the heavy water, slowed down, and induced fission reactions when they reached the first layer of uranium. These fissions released more neutrons, which again slowed down as they moved outward and encountered the next layer of heavy water. Which of course caused still more fissions in the outermost layer of uranium. Because Heisenberg's team knew roughly how many neutrons the radium-beryllium ball at the center released each second, they could measure the neutrons escaping the sphere's surface and calculate a "neutron multiplication" factor, a measure of how many extra neutrons the setup produced. In June 1942 they got a multiplication of 13 percent—the first positive neutron production in history, and step one in a chain reaction. Shortly after this, Heisenberg reported the results of his experiment to several Nazi leaders, including Albert Speer.

Unfortunately for Heisenberg, a series of accidents soon destroyed the L-IV apparatus. Powdered uranium is pyrophoric, meaning it can react with the oxygen in the air and ignite spontaneously. (It's a chemical reaction, mind you, not nuclear, but it's still a doozy.) Heisenberg's lab had learned this fact the hard way a few months earlier. An assistant had been spooning powdered uranium into a funnel to fill one of the shells. He suddenly heard a "dull thud." He paused to listen, and a twelve-foot flame exploded out of the funnel, knocking him backward and scorching his hands. He tried smothering the fire with his coat, but a nearby box of uranium powder caught fire as well, and another assistant had to rush it outside and smother it with sand before evacuating the room. To the researchers' horror, the metal was still smoldering a day later, so they dumped the box in a bucket of water—which quenched the fire, but taught them an unfortunate lesson.

The second accident unfolded more slowly, but the pyrotechnics proved even more spectacular. Heisenberg's team normally kept the sphere in a tank of cool water, and one afternoon in late June an assistant noticed a stream of bubbles in the tank. He captured some and determined that they were hydrogen, a bad sign. Apparently a leak had developed in the sphere, and the uranium inside was stripping the H_2 off the O and producing explosive gas. Around 3:15 p.m. the bubbles stopped, so they winched the sphere up to take a peek inside. The fellow assigned to remove the cover was the same hapless assistant who'd burned his hands before, and he had no better luck this time. As he unscrewed the cover he heard a hissing noise, an inrush of air. Two seconds of ominous silence followed—then another volcano, with burning uranium playing the part of lava. The poor fellow covered his head and ran.

The fun wasn't over. Remembering the lesson from the first incident, the team decided to douse the burning uranium with water. This would cut off the flames' oxygen supply and put the fire out. Unfortunately, while uranium powder can ignite in air, it can also react explosively with water. (It's fun stuff.) Unaware of this, the assistants dunked the apparatus back into the tank and waited.

At this point Heisenberg ducked his head into the lab. *Everything okay, boys? Yeah, chief, great.* The place couldn't have smelled good — burning metal has a distinct odor — but Heisenberg said okey-doke, and wandered off to teach an evening seminar.

In the middle of the seminar, around 6 p.m., an assistant interrupted with a knock. *You might want to take a look at this, chief.* Heisenberg ran back to find the tank of water seething with bubbles. Two assistants stood next to it, staring, wondering what to do. Suddenly the sphere shuddered, visibly bucking. Without even wasting time to yell, everyone dove for the hallway.

The sphere exploded with such force that a hundred metal rivets holding the aluminum shells together were ripped apart, shattering completely. The flames reached twenty feet high this time, licking the ceiling. The scientists ran for their lives, and the local fire brigade came roaring up a few minutes later. Despite their best efforts, the firefighters couldn't put the inferno out, only confine it. The sphere continued to smolder for two full days afterward, forming what one witness called a "gurgling swamp" of singed metal and polluted water.

Still, because no one had died, the firefighters had a good laugh about the whole thing. One of them who knew Heisenberg slapped the physicist on the back and congratulated him on finally producing an "atomic explosion."

Heisenberg had plans to transfer his lab to Berlin after the L-IV experiments, so the mishap provided a convenient stopping point. But no one could deny what he'd achieved in Leipzig: the first positive neutron multiplication in history, and arguably the first real, if small-scale, nuclear chain reaction.

Rumors of the accomplishment spread to other members of the Uranium Club, then throughout Germany. Scientists in neutral countries like Switzerland heard next, and they in turn told colleagues across the world. You can imagine what happened then. However precise they strive to be in their work, scientists are no less prone than laypeople to embellish hot gossip. And before long, several nuclear scientists were conflating the 13 percent neutron multiplication with the later reactor fire, as if some sort of atomic explosion really had taken place. Similarly, a few burned hands and some singed clothing metamorphosed into grave injuries.

By the time the scuttlebutt reached the United States, Werner Heisenberg had achieved a full-scale nuclear reaction in his lab, with the resulting deaths of several brilliant young physicists. And because the United States lacked any sort of scientific espionage program, officials had no way of determining the truth of this; as a result, they believed the falsehoods. One top scientist in Chicago, after sitting down and plotting out Germany's apparent rate of progress, calculated that Hitler could have an atomic bomb in hand in six months. If nothing else, he reasoned, a working nuclear reactor would allow the Reich to mass-produce radioactive species for dirty bombs to attack European cities.

The rumors about the L-IV fire had two effects on the Allied nuclear community. First, American physicists looked at their own, so-far feeble attempts to build a bomb and grimaced. They simply couldn't dillydally any longer. So in a scientific declaration of war, they founded the Manhattan Project on December 6, 1941—just hours before the raid on Pearl Harbor pulled the rest of the country into war. Second, officials began scheming about ways to strangle the Nazi atomic bomb program. One obvious way involved Werner Heisenberg: remove him from the project, and it would surely stumble. Another chokepoint was heavy water. The Nazis required it for their research, and there was only one firm in the world producing it at the time. A plan quickly emerged to reduce that number to zero.

CHAPTER 18

Off to War

Although he didn't know it at the time, Joe Kennedy's life took a deadly turn on May 15, 1942. That afternoon a British Spitfire returning from a reconnaissance run over the Baltic Sea passed by a small island off northern Germany. A photographer in the plane happened to glance down as they were crossing a cape called Peenemünde on the northern tip, and he noticed some circular embankments there. They seemed awfully large, so he snapped several pictures. A few days later, those pictures landed on the desks of British intelligence agents. The name Peenemünde was not unfamiliar to them: the Allied underground had recently passed along rumors that the Nazis were shipping expensive, high-end vacuum tubes there. But the circular embankments baffled the agents. They decided to expand surveillance of Peenemünde to figure out what was going on there.

If the British had only listened, a spy in Berlin had already told them exactly what was going on—and it was ominous indeed. In late September 1939, an aerodynamics expert named Wernher von Braun had gathered dozens of top scientists and engineers in Peenemünde for a conference. Von Braun would go on to become a legendary rocket scientist in the United States, the architect of the Apollo moon mission. Back then, he was just an ambitious twenty-seven-year-old Nazi engineer, and at the three-day conference—jokingly titled *Der*

Tag der Weisheit (The Day of Wisdom)—he and his colleagues swapped ideas about several revolutionary new weapons.

The Day(s) of Wisdom were top secret, of course, but word of them soon trickled out to Paul Rosbaud, the publisher who'd rushed the first uranium fission paper into print to warn the Allies. Since then, he'd adopted a nom de guerre, the Griffin, and had become the top scientific spy in Axis territory. He channeled information about all manner of topics to Allied agents, often by encoding messages in books whose printing he oversaw. After hearing about von Braun's little conference, Rosbaud began gathering string about it. He was something of a scientific bon vivant, and would invite colleagues to lavish dinners with lovely wines. None had any idea they were betraying secrets. They didn't even know their friend traded in secrets. They simply enjoyed his company and were happy to pass along bits of gossip about von Braun and the new rocket weapons at Peenemünde. Everyone agreed that the British were in for a hell of a surprise.

In a huge gamble, Rosbaud eventually snuck up to Peenemünde to poke around. Geographically, Peenemünde sits on an island near the present-day border between Germany and Poland. Part of it was once a beach resort; the rocket-testing grounds occupied the northern half. Although modest in size (around ten square miles), its shape stood out ominously on a map: it looked like the bald head of a wraith, its mouth gaping wide, screaming straight at London.

For ten days the Griffin haunted the woods around Peenemünde, watching and listening. The Nazis had restricted access to the island, but locals had seen strange vapor trails arching across the sky, followed by sudden explosions. Unbeknownst to them, they were watching the birth of the dreaded *Vergeltungswaffen*, or vengeance weapons, the V-1s and V-2s that would so terrorize London. At the end of World War I, the Treaty of Versailles had forbidden Germany from stockpiling certain weapons, and in a topsy-turvy way, the ban benefitted the German military in the long run, since it forced them to innovate and invent new ones. The V-weapons grew directly out of such programs. The V-1 was essentially a 2½-ton, 27-foot-long drone airplane packed with explosives. It proved difficult to aim, but had a crude autopilot to steer it toward its target. The V-2s were even bigger and badder, stretching 44 feet and weighing 14 tons. As the first true ballistic missiles, they didn't fly so much as whoosh straight upward and kiss the edge of space before plummeting back down. Perhaps most frightening of all, because the V-2s moved at supersonic speeds, people never heard a thing before they smashed home. If you heard the explosion, you'd already survived.

After scouting Peenemünde, Rosbaud returned to Berlin in late August 1941 and wrote a report for British intelligence; it included a description of the cigar-shaped V-2 and its components. Virtually no one read it. Who'd ever heard of Peenemünde? But in 1942 the photographs of the mysterious circular embankments gave the Griffin's report new impetus. Over the next two years the nature of these embankments would become the subject of intense, vitriolic debate within the intelligence community—debates that would eventually send Joe Kennedy to his doom.

Joe had had a frustrating war up to this point. He enjoyed the social side of law school, filling his days and nights with plenty of dates and hijinks. (He reportedly kept a "cretinous alligator" named Snooky in his bathtub, and would occasionally toss it into his roommates' tubs, while they were bathing, as a joke.) But law school itself bored him and, sensing his destiny lay elsewhere, he dropped out after his second year to join the navy. Specifically, he wanted to enroll in flight cadet school, which he considered the most dashing program in the military: "I've always fancied the idea of flying," he wrote to a friend. "I've never fancied the idea of crawling with rifle and bayonet through the European mud." The suggestion appalled his father, who thought flying far too dangerous. Flight school also seemed unlikely to provide the quick advancement that a Kennedy boy deserved. But Joe Junior proved every bit as bullheaded as his namesake and signed up for a flight cadet program in June 1941.

Over the next few months, Joe proved his father right about flight school being a bad idea. For one thing, Joe was a terrible pilot. He struggled to control his plane in the air and didn't really retain things from one day to the next. He also lacked a natural instinct for the craft, jerking the plane around and relying too much on his instruments. As one historian noted, "A talented pilot—a natural—*sees* and *feels*; Joe Jr. spent his time trying to *remember* what he was supposed to do next." His instructors didn't trust him with a solo flight for several weeks, and he eventually graduated seventy-seventh in a class of eighty-eight.

Even worse for Joe, his brother Jack's military career got off to a soaring start. In reality, the military never should have accepted Jack, since his health was about on par with Irène Joliot-Curie's. (He'd caught scarlet fever at age two and remained perpetually sickly after that; Catholic priests read him Last Rites three separate times during his life.) But Kennedy Senior pulled some strings, and after a sham medical exam Jack joined naval intelligence. Unlike flight school, this proved a quick route to advancement. Pretty soon Jack outranked Joe, despite having joined the navy four months after him. The thought of having to salute Jack, even in theory, left Joe sputtering.

Temperamentally, the two brothers had always been different. Whereas Jack was sickly, Joe was the soul of vigor. Joe was considered brighter as well: Kennedy Senior had once dismissed Jack's prospects by saying that he "wouldn't get very far." In sports, Joe argued aggressively with refs and umpires, and he loved smashing into people on the football field. Jack, meanwhile, was more retiring; he was a wily quarterback and later became a cheerleader. (As a youth he'd also served as a batboy for the Boston Red Sox; one of his favorite players was Moe Berg.) Both Kennedy boys proved popular with their peers, but as one historian noted, "Joe swaggered while Jack charmed."

Still, being just twenty-two months apart in age, the boys couldn't help but clash over certain things. With his swagger and arresting blue eyes, Joe had no trouble attracting women; in college he'd gone on dates with the likes of Ethel Merman and Katharine Hepburn. Joe wasn't above scooping up Jack's dates, too, just to stick it to him: he'd plop down next to Jack's girls, throw some money around, then tell his kid brother to scram. They nearly came to blows over dolls several times—and didn't in large part because Joe would have walloped Jack. Their senior theses at Harvard turned into another competition. Kennedy Senior believed that publishing a book gave a young man gravitas. So he'd taken Joe's *cum laude* thesis on the Spanish Civil War, spiced it up with some of his letters from Valencia and Madrid, and peddled it as a sort of cerebral adventure story, à la Lawrence of Arabia. Editors didn't bite. Jack, meanwhile, wrote his Harvard thesis on the buildup to World War II, and this time Kennedy Senior succeeded in getting it published, as *Why England Slept*. Joe remembered Jack throwing the thesis together with the help of five stenographers the week it was due. The book nevertheless sold eighty thousand copies, in part because Kennedy Senior purchased them in bulk. But father again knew best: the book gave Jack prestige, to Joe's frustration.

Now Jack was surpassing him in the military. But Joe finally caught a break during the summer of 1942. By this point he'd graduated to flying larger, bulkier planes called PBM Mariners; they topped out at 211 miles per hour. Giving an iffy pilot a larger plane might not sound like a great idea—if someone struggles to back up a sedan, you don't hand them the keys to an eighteen-wheeler—but flying larger planes actually required less brio and natural skill. Pilots relied on instruments instead, which played to Joe's strengths: he'd always flown more with his head than his hands, and he didn't mind all the procedures and checklists the way some hotshots did. By the end of the summer, he was even instructing other pilots. He soon got shipped to Puerto Rico for his first assignment, hunting submarines, and earned a promotion there as well, drawing even with Jack.

Only to fall behind again that autumn, when Jack was promoted a second time. Worse, Jack soon transferred out of naval intelligence to become a PT boat jockey, one of the studliest jobs in the war. PTs were basically mobile torpedo units: they'd zoom up to Japanese ships and unload on them, then dart away. The job rewarded reckless daring, and because Jack would be risking his neck in the South Pacific, he was virtually guaranteed a medal. Meanwhile, Joe found himself flying circles in the Caribbean, hunting for nonexistent German subs. Flying out of Puerto Rico was about as unglamorous a job as an ambitious young Kennedy could hope for, and it left Joe seething, once again, over his status compared to his kid brother.

A general note: We didn't have room in the book for all the great pictures out there, so I've made them available on my website instead. Visit http://samkean.com/books/the-bastard-brigade/extras/photos/

Polyglot Moe Berg, a former Major League Baseball catcher, would become America's first atomic spy. *(Photo courtesy Baseball Hall of Fame)*

No matter how cluttered things got, Moe Berg never let anyone touch his "live" newspapers. *(Department of Rare Books and Special Collections, Princeton University Library, Moe Berg collection)*

Lou Gehrig (right), Babe Ruth (center), and Moe Berg (next to Ruth, staring at the camera) traveled to Japan in 1934 on an all-star baseball team. Ruth starred on the diamond, while Berg took the opportunity to spy. *(Photo courtesy Charlie A. Barokas Collection)*

A young Irène Joliot-Curie works in the lab with her mother, Marie Curie. *(Photo courtesy Wellcome Library, Wellcome Images)*

Irène Joliot-Curie and her husband, Frédéric, in 1935, the year they shared the Nobel Prize. *(Photo courtesy Bibliothèque Nationale de France)*

Lise Meitner and Otto Hahn codiscovered nuclear fission in December 1938. *(Chemist Otto Hahn with Physicist Lise Meitner in the Laboratory, OPA at National Archives, author unknown, PD-1923)*

Coach Boris Pash was already a two-war veteran by the time he joined the staff at the legendary Hollywood High. *(Photo courtesy Hollywood High)*

Dutch physicist Samuel Goudsmit as a young man. He would later lead the Alsos atomic espionage mission in Europe. *(Photo courtesy AIP Emilio Segrè Visual Archives, Crane-Randall Collection)*

Dozens of world-renowned physicists gathered in Ann Arbor in the summer of 1939 for the University of Michigan's annual physics camp, including Samuel Goudsmit (far left), Werner Heisenberg (center), and Enrico Fermi (second from right). *(Photo courtesy AIP Emilio Segrè Visual Archives, Crane-Randall Collection)*

Joe Kennedy Jr., his sister Kathleen, and his brother John (the future U.S. president) make their way to the House of Commons in London in September 1939 to hear the British prime minister ask for a declaration of war. *(Photo courtesy John F. Kennedy Presidential Library & Museum)*

Lovelorn German physicist Walther Bothe, who bungled the graphite experiments. *(Portrait of Walther Bothe, Mondadori Publishers, author unknown, PD-1996)*

Pathetic, striving German physicist Kurt Diebner, who helped found the Uranium Club. *(United States Department of Energy)*

German physicists Werner Heisenberg *(left)* and Carl von Weizsäcker were best friends and fellow members of the Uranium Club. *(Photo courtesy Universitätsarchiv Leipzig, DC 2280)*

CHAPTER 19

Brazil and Beyond

Like Joe Kennedy, catcher Moe Berg whiled away the early days of the war in an obscure theater, South America.

Berg coached for the Red Sox in 1940 and 1941, earning a tidy $4,000 a year ($70,000 today). But as the seasons dragged on he found himself increasingly restless. Europe was in chaos, ravaged east and west by war—and here he was, still cracking jokes with relief pitchers in the bullpen. As a Jew, Berg felt a natural antipathy for the Nazis, and even more than that, he hated the persecution of intellectuals. The Nazis jailed professors and harassed scientists, smashed printing presses and burned books. It infuriated him.

In January 1942, Berg finally got a chance to do his part in fighting fascism. The Office of Inter-American Affairs, a government agency run by Berg's friend Nelson Rockefeller, aimed to promote U.S. interests in Latin America during the war, work that took many forms. In one case, OIAA officials brokered a deal on behalf of Ecuador to clear Peruvian refugees out of the Galápagos Islands, which the U.S. Navy then got to use as a base. OIAA also aimed to undermine the widespread popularity that Nazism enjoyed in South America. The Nazis had spent years stirring up resentment on the continent by spreading rumors about American soldiers raping women and stealing meat and sugar and kerosene intended for locals. Rockefeller and his staff fought back with pro-American puffery. They arranged to exhibit South American painters in New York, and leaned hard on Hollywood studios to re-edit romance flicks so that the lovers jetted off to, say, Rio instead of Paris at the end. Because sports were popular down south, Rockefeller also suggested hiring athletes like Moe Berg as goodwill ambassadors and sending them on tours.

In addition to battling the Nazis, Berg wanted to join OIAA for another, more personal reason. By the winter of 1941 his father Bernard was bedridden with incurable cancer, but he remained as stern and unforgiving as ever. Just before the attack on Pearl Harbor he lapsed into unconsciousness for a few days, and after coming to and hearing the news, he struggled to one elbow. "Where are the boys?" he demanded. He wanted his sons out fighting for their country. Berg's sister told him they were at home. "And what are they doing there?" he growled.

In joining OIAA, Berg hoped to win his father's approval at last. But before he could tell him anything, Bernard lost consciousness again and passed away. Berg once told a friend that seeing his father in the stands at one of his games would have given him more joy than breaking all of Babe Ruth's home run records. But Bernard never did attend a game, and he would now never know that his son had joined the war effort. Although he'd played his last Major League game in 1939, Berg had never officially retired from baseball. He did so the day his father died in mid-January 1942, determined to leave that life behind and make Bernard's ghost proud.

While OIAA worked out the details of Berg's trip—which took a mysteriously long time—another agency asked him to deliver a radio speech to the people of Japan in their native language. Berg agreed, and gave a heartfelt if rambling talk. He evoked memories of his two trips to the island, and lamented the fact that two countries who both loved baseball were now enemies. More to the point, he scolded Japan for overrunning and attacking other countries—"You have lost face and are committing national seppuku"—and urged the

Japanese people to overthrow their government and embrace democracy. The speech made no difference, of course, and the mention of baseball arguably backfired: within a year Japan had banned this decadent American sport. Later, when American troops began invading Japanese-held islands in the South Pacific, Japanese soldiers would scream the vilest insults they could imagine about Babe Ruth, convinced that this would crush the morale of American troops.

After the speech, Berg continued to cool his heels in Washington, waiting month after month to deploy. Growing frustrated, he eventually dug out the twenty-three seconds of film he'd recorded in Tokyo in 1934 and began shopping it around to various military officials. It was essentially his espionage audition tape.

Berg finally left for South America in late August 1942, and spent the next six months bouncing around to military bases in Panama, Peru, the Galápagos, Costa Rica, Trinidad, Aruba, British Guiana, Dutch Guiana, Guatemala, and Brazil. He earned $22.22 a day ($330 today), and traveled mostly on small planes and in the back of jeeps. (He had to bum rides because, as a native New Yorker, he couldn't drive.) Rather than his standard black suit and tie, Berg made a concession to the equatorial heat and, for the first and last time in his life, wore khakis and no suit coat. It wasn't the most flattering look: he appears sweaty and pudgy in pictures, well above his playing weight.

In addition to spreading goodwill, Berg served as a sort of roving health inspector for OIAA, and he spent much of the mission fretting over those age-old vices of soldiers: booze, cards, and especially sex. Given the hordes of young men, prostitution flourished near U.S. bases, and the rates of syphilis and other venereal diseases reached pandemic levels in some spots, as high as 66 percent among local women. Berg worried long and hard over how to keep American boys "out of the clutches of these girls" (as if the women were solely responsible). Naïvely, he suggested that the military could eliminate most hanky-panky by distracting the troops with ice cream, movies, fishing gear, badminton, and horseshoes. He figured that books would also help, and made a rather poetic plea to this effect: "Let the servicemen see the better things," he declared, "and they may choose not to follow the worse." Elevate the young men's spirits, in other words, and they'll shun whoring and drinking and gambling. A nice sentiment, if doomed.

In between visits to bases, Berg took some daring field trips, often rising at 5 a.m. to fly to obscure locations. He tracked snakes, monkeys, jaguars, and iguanas through the jungle. He observed a leper colony on an offshore island. He hunted for submarines from the glass nose cone of an airplane. (Much like Joe Kennedy, he saw nothing of note.)

Even these adventures, though, left Berg dissatisfied. To be sure, South America was strategically important for the Allies, given its location and abundance of rubber and other natural resources. In fact, President Franklin Roosevelt had woken up in the middle of the night after Pearl Harbor and, despite all the other things he had to worry about, gave orders to deploy U.S. marines to eastern Brazil. He did so to protect the airports there, which the military would need to bounce troops and supplies over to North Africa and eventually Europe. U.S. officials called this scheme the Trampoline to Victory.

But however strategically important, South America wasn't a battlefield, and the war would not be won there. Plenty of baseball stars had already joined the military and were doing great things—Ted Williams, Hank Greenberg, Bob Feller, Pee Wee Reese. Meanwhile Berg was stuck on a different continent, interviewing knucklehead sailors about where they'd caught the clap. So in addition to spreading goodwill, Berg also went a bit rogue in South America, and investigated matters he had no authority investigating. He sized up various political leaders for the State Department, met with a former president of

Brazil, and inquired about a series of assassination attempts on an unpopular general. He then wrote a secret report when he returned to Washington in February 1943, hoping to impress people there with his acumen and insight.

In the spring of 1943 Berg also followed up on the espionage audition reel he'd sent out several months before. His friends would later claim that military leaders used this film to plan the famous Doolittle Raid on Tokyo in April 1942, the first American raid to strike the Japanese homeland. Given the timing of events, that seems impossible: the Doolittle raid occurred before Berg ever shopped his movie around. But film footage of Tokyo was rare then, and military officials eagerly blew up stills from it to pinpoint docks, warehouses, factories, gas tanks, and power plants. As one officer told Berg, "We were able to get more good from your movies than would have been accumulated over several months of looking through texts and travel magazines." The fact that they relied on travel magazines should tell you something about the sophistication of military intelligence back then. Berg's climb to the rooftop in Tokyo made him look like a Green Beret in comparison.

Ultimately, given his age, the military couldn't take Berg. But there was an agency that could. In fact, given its freewheeling and even reckless approach to intelligence work, Moe Berg would prove a perfect fit for the Office of Strategic Services.

At the start of World War II the United States had no civilian agency dedicated to gathering foreign intelligence. Not that Americans never spied: the army and navy both had intelligence branches, and even private companies like General Electric sponsored corporate espionage. But the genteel Ivy Leaguers who ruled the federal government tended to view such activity as immoral, even dirty. As Roosevelt's secretary of war, Henry Stimson, once said, "Gentlemen don't read each other's mail." This squeamishness put the United States at a disadvantage compared to Great Britain, Germany, and Russia, all of which had sophisticated intelligence bureaus and happily spied on adversaries and allies alike.

Pearl Harbor finally forced the U.S. government to admit its shortcomings and establish the Office of Strategic Services. Most people know it today as the precursor to the Central Intelligence Agency, but OSS's mandate was broader than that. In addition to espionage, it carried out paramilitary operations overseas and helped pave the way for the U.S. military's Special Forces. In many cases, as with Moe Berg's adventures, the espionage and the extralegal activities went hand in hand.

OSS was primarily shaped by two men, director William "Wild Bill" Donovan and chief scientist Stanley Lovell. Donovan first won fame during World War I for leading a spectacularly idiotic assault. He commanded the 69th Infantry of New York, the famous "Fighting Irish," who were trying to conquer a German fortress in the Argonne Forest in October 1918. During an intense shootout one day, Donovan received orders to fall back. After considering his options, he ordered his men to charge instead. When the Fighting Irish hesitated, he screamed, "What's the matter with you? You want to live forever?" and charged off alone, confident his men would follow. They did.

The Germans stopped them cold, and a machine-gun bullet shattered Donovan's knee. But he once again refused orders to evacuate, and spent the next five hours hobbling around and preparing his men for the inevitable German counterassault. When it came, he rallied the Fighting Irish and drove the Huns back into the fortress in a rout, all but winning the battle single-handedly. Had the assault failed, Donovan would have been court-martialed (assuming he even lived). As it was, he earned the Medal of Honor and returned home one of the most highly decorated soldiers in American history.

When World War II rolled around, Donovan was working in a New York law firm. He happened to have attended law school at

Columbia with Franklin Roosevelt, and Roosevelt sent his old chum to England in July 1940 to provide a more accurate picture of events there than the defeatist Joseph Kennedy Sr. could. (Kennedy took this as a slap in the face, exactly as FDR intended.) Although Donovan agreed that things were grim, he emphasized the grit of the British people and singled out Winston Churchill—who wasn't even prime minster yet—as a stupendous leader. The assessment bucked up FDR's spirits and helped forge the Churchill-Roosevelt alliance that would ultimately help defeat Hitler.

Donovan parlayed his field trip to England into a job as Roosevelt's coordinator of intelligence, and from there he founded OSS and became its chief. But while the role made sense on paper—Donovan clearly had the vision and drive to see OSS succeed—Wild Bill also lacked pretty much every other skill necessary to run a government agency. Even those who adored him admitted that he had "abysmal" if not "atrocious" administrative skills, and he simply didn't have the patience or fortitude to manage people. As a result OSS became one of the most poorly run agencies in American history—the bureaucratic equivalent of Prohibition. Employees used to laugh over a line from Macbeth that perfectly summed up the enterprise: "Confusion now has made his masterpiece."

Nowhere were Donovan's flaws more evident than in his hiring practices. Needing to throw together an agency quickly, he turned to his circle of friends in New York and hired bluebloods by the dozen: the OSS roster was lousy with Mellons, Du Ponts, Morgans, and Vanderbilts. Columnists joked that the agency's initials actually stood for "Oh So Social." In Donovan's defense, hiring aristocrats did make sense on some level: they usually spoke several languages and knew Europe well. But holidays on the Riviera were a far cry from war. As one reporter noted, "Knowing how to speak French in a tux didn't necessarily prepare recruits for parachuting into enemy territory or blowing up bridges." More than a few heirs and heiresses suffered "dramatic mental crackups" in the field.

Even more than aristocrats, however, Donovan loved misfits, and he staffed OSS with a bizarre array of talent. There were mafia contract killers and theology professors. There were bartenders, anthropologists, and pro wrestlers. There were orthodontists, ornithologists, and felons on leave from federal penitentiaries. Marlene Dietrich, Julia Child, John Steinbeck, John Wayne, Leo Tolstoy's grandson, and a Ringling circus heir all pitched in as well. Observers sometimes referred to OSS as "St. Elizabeths," after the well-known Washington lunatic asylum, and they weren't just saying that. One top official there admitted that "OSS may indeed have employed quite a few psychopathic characters." Donovan once said, "I'd put Stalin on the OSS payroll if I thought it would help defeat Hitler." No one knew whether he was kidding.

To be fair, Donovan did hire some brilliant misfits as well, including chief scientist Stanley Lovell. When Donovan first interviewed Lovell, he asked him to become the OSS equivalent of Dr. Moriarty, the nemesis of Sherlock Holmes. It's more accurate to think of Lovell as Q from the James Bond franchise: his job basically consisted of puttering around in a lab and thinking up cool spy shit. To take a few examples, he and his labmates developed bombs that looked like mollusks to attach to ships. They crafted shoes and buttons and batteries with secret cavities to conceal documents. They invented pencils and cigarettes that shot bullets. They devised an explosive powder called Aunt Jemima that had the consistency of flour and that could be mixed with water and even baked into biscuits and nibbled on without any danger; only when ignited with a fuse did Aunt Jemima detonate. Like overgrown toddlers, Lovell's team also developed several feces-based weapons. One, called caccolube, destroyed car engines far more thoroughly than sugar or sand when dumped into gas tanks. Another weapon involved creating artificial goat turds to bombard North Africa with, in the hope of attracting flies that spread diseases. (They called it Project *Capri*cious.) Yet another project required synthesizing what was essentially eau de diarrhea, a compound that, as

Lovell said, "duplicated the revolting odor of a very loose bowel movement." They then hired small children to dart out and squirt it onto the pants of Japanese officers in occupied China. Lovell dubbed it the "Who, Me?" bomb.

And those weren't even the crazy ideas. After hearing that Hitler and Mussolini would be holding a summit at the Brenner Pass between Austria and Italy, Lovell devised a scheme to dump a vial of caustic liquid into a vase of flowers in the conference room. Within twenty minutes this volatile liquid would evaporate, turning into mustard gas and frying the corneas of everyone present. To really add some punch, Lovell suggested contacting the pope ahead of time and having him prophesy that God would strike the fascists blind for violating the Ten Commandments. When the mustard gas fulfilled this "prediction," the citizens of Germany and Italy would surely revolt, he argued, and take down the fascists from within. (Alas, the summit location was changed at the last second, and the plan never went into effect.) Lovell also developed what he called the "glandular approach" to winning the war. Drawing on some dubious Freudian theory, Lovell declared that Hitler straddled the "male/female gender line" and therefore might easily be pushed toward one sex or the other. Accordingly, Lovell isolated several feminine hormones to inject into beets and carrots in Hitler's personal vegetable garden. He hoped that Hitler's breasts would swell, that his mustache would fall out, that his voice would rise to a humiliating soprano register. The plan got far enough for Lovell to bribe one of Hitler's gardeners, but ultimately nothing came of it. As Lovell later admitted, "I can only assume that the gardener took our money and threw the syringes and medications into the nearest thicket."

The stories go on and on. But the craziest, nuttiest, most unbeliev-able thing about OSS was this: that often as not, its schemes worked. Whatever his faults as an administrator, Wild Bill Donovan pos-sessed a rare combination of physical courage and mental daring. As film director John Ford—another OSS recruit—once said, "Bill

Donovan...thought nothing of parachuting into France, blowing up a bridge, pissing in Luftwaffe gas tanks, then dancing on the roof of the St. Regis Hotel with a German spy." A man like that couldn't help but inspire people. And for every twenty of Lovell's cockamamie ideas, one or two worked brilliantly, seriously disrupting Axis missions. In fact, given the chaos of the world then, perhaps only something as haphazard as OSS could have succeeded. The agency ran a dizzying array of covert operations all over the world, gathering data on everything from troop movements to nuclear bombs. It also made brilliant use of several eccentrics who otherwise might not have contributed to the war effort, including catcher Moe Berg.

CHAPTER 20

Baja Days

Like Moe Berg, Boris Pash had Latin American sports on his mind in the summer of 1942. But while Berg's interest in them was wholesome and innocent, Pash was playing a wily game of espionage.

In 1940 Coach Pash had quit Hollywood High School and moved to San Francisco to help run the U.S. army's intelligence division there. Although Major Pash's jurisdiction was massive—the seven westernmost U.S. states, plus the Alaska territory—the assignment was a fairly sleepy one until the attack on Pearl Harbor. After that, the region remained on high alert for several months, and for good reason. Most people don't realize it today, but Japan actually attacked U.S. territory a second time shortly after Pearl Harbor, on February 23, 1942, when a submarine surfaced near Santa Barbara, California, and shelled an oil refinery. The assault did little damage, and no one died. But for a country still reeling from Pearl Harbor, the incident looked dire. Japan could seemingly strike from the air or sea at any moment.

Given his interactions with Japanese families at Hollywood High— not to mention his familiarity with the Japanese military after it began drafting his ballplayers—Pash was assigned the task of hunting for Japanese insurgents along the Baja peninsula of western Mexico. The military especially feared a surprise attack on the naval yards in San

Diego, just north of Baja. Now, hunting for Japanese insurgents in Mexico might sound daft, but more than 3,600 Japanese people lived on Baja in the early 1940s, descendants of farmworkers who'd migrated from Japan for jobs in the cotton industry there. In fact, the Mexican government had already rounded most of them up in January 1942 and shipped them off to Mexico City and Guadalajara, inland sites where people of supposedly dubious loyalty could do less damage. (Similarly, the Canadian government evicted 2,500 Japanese people from Vancouver, and the Roosevelt administration relocated and interned 110,000 over the course of the war.) Despite the relocations, Pash's office continued to worry about subversive activity in Baja. So in the summer of 1942 he deployed a half dozen agents to the area, instructing them to scour the peninsula "by auto, burro, horse, boat, and on foot."

Every so often the agents—who went by code names like "Z" and "Brush" and "Strong-Arm"—mailed off reports on their findings. They usually addressed them to "Dr. Bernard T. Norman," an alias Pash had cooked up, and at first glance the reports seem laughably incompetent. The most prolific letter writer, "Carlos," kept getting stranded in small towns in Baja without transport out; in one case he tried buying passage on a ship hauling dynamite, but the captain rebuffed him, forcing him to waste three weeks there doing nothing. Whenever Carlos did wash up somewhere new, instead of scouting around for subversives he'd spend his days fishing or hunting on horseback in the countryside. He'd also wander down to the local ball diamond to catch a few innings. Carlos in fact seemed obsessed with baseball: in every single village he'd count how many ballplayers and ball fields and baseball gloves there were; he even tallied umpires. We know all this because he vomited every last detail of his "findings" onto Pash, blathering on and on about the baseball-playing habits of Baja Mexicans. Today these letters seem both monotonous and exasperating, which is quite a literary feat.

And that's exactly what Pash intended. The mail service in Baja then was spotty and unsecured, and anyone snooping through the letters would have found them vacuous. But among Pash's personal papers, there's a decipherment key that makes sense of everything—for the letters were written in code. In prattling on about the size and location of "baseball diamonds," Carlos was really providing intel on airplane runways. "Gloves" were planes, "bats and balls" were gasoline depots, and "umpires" were potential saboteurs. Other code words (e.g., "handball courts") referred to submarine docks or caches of oil and cement. Similarly, the fishing and hunting trips were perfect opportunities to scout out beaches and other areas where enemy troops could invade or hide. All the mindless blather had a serious purpose.

To be honest, this surveillance didn't amount to much. All the "baseball diamonds" and "bats and balls" in Baja were for civilian use, and despite the persistent rumors of Japanese subs surfacing off Mexico, no U.S. agent ever saw one. Indeed, there were probably more U.S. spies hunting Japanese subversives in Baja than there were actual Japanese subversives. (One agent, whose job entailed stopping local tuna boats and interrogating the crews, was chagrined to discover that the captains of two of them were also American military personnel working undercover.) What's more, some of Pash's sources of information seemed less than reliable—lazy, indiscreet con men who were taking money from both sides and were more intent on settling personal scores than providing good intelligence: they'd snitch on each other for visiting "cat houses" while their wives were away or for smuggling booze across the U.S. border.

Still, if the Baja campaign didn't exactly win the war, it did change the direction of Boris Pash's life. He loved trading in secrets and delighted in outwitting people with codes and other stratagems: in addition to managing agents, he reportedly experimented with disguises during this period, dressing up in wigs and employing "voice-altering devices" to conceal his identity. Although a simple soldier and teacher before the war, intelligence work enthralled him, and he wanted more. Luckily for him, a new scientific venture needed a man with his exact skills. It was called the Manhattan Project.

Although founded before Pearl Harbor, the Manhattan Project had drifted along for months without much energy or purpose—little more than a series of meetings and studies by government officials, the main outcome of which was to suggest further meetings and further studies, ad infinitum. This lack of urgency infuriated nuclear physicists, especially the refugees from Hitler's Europe, who knew that Germany had been working on fission for nearly three years by then.

This lethargy ended the day General Leslie Groves took command of the project in September 1942. Although practically synonymous with the Manhattan Project today, Groves originally loathed the idea of running it. Known as a brilliant but ruthless manager—simultaneously the best construction foreman and the biggest asshole in the military—Groves was in charge of all army construction within the United States and on offshore bases at the war's outset, overseeing a budget of $600 million per month ($10 billion today). Running the Manhattan Project—whose entire budget was only $100 million at the outset—was effectively a demotion. It turned out that Groves had pissed off one too many people in the Pentagon (whose construction he'd personally overseen), and the Manhattan Project was his punishment. His superiors figured he had no chance in hell of building one of those nuclear whatsits before the war's end, and the failure would teach him some humility.

Things didn't quite work out that way. Roused to anger, Groves threw himself into the task, and in one of his first moves as director he named physicist Robert Oppenheimer, a professor at the University of California at Berkeley, as head of the weapons design lab at Los Alamos. In hindsight, hiring Oppenheimer looks like a brilliant move, but at the time it was radioactively controversial, among both scientists and military officials. Rather snootily, several top physicists suggested to Groves that Oppenheimer wouldn't command much respect as a leader since he lacked a Nobel Prize. More to the point, he was a theoretical physicist, not an experimentalist, and he seemed as clueless about administrative matters as OSS chief Wild Bill Donovan. "He couldn't manage a hamburger stand," one scientist complained to Groves. Groves didn't care. He considered himself a shrewd judge of character, and against all available evidence he decided that Oppenheimer would thrive as head of Los Alamos.

The military's objection to Oppenheimer was more grave. Having grown up rich and politically naïve—a classic limousine liberal—Oppenheimer had always vaguely supported whatever causes were fashionable among the Berkeley elite. In the 1930s that meant communism, and Oppenheimer had spent the decade befriending Stalinists and raising thousands of dollars for the Spanish Civil War. More worrisome, he surrounded himself with student radicals who worked on fission research by day and met secretly with Communist Party officials by night, to spill everything they knew. In short, Oppenheimer was a gargantuan security risk. So much so that, despite being head of the most secret laboratory of World War II, the army refused to grant him a security clearance to work there—which was a bit awkward. Groves finally had to intervene and arrange for a temporary pass so his tenure could begin.

Despite his confidence in Oppenheimer, Groves felt duty-bound to investigate the physicist's ties to communism, and he soon found the perfect person for the task: a dogged, clever intelligence officer already living in the Bay Area—someone who, ever since fighting against the Red Army as a teenager, had loathed communists and all they stood for. Boris Theodore Pash.

CHAPTER 21

V-1, V-2, V-3

In the spring and summer of 1942, Wernher von Braun scheduled three V-2 launch trials at the Peenemünde rocket-testing grounds in northern Germany. All three ended in failure. The body of the first rocket blew up during a preliminary check of the combustion chamber. The second wobbled dangerously on takeoff and crashed into the Baltic Sea a mile away. The third lifted off just fine, but suffered an electrical short and exploded in midair like a fourteen-ton bottle rocket—an entertaining but frustrating development.

On October 3, however, the V-2 finally took flight. The rocket was actually pretty sluggish on the launch pad; as one former slave at Peenemünde remembered, "V-2s rise slowly, as if pushed up by men with poles." But they acquire momentum fast, and can reach 3,500 miles an hour at top speed. In anticipation of the launch that day, von Braun's crew had decorated the fuselage with a painting of a comely woman straddling a crescent moon. The rocket didn't fly that far, but it did land an incredible 118 miles away.

Von Braun's staff filmed this and other successful launches over the next few months, and Nazi military leaders began passing the reels around like snuff films, ecstatic at the possibilities. Propaganda minister Joseph Goebbels squealed, "If we could only show this film in every cinema in Germany, I wouldn't have to make another speech or write another word. The most hardboiled pessimist could doubt in victory no longer." Hitler was less enthused. In a rare display of common sense, he pointed out that the rockets seemed difficult to aim— the October 3 launch had flown far, yes, but had also missed its target by two full miles. Surgical strikes seemed all but impossible.

The Nazis nevertheless poured billions of Reichsmarks into the development of the V-weapons over the next several months. And although no one realized it at the time, the rockets' inaccuracy would soon prove their greatest asset. Because they were impossible to aim, they could land anywhere within a huge radius. No one on the ground ever felt safe as a result, no matter how far they lived from potential targets. The vengeance weapons, in other words, were supreme weapons of terror—perfect for pummeling the 360 square miles of London and the millions of people who lived there.

So while testing continued at Peenemünde, the German army began toiling away in northern France, just south of London, to build launch facilities for the V-weapons. In the meantime, von Braun's engineers began drawing up plans for an even more ambitious weapon, the enigmatic V-3.

CHAPTER 22

Letters

After a year of chaperoning frat boys at Harvard, Samuel Goudsmit was finally doing something useful. In November 1941 a colleague down the road at the Massachusetts Institute of Technology had recruited him to work on radar there. As someone whose interests ranged far beyond physics, Goudsmit enjoyed the eclectic mix of people in his new lab. There were lawyers and soldiers and intelligence analysts running about, even a husband-and-wife team on loan from Disney to illustrate a book about radar—not for children, but for slow-witted generals. Best of all, Goudsmit felt he was really contributing to the war at last. He was feeling so confident, in fact, that he took a foolish gamble and nearly got himself arrested.

As so many scientific troubles did during the war, it all started with Werner Heisenberg. In October 1942, an exiled Austrian physicist in New York, Victor Weisskopf, received a letter from a colleague in Switzerland that mentioned two things. First, that Heisenberg would be giving a lecture in Zurich soon. Second, that Heisenberg had accepted a new job at the Kaiser Wilhelm Institute in Berlin. This was mostly just chitchat, but as the politically savvy Weisskopf read and reread the letter, his face creased with worry. KWI was a state-run (i.e., Nazi-run) entity with a strong nuclear physics program.

Rumors had already been swirling about Heisenberg's "atomic explosion" in Leipzig, as well as the German interest in heavy water. Heisenberg's move to KWI, then, could mean only one thing: Germany was ratcheting up its nuclear bomb program.

Just before Halloween, Weisskopf showed the letter to another émigré who worked at MIT. He agreed that it sounded dire. Unsure what to do, they discussed the matter exhaustively, and as the night passed, a plan took shape. The letter had mentioned Heisenberg giving a lecture in Zurich in early December. Switzerland was a neutral country, offering free passage to both Axis and Allied citizens. Perhaps they could send a scientist there to question Heisenberg—feel him out about German progress. Why not?

Soon the idea got a little bolder. Instead of just sounding Heisenberg out, what if they could disrupt the German project somehow? Maybe by delaying Heisenberg's return to Germany. Or heck, maybe they could delay him permanently—detain him and prevent him from returning at all. It was a stupid idea, of course—far too risky. But after another minute or two, it didn't seem so ridiculous anymore. Why couldn't they detain him? This was war, after all. Anything goes. If they planned properly, Heisenberg wouldn't be in any real danger, and getting him away from Germany would derail the entire Nazi bomb project. The more they egged each other on, the more clever the idea seemed, and the very next day Weisskopf wrote a letter to Robert Oppenheimer.

The letter relayed the gossip about Heisenberg's new job and what that portended, then mentioned his upcoming lecture in Zurich. With startling bluntness, Weisskopf added, "By far the best thing to do... would be to organize a kidnapping." The Germans wouldn't hesitate to kidnap *you*, he pointed out to Oppenheimer. And while the plan certainly carried risks—they would be violating the neutrality of Switzerland, and whoever approached Heisenberg might be seized as a spy if things went south—those dangers seemed negligible weighed against the possibility of a Nazi atomic bomb. Despite the risk of capture and torture, Weisskopf volunteered to go to Zurich himself. "It is evident," he concluded, as if this were a simple mathematical proof, "that the kidnapping is by far the most effective and safest[!] thing to do."

By this time Oppenheimer was running the Los Alamos weapons lab and had no time for harebrained nonsense. He'd in fact already heard the news about Heisenberg through other channels, and his response to Weisskopf was filled with bland bureaucratic assurances about how the "proper authorities" would be notified. In what seemed like a polite brush-off, he added, "I doubt whether you will hear further of the matter."

In private, though, Oppenheimer thought the kidnapping idea wasn't half bad, and he forwarded it to the head of wartime scientific research, Vannevar Bush. In passing the proposal along, Oppenheimer noted that he wasn't necessarily endorsing it, and he pointed out that Weisskopf would obviously be in over his head; they'd need a professional spook. Still, the visit to Zurich "would seem to afford us an unusual opportunity," he wrote.

Bush in turn put Oppenheimer off with more blandishments. But he secretly liked the idea, too, as did the military honchos he mentioned it to. Ultimately, those officials vetoed the plan—not because it was dangerous and illegal, but because of game theory considerations. If we contact Heisenberg, they reasoned, the Nazis will know that *we* know about their atomic bomb program. This would in turn imply that we have a similar program, which would put us at risk. What's more, the Nazis would redouble their efforts, and since we're already lagging behind, that doesn't seem wise. Incredibly, though, no one dismissed the idea outright, and the plan to shanghai a Nobel Prize–winning physicist continued to percolate in their minds.

Not that Victor Weisskopf knew any of this. For security reasons the people making such decisions never deigned to explain their reasoning to a peon like him. Beyond Oppenheimer's letter he heard nothing back about his proposal, and after a week passed, he grew fretful. The lecture in Zurich was coming up soon—days were ticking off the calendar—and bureaucratic inaction seemed likely to ruin this opportunity.

Unable to sit still, Weisskopf and his buddy decided that the British, who had the most to lose from an atomic Hitler, might be more open to their scheme. So they turned to their fellow émigré Samuel Goudsmit, who knew a few Brits working on radar at MIT. Goudsmit thought that kidnapping his old friend Werner was a smashing idea. His British contacts gave him a name to write to and encouraged him to slip the phrase "tube alloys" into the letter, that being the British code word for atomic bombs. Goudsmit did so and rushed to the mailbox on November 7, convinced that the letter would spur people into action.

It did—but mostly on this side of the Atlantic. Seeing the phrase "tube alloys" panicked British officials. How had this random physicist learned their most secret code word? They alerted Vannevar Bush, who opened an investigation into the breach. Humorless intelligence agents swarmed Goudsmit and demanded to know where he'd heard the term. Even more suspicious, Goudsmit had mentioned in the letter that he occasionally came across other tidbits of intel. Would the British like to hear those as well? You can imagine what the American agents thought of that. Goudsmit sputtered out an explanation and apologized for his clumsiness. More hard questions followed, and only begrudgingly did the agents clear him. He walked away from the incident chagrined, and with a clear understanding of his place within the wartime scientific hierarchy—at the bottom.

And yet, similar to Boris Pash, this brush with the clandestine side of science intrigued Goudsmit. He'd always adored detective novels and fancied himself an amateur sleuth. (When thieves had stolen equipment from his lab in Ann Arbor, he'd pulled out a fingerprinting kit and dusted for clues himself.) The idea of plotting war stratagems excited him even more. So despite his dressing-down, Goudsmit wrote another letter to intelligence officials offering his services. He pointed out that he spoke several languages, a useful asset abroad, and that he knew scientists in Italy, Holland, Belgium, and France. "I think there are even some German physicists who still believe I am their friend," he added. The tone of the letter, frankly, was a tad pathetic. *Please keep me in mind—I'll do anything!* And perhaps not surprisingly, no one bothered answering him. After his bungling attempt to arrange the kidnapping, it seemed that the intelligence community wanted nothing to do with Samuel Goudsmit.

Another letter soon banished his fantasies about espionage anyway. In the summer of 1942 the Reich had started deporting Jews from the Netherlands, and all communication between Goudsmit and his parents ceased. For months he'd been writing to friends in Holland, begging for help in tracking them down. This included Dirk Coster, the physicist who'd helped Lise Meitner escape Berlin.

Coster then turned around and, in late 1942 or early 1943 (the dates are uncertain), asked the most powerful German citizen he knew for help on Goudsmit's behalf—Werner Heisenberg. After all, Coster reasoned, weren't Goudsmit and Heisenberg friends? Hadn't Heisenberg dined at the Goudsmits' home? Ironically, then, the man Goudsmit had been plotting to kidnap was now, unbeknownst to him, the best hope to save his parents.

Meanwhile, in March 1943, Isaac and Marianne finally got a letter through to their son. Seeing their names on the envelope must have brought a surge of hope to Goudsmit—followed by despair, when he noticed the postmark. He had no idea how long the letter had been delayed in transit, nor how exactly it had reached him—a merciful German soldier? The Dutch underground? It had been routed through Portugal, he saw, but the original postmark was from a city called Theresienstadt in the former Czechoslovakia. It stood right next to a massive concentration camp.

CHAPTER 23

Operation Freshman

Winston Churchill once called *heavy water* a "sinister term, eerie, unnatural, which began to creep into our secret papers" in 1940 and 1941. Perhaps in deference to this squeamishness, British officials began referring to D_2O as "Juice" instead. And as spies confirmed that the Germans were buying up every ounce of Juice they could find, intelligence agents began arguing over what to do about it.

Some advocated doing nothing, citing the same game-theory considerations that had recommended against kidnapping Werner Heisenberg. Yes, the Allies could no doubt disrupt German access to heavy water, perhaps by damaging the Vemork plant in Norway. But that would tip the Nazis off to Allied interest in this eerie, unnatural liquid. The Nazis would then conclude that the Allies were also working on a bomb, which would spur them to work even harder on theirs.

A dissenting group argued that action was imperative, especially given the Nazis' head start. They wanted to bomb Vemork immediately. Norwegian agents, however, recommended against this approach. The mountainous terrain and unpredictable weather near Vemork would make bombing runs tricky. The cells that produced heavy water were located in the basement of the plant anyway, which made them difficult to destroy with bombs. And Vemork did more than produce heavy water: it supplied power and fertilizer for the region, and both of those operations would suffer far more damage in any raid.

In the end, the British settled on a third, riskier option: sending a team of commandos into Norway to infiltrate Vemork and sabotage the heavy-water cells. Responsibility for this mission fell to an agency called the Special Operations Executive, known informally as the Ministry of Ungentlemanly Warfare. Their job, as one historian put it, was to "singe Hitler's mustache" by carrying out sabotage throughout Europe, and nothing would burn the Führer more than taking out Vemork—"the first attempt in history to strike a blow against atomic weapons."

To disable Vemork, the Ungentlemen first had to understand how heavy-water production worked there. It started on the fifth floor of the hydrolysis building, where local river water was pumped in through pipes and collected in vats. An electric current then zapped the H_2O and split off the hydrogen for use in fertilizer. Heavy water was a side effect of this process. Molecules of heavy water break down less easily than molecules of regular water, so if you zap a tank of water long enough, the leftover liquid contains a higher percentage of heavy water than before. You can now run this slightly enriched water through another zapping tank and enrich it further. After six such stages, running from the fifth floor to the first, Vemork could increase the concentration of D_2O from 0.000002 percent to 13.5 percent.

To raise the percentage higher still, Vemork employed three more stages of filtration cells. These cells were engineering marvels, a warren of pipes, condensers, seals, valves, and asbestos diaphragms, with all sorts of feedback loops to goose production even more. They were complicated as heck and incredibly wasteful: for every 100,000 gallons of river water that Vemork consumed, just one gallon of heavy water dribbled out at the far end. But that gallon was 99.5% pure reactor Juice.

The question for the Ungentlemen was how to sabotage this process. It turned out that the head engineer at Vemork—who knew what the Nazis used heavy water for—had already been sabotaging things himself, by pouring castor oil into the production lines. This caused the water to foam up and disrupted its flow through the filtration cells, leading to backups and delays. The engineer had helped build the plant years earlier, and it broke his heart to foul up his own machinery, but he felt he had no choice. The British cheered the engineer's cleverness—imagine stopping an atomic bomb with veg oil!—but he couldn't disrupt things often or he risked being caught and shot. Such measures were only temporary anyway: after a little cleaning, production started up again. The British needed something more permanent, and violent. The last stage of processing involved filtering the Juice through eighteen ultraprecision cells in the basement of the plant. The British therefore decided to send commandos in to blow those cells to smithereens. They called the mission Operation Freshman.

With that plan in mind the British began selecting saboteurs. A few Norwegian expats advising the British government suggested sending in a crew of eight soldiers who'd grown up near Vemork; they were expert skiers and knew the terrain well. The British overruled them. Norwegian troops had no experience with sabotage or clandestine warfare, and on a precision mission like this, familiarity with explosives trumped familiarity with terrain. Rather than skiing in, the commandos could approach on foot instead, or on special folding bicycles. Eight soldiers seemed too few anyway, leaving no margin for error. The British thought thirty sounded more like the thing, and they drew them exclusively from the ranks of their own Ungentlemanly troops.

The thirty commandos trained on an estate outside of Cambridge. To mimic the approach to the plant, they practiced walking over tall grass in snowshoes, that being the closest approximation to snow they could find. (To conceal the reason for all this tramping about, their supervisors spread a story about training for an endurance competition against the Yanks, to win the coveted, and fictitious, Washington Cup.) To practice the attack itself, carpenters constructed an inch-by-inch replica of the plant's basement using microfilmed blue-prints smuggled out of Vemork in toothpaste tubes. The model was accurate down to the lock on each door, which the commandos learned to pick with the help of a master thief furloughed from a local prison. The soldiers drilled endlessly on navigating the mock-up's passages and laying the explosives just so, until they could complete the task in pitch blackness. Someone then came up with a felicitous final touch. After the commandos had destroyed the cells and escaped Vemork, the Nazis would of course pursue them. So at an easily discoverable spot the raiders would drop a map (drawn on a silk pocket handkerchief) with Vemork circled in blue and a false escape route plotted to western Norway. Meanwhile, the troops would be hightailing it a hundred miles east, to neutral Sweden.

As a concession to the Norwegians, the British recruited four Norse soldiers as an advanced reconnaissance team. These four, all drawn from central Norway, would parachute in a month ahead of time to prepare a landing strip, scout Vemork's defenses, and escort the British commandos there for the attack. This prep work was essential because the area around Vemork was rugged beyond belief, a 3,500-square-mile plateau populated mostly by reindeer and swept by gusts of wind strong enough to knock grown men on their keisters. One historian described it as "one of the most godforsaken places on earth—the largest, loneliest mountain range in northern Europe." Locals said the cold descended so quickly at night sometimes that flames froze in place. Even the plateau's name, Hardanger, looked scary to English speakers, a seeming portmanteau of *hard* and *danger*.

For maximum secrecy, parachute drops took place only at night, and for maximum visibility, night drops took place only around full moons. By the mid-October moon, the advance team of four Norwegians—code-named Grouse—was ready to deploy. In cooperation with British intelligence, the BBC altered its usual broadcast on October 18. Instead of saying, "This is the news from London," the announcer began, "This is the *latest* news from London." This one word alerted undercover agents near Vemork that the commandos would make the drop that night.

The Grouse team had to parachute down onto a scary, unforgiving landscape studded with boulders and crisscrossed by gorges you could disappear into and never be heard from again. The terrain compounded the danger, since British pilots were used to navigating by rivers and roads, whereas Hardanger offered only a confusing mess of black-and-white "tiger stripes" of rock and snow, all of which looked identical. Sure enough, although all four Grouse men landed safely, they were dumped thirty miles from their rendezvous point, with their equipment strewn all over kingdom come. They wasted two days tracking it all down, and as soon as they had, the Hardanger plateau clobbered them with a blizzard.

After that cleared, they began dragging their hundreds of pounds of food and equipment toward Vemork, a job made slightly easier when one of them discovered an old sled half buried in the snow. (Miraculously, he recognized it as his old childhood sled; two years earlier he'd loaned it to Norwegian resistance fighters, who'd apparently abandoned it.) But even the sled didn't help much: the landscape was uneven, with snow reaching past the soldiers' knees, and the wind and elevation made breathing difficult. After fifteen days of toil they finally arrived in the vicinity of Vemork, and broke into a log cabin to rest. Many such cabins dotted the plateau, but they were mainly for summer use, with thin walls and cracks that the snow whistled right through.

After scouting Vemork, the Grouse team tried establishing radio contact with London, a job that, given the weather, proved virtually impossible. Their batteries had frozen solid, and their antenna kept snapping in the wind; they had to jerry-rig something from ski poles and abandoned fishing rods. When they did finally reach London, they had nothing but bad news to report. Since taking over Vemork, the Germans had planted land mines, ringed the place with barbed wire, and emplaced several machine guns. Casualties seemed likely.

Nevertheless, plans went forward. On the afternoon of November 19, the Grouse radio operator sent a brief weather report indicating clear skies. London responded with the seeming non sequitur "Girl." The operator's heart no doubt began racing; Operation Freshman would begin that night.

Rather than parachute thirty troops onto boulder-strewn terrain and risk broken ankles, the British opted to deploy them with gliders, an idea they stole from the Nazis. Because gliders lacked engines, they couldn't take off or fly on their own. A prop plane would tow them to the drop site instead using a 350-foot rope that was severed in midair. At this point the sixty-five-foot gliders would coast to the ground in silence, which masked their approach. But if you're picturing a calm, controlled descent, think again. The gliders were made of plywood and weren't exactly renowned for aerodynamic stability: to prevent people from slipping, the fuselage had a corrugated-iron floor specially designed to drain off vomit. And while German gliders had deployed troops with great success on the soft plains of Belgium, the hard danger of the Hardanger plateau offered far more opportunities for things to go awry.

Throughout November 19, the Grouse team sent periodic reports to London on the local wind speed and visibility, but by the time the British decoded these messages, fickle clouds and cold fronts had remixed the weather there. Feeling antsy, the British gave the go-ahead anyway, and the first plane-and-glider combo lifted off from northwest Scotland at 6:45 p.m., with the second following fifteen minutes later. Almost immediately winds began to buffet them, and by the time they reached the coast of Norway a storm was threatening to erupt. One observer later described the weather as "sticky with cloud."

Down on the ground, the Grouse team prepared the landing site, arranging several red-beamed flashlights in an L shape and firing up a radio beacon (called Eureka) for the planes to home in on. Not long afterward they heard the drone of motors through the clouds. After a cruel month alone on the plateau, their spirits leapt — "Here they come!" one cried. "Up with the lights!" They began blinking the red beams on and off, their eyes straining for a glimpse of the plane above.

Then the drone faded — the plane was sailing past them. The men frowned, but weren't overly worried. Perhaps the pilot just wanted to take a different approach; a second plane should be coming along anyway. Sure enough, the drone returned a few minutes later, and they resumed blinking the flashlights — only to hear the noise melt away again. When this happened a third time, the Grouse boys swallowed hard.

Finally, after more near misses, the noise faded one last time and never returned. At midnight the Grouse team trudged back to their flimsy cabin and radioed London, desperate to know what had happened. They soon regretted asking.

Up in the air, the pilots were lost. The receivers for the Eureka beacon had failed in both planes, and even when the sheets of clouds broke up a bit, giving them a chance to scan for the red L, they saw nothing but tiger stripes below. Twenty minutes passed, then an hour, then two, as the planes and gliders circled and recircled the plateau.

Eventually the pilots risked running out of fuel and had to turn back — at which point everything went to hell. Ice had been accumulating all night on the towlines holding each glider, and the ropes began to strain under the excess weight. One finally snapped as they neared the Norwegian coast, and the towing crew could only whip their heads around in horror as the glider began plummeting. It torqued into a spin at 200 miles an hour and disappeared into the clouds. The plane circled and searched to no avail, and when it finally returned in England at 3 a.m., the crew reported the glider lost at sea.

The towline of the second plane snapped as well, this time over land. Hoping to fix the glider's position, this crew began circling the area, straining for a glimpse. Unfortunately, the pilot pushed things too hard in the heavy cloud cover and clipped the side of a mountain. The plane went tumbling down the slope and exploded after a few rolls, killing everyone aboard.

Its glider, meanwhile, crash-landed. Of the seventeen people aboard, three died instantly and six sustained critical injuries. After clearing everyone out of the wreckage, the commanding officer sent two men who could still walk to a local farmhouse to determine their whereabouts; perhaps they could still make a break for Sweden. Alas, the farmer told the soldiers they were four hundred miles from the border. He then added that he would have to alert the local Nazi command to their presence: the Germans were bound to hear about the crash, and he'd be shot as a traitor if he stayed silent. The soldiers returned to the crash site and informed their officer. He thought his men could probably fight off the first Nazi patrol, but more Germans would inevitably follow, and several troops needed medical care anyway. So when the Nazis arrived, the survivors surrendered. While they were being driven away in trucks, one captive playfully flashed a V-sign to the Norwegians who had gathered to watch.

What happened next came out only years later, during a trial for wartime atrocities, and because of contradictory testimony the truth remains somewhat obscure. Apparently the Nazi military commander of Norway had his men question the British about their mission, but they refused to say anything more than their name, rank, and serial number. The commander now faced a dilemma. In October 1942, Hitler had issued a secret *Führerbefehl* (Führer command) to shoot all foreign saboteurs on sight; the Germans privately called them "pirates." And given the explosives discovered in the glider, these men were clearly saboteurs. But they were also wearing army uniforms. This technically made them prisoners of war, and the commander later claimed that every fiber of his being resisted the idea of executing them.

After wrestling with his conscience, the commander called up the field marshal of the Nazi armed forces, Wilhelm Keitel. (In a bizarre coincidence, Keitel was the uncle of Robert Oppenheimer's wife Kitty, making him related to Oppenheimer by marriage.) Must we shoot them? the commander asked. Keitel gave a vague answer but didn't contradict the Führer's orders. Still unable to decide, the commander played Pontius Pilate, washing his hands of the matter and turning the commandos over to the Gestapo, who had fewer scruples about men in uniform.

After briefly imprisoning the fourteen men, the Gestapo rounded them up and herded them into a mountain valley. There, guards grabbed one man and frog-marched him toward a concrete ammunition shed. As soon as he was out of sight, the valley exploded with gunfire. The remaining thirteen men, still assuming that their uniforms would protect them, looked around in bewilderment. *What are they shooting at?* The guards returned for another man, and a minute later gunshots rocked the valley again. All fourteen were eventually marched off to the shed, where a firing squad awaited. Upon arrival, each commando was ordered to remove his army jacket, to preserve the fiction that the Germans weren't shooting prisoners of war. Some refused and were shot anyway. One held up pictures of his wife and children, begging for his life; the reply was a barrage of bullets. One man was too injured to stand, so they propped him against a rock before taking aim. After each man fell, the commanding officer walked up and shot him in the head. All fourteen bodies were then tossed into a ditch.

When word of the atrocity reached Berlin, officials were furious. Not because it violated every international standard of humane warfare dating back several centuries. No, they bemoaned the lost intelligence. *(Interrogate first,* then *shoot, you imbeciles.)* Now the Gestapo would never know the saboteurs' target.

Luckily for the Gestapo, it soon got a second chance.

Despite what the crew returning to England had said, the other glider hadn't crashed into the sea. It landed short, splitting in half and killing six men on impact. Four others were critically injured, with fractured skulls and shattered limbs; two of them ended up frozen to the ground by their own blood. Here too the commanding officer judged the situation hopeless and surrendered to local Nazi patrols.

At least the men executed by firing squad had died quickly. Not so with this group. The German officer in charge—known as the Red Devil—hustled away the four critically injured soldiers and ordered a doctor to snuff them out with morphine injections. One died straightaway, but when the other three realized what was happening—the doctor had lied and told them they were receiving booster shots for typhus—they tried to resist. Big mistake. This enraged the Red Devil, who strangled one soldier with a belt. The doctor, wanting to be merciful, tried injecting air into the veins of a second, to kill him with an embolism. It didn't work, so the Red Devil smashed the man's Adam's apple with his boot. The last British soldier was pushed down a flight of stairs and shot in the back. The Germans then dumped all four bodies into the sea.

The five uninjured soldiers ended up in Oslo for interrogation. These men also refused to say much, so the Nazis bound their hands in barbed wire, blindfolded them, and ordered them into a truck, supposedly for transport to a POW camp. After driving several hours, the truck stopped and the blindfolded soldiers stumbled out, their wrists no doubt shredded. Suddenly someone yelled "Achtung!" A blast of rifle fire, and all five men slumped to the ground. They were discovered after the war in a pit, still wearing their uniforms.

The only consolation the British could take from Operation Freshman was that none of the commandos had revealed the mission's goal. But in the end, all their bravery in staying silent went for naught. In digging through the wreckage of one glider, the Nazis discovered a map on a silk handkerchief. The commandos had intended it as a red herring, to fool the Nazis into thinking they'd flee west instead of east. But the Germans ignored the supposed escape route and zeroed in on the site circled in blue—the Vemork heavy-water plant. They now knew the saboteurs' target.

Freshman was a disaster on every level. Thirty top British commandos died. The map betrayed their target to the Germans, who then reinforced the defenses at Vemork (more barbed wire, more land mines, more guns). Worst of all, the Allies had tipped their hand about their interest in nuclear fission. They might as well have sent Hitler a postcard about Los Alamos.

Still, if Freshman left most people in despair, it steeled others. The Norwegian expats who'd been advising British intelligence had always hated Freshman: it was unwieldy, and involved troops unfamiliar with the terrain. (Who the hell ever dreamed that folding bicycles would work there?) Within a day of hearing about the disaster, two Norwegian officers began sketching out an even more audacious plan, called Operation Gunnerside. They were confident that they knew better than the British how to cripple Vemork.

CHAPTER 24

The Italian Navigator

After the lovelorn Walther Bothe bungled his graphite experiments, the Nazi atomic bomb project focused on using heavy water to study nuclear reactions. The American project, meanwhile, had always considered graphite the superior option, and their confidence in this carbon compound paid off handsomely in December 1942.

The Americans called their first nuclear reactor a "pile," and for once this wasn't an empty code name. It really was just a huge pile of graphite, 771,000 pounds of slippery black bricks stacked on the floor of an unused squash court at the University of Chicago. It took technicians seventeen days of around-the-clock labor to arrange them all, and the end result was an egg-shaped mound twenty-five feet wide and two stories tall. Although the court was unheated, the men building the structure worked up a good sweat, and many stripped off their shirts to stay cool—to their immediate regret. With all the graphite dust in the air, "We found out how coal miners feel," remembered one. "We looked as if we were made up for a minstrel show. One shower would remove only the surface graphite dust. About a half hour after the first shower the dust in the pores in your skin would start oozing."

The outermost bricks in the pile were solid hunks of graphite, but most bricks toward the center had holes drilled into them, which were filled with five-pound slugs of uranium. Each of the 18,000 slugs was capable of firing off neutrons in any direction. These neutrons slowed down when they entered the surrounding graphite, and the reduced speed allowed fission to take place when another slug of uranium absorbed them a fraction of a second later. You can think about the slugs as nuclear raisins suspended in a loaf of graphite, with each little bit of fruit contributing to the overall chain reaction.

Once the graphite and uranium were stacked in place, the pile would have gone critical by itself if not for a third component, cadmium. Cadmium is a particle sponge, gobbling up neutrons by the billions. So during the pile's construction, workers occasionally left horizontal shafts between the graphite bricks, into which they slid thin wooden beams with cadmium slabs affixed to them. These were padlocked into place to prevent the pile from going critical prematurely.

Everything finally came together on December 2, 1942, one of those Chicago days when it's so cold it hurts to breathe. The United States had been at war for a year at that point, largely in North Africa and on islands like Guadalcanal in the South Pacific. That very morning, the State Department announced that two million Jews had already died in Europe.

Forty-two people packed into the squash court that afternoon wearing hats, scarves, and gloves; most stood on a balcony overlooking the pile, but several technicians puttered about on the floor below. One of the men down there held an axe. They called him the SCRAM (safety control-rod axe man), and his job was to prevent this first nuclear chain reaction in history from deteriorating into the first nuclear meltdown in history. If the atomic pile got out of hand, he'd swing the axe down executioner-style and chop a nearby rope. This would release a Damoclean rod of cadmium hanging from the ceiling, which would plunge down a hole into the heart of the pile and kill the nuclear reaction. Should the SCRAM fail, another trio of scientists—jokingly called the "suicide squad"—stood ready on the balcony with buckets of cadmium-laced water, which they planned to hurl like slop onto the pile as a last resort. Both safety measures were crude, but then again, so was everything about the pile.

The critical experiments started around 2 p.m. Enrico Fermi, the Italian physicist who'd once sprinted through the halls of his institute in Rome, was in charge. He ordered the padlocked cadmium rods removed one by one while a radiation detector clicked away nearby. At each step Fermi made a few calculations with his slide rule and checked the readings on some instruments. (He'd been reading *Winnie-the-Pooh* to boost his language skills, and named each instrument after a different character: Pooh, Roo, Tigger.) Ultimately, he was trying to determine what's called the "k factor." The k factor governs whether a chain reaction fizzles or goes critical. Each time a uranium atom fissions, it releases neutrons. Sometimes those neutrons fly off and cause more fissions, sometimes not. The k factor measures how often on average they succeed. So if it takes four uranium fissions to produce one additional split, then $k = \frac{1}{4}$, or 0.25. If it takes just two fissions to produce an additional split, then $k = \frac{1}{2}$, or 0.5. The key threshold is $k = 1$. After this point each split produces more than one new fission, and the reaction becomes self-sustaining.

With all but one cadmium rod removed, k had inched to within striking distance of 1. But rather than yank the last rod out and vault to criticality, Fermi milked things a bit. He ordered it withdrawn six inches. Then he checked Roo and Tigger, did a little calculation, and nodded. Getting warm. The rod was withdrawn six more inches, and Fermi did another calculation. Really close now—try six more. All the while, the axe man and the suicide squad stood at their posts, their fingers itchy. The radioactivity detector was meanwhile click-click-clicking more frantically with each step, eventually buzzing so fast that, by some accounts, it sounded like "a burring drill." Fermi continued to calculate. Six more inches, please.

Finally, at 3:53 p.m., Fermi looked up and grinned. K has reached 1.0006, he announced. It must have been an odd moment. The pile didn't shudder or change visibly; it certainly didn't start glowing or anything. Yet the graphite in Fermi's pencil had just proved that the graphite in Fermi's pile had crossed a new threshold in history—they were standing next to the world's first nuclear reactor. A bomb was still a long way off: among other things, bombs require either enriched uranium-235 or plutonium. But Fermi had just narrowed the gap considerably.

There was no point in going further that day, so his assistants slotted the cadmium rods back in and let everything cool down. To celebrate, the assembled scientists broke out some Chianti and sipped from paper cups. The bottle had the traditional straw jacket on it, so they passed it around and everyone signed it.

Soon afterward the top Manhattan Project official in Chicago phoned another top official in Washington. "Jim, you'll be interested to know," he said, "that the Italian navigator has just landed in the New World."

The other official perked up when he heard the code. "Is that so? Were the natives friendly?"

"Everyone landed safe and happy."

The success of Fermi's pile kicked the Manhattan Project into high gear. Now that self-sustaining chain reactions were possible, General Leslie Groves could start building industrial plants to enrich uranium in earnest. With a working pile, scientists could start breeding plutonium as well. Both developments gave Manhattan Project officials confidence.

On the other hand, Fermi's success also revved up the fear of a Nazi nuclear bomb. Because if a motley crew of Americans could build a reactor, surely geniuses like Heisenberg and Hahn already had. In fact, within days of December 2, one Manhattan Project physicist—a future Nobel Prize winner, no dummy—grabbed some chalk in his office and "proved" on a blackboard that, given the rate of American progress, the Nazis would have the Bomb no later than December 1944, probably sooner. No one had any reason to doubt him.

PART IV

1943

CHAPTER 25

Secret Messages

British physicists had been trying to evacuate Niels Bohr from Denmark for years, partly for his own protection and partly to exploit his genius for the Allied war effort. To this end, a member of the Danish underground visited him at his home in Copenhagen in January 1943 with an offer to smuggle him out.

Wary of a trap, Bohr refused to consider the offer without written confirmation from the British. Perhaps a little elaborately, the agent relayed Bohr's request by writing a letter, shrinking it onto microfilm, and concealing copies beneath the stamps of several postcards, which he mailed to a second agent in Stockholm. After peeling the microfilm off the stamps, this agent locked himself in a room and used a projector to magnify the message, which he then dispatched to the British.

As if trying to one-up the Danes, the British responded in even more elaborate fashion. In early February the Copenhagen agent received a phone call from a contact at a bank saying that someone had left a set of old iron keys there. After retrieving them, he clamped one of the keys—marked 229—into a vise and began filing it at a specific point. A sixth of an inch down, a tiny cavity appeared. Inside were three half-millimeter-square specks of microfilm. At this point he tied a kerchief around his nose and mouth to avoid accidentally sneezing or coughing and sending the specks flying. He then floated them out with water and read them with a 600-magnification microscope borrowed from a doctor friend. (All three specks held the same message, but some words inevitably got blurred during the shrinking process, and only by comparing all three versions could he make them out.) He passed the message to Bohr, then had Bohr bury the keys in his garden.

The message came from James Chadwick, the nuclear physicist who'd discovered the neutron years before. He knew Bohr well and begged him to come to England. "I have in mind a particular problem in which your assistance would be of the greatest help," he wrote. (Charles Darwin, a physicist and the grandson of the naturalist, also solicited Bohr's help.) Given Chadwick's interest in neutrons, Bohr deduced that Britain had started an atomic bomb program.

But this news didn't stir Bohr as deeply as you might expect. Bohr was the scientist who'd first discovered, back in 1939, that uranium-235 was the key isotope for fission: it alone drove nuclear chain reactions, not the vastly more abundant uranium-238. Cut off as he was in Denmark, he knew nothing of recent advances in uranium enrichment and therefore dismissed the idea of enrichment as fantastical. To make a single bomb, he famously declared, you'd have to turn your entire country into a factory. Consequently, the offer to work in England seemed futile: "I have to the best of my judgment convinced myself," he wrote in his typically tortured syntax, "that in spite of all future prospects any immediate use of the latest marvelous discoveries of atomic physics is impracticable." This response, naturally, was shrunk onto microfilm, wrapped in metallic foil, hidden in a false tooth in a courier's dentures, and smuggled to England.

The British persisted in pleading with Bohr, and Bohr later passed along some rumors about German scientists buying up uranium and heavy water. But the British simply couldn't dislodge the Great Dane from Copenhagen. Beyond the impracticable physics of nuclear bombs, Bohr also felt obligated to stay in Denmark and protect his fellow citizens. As with Heisenberg and Joliot, he felt his country needed him, and thanks in part to Bohr's influence—and his underground resistance work—Denmark did enjoy relative freedom compared to other conquered countries in Europe.

In reality, however, Denmark owed most of its precious freedom not to resistance fighters like Bohr but to other factors. For one thing, the starving German populace needed the country's beef, cheese, and butter. For another, the Danes had a kind of—if one can even say this—Nazi guardian angel in Berlin. This savior was Ernest von Weizsäcker, the second-ranking officer in the Reich foreign ministry and the father of the patrician physicist Carl von Weizsäcker. Weizsäcker père certainly participated in despicable deeds, but he had more humanity in him than most Nazis and had fond memories of Copenhagen from his days as a diplomat there. He therefore defied Hitler's orders to round up and deport Danish Jews. Out of deference to his son, Weizsäcker also protected both Bohr and his physics institute from harassment.

Alas for Bohr, the elder Weizsäcker had grander ambitions than saving Denmark. After months of scheming, Ernest accepted an offer in March 1943 to become the Nazi ambassador to the Vatican. On paper this was a curious decision. Given his exalted position in Berlin, moving to the tiny papal state was a de facto demotion. It also played right into the hands of Weizsäcker's boss, the odious Joachim von Ribbentrop. Like many top Nazis, Ribbentrop preferred to surround himself with toadies, and he happily bid adieu to the talented, humane Weizsäcker. But Ribbentrop never suspected his underling's real motivation for leaving. Weizsäcker had grown disillusioned with Hitlerism, and as the Wehrmacht got bogged down in North Africa and Russia, he wanted to end the war quickly and with honor. Although puny in size, the Vatican had worldwide reach and outsized moral standing. (As a matter of fact, in one of the most head-scratching incidents of the war, Pope Pius XII broadcast a speech in February 1943 warning humanity about the menace of nuclear bombs. No one has ever quite figured out how he learned about this top-secret research so early.) As ambassador to the Vatican, Weizsäcker hoped to open secret negotiations between the Holy See and Italy, to convince the Italians to lay down arms. This would deprive the Reich of its strongest ally in Europe, at which point, he calculated, the Reich would quit the war as well.

All of this was treason, obviously. Had Ribbentrop caught even a whiff of it, he would have had Weizsäcker arrested and executed. But Weizsäcker thought the gamble worth taking and moved to Rome in April 1943. This would leave Denmark vulnerable, but ending a world war was a higher priority. Besides, the political situation in Denmark seemed stable. He saw no reason to worry about the fate of Niels Bohr.

CHAPTER 26

Operation Gunnerside

The Grouse team was indignant. Despite the Operation Freshman fiasco, the British Ministry of Ungentlemanly Warfare still wanted to take out the heavy-water plant at Vemork and needed the quartet's help. A few days after the failed mission, they radioed the crew with orders to sit tight and make do, despite the approaching winter. "How? With what?" one commando remembered thinking. "We had no food left. We had nothing."

You couldn't say the weather on the plateau never changed—sometimes the blizzards blew in from the north, sometimes they blew in from the south—but it was always harsh. With no dry wood for fires, the Grouse team had to bundle themselves up in everything they owned: long underwear, stocking hats, sweaters, parkas, windproof pants, multiple pairs of socks and gloves. They still felt like frostbitten icicles, and they spent most nights shivering inside their sleeping bags. And as bad as the cold was, the hunger was worse. Aside from handfuls of oatmeal and other goods they had nothing left to eat, and they began breaking into nearby cabins to scrounge whatever they could—moldy turnips, cod-liver oil, fermented fish, even dog food. Finally they got so desperate for meat that one man risked stealing a sheep from a local rancher. He then made a succulent stew, letting it steep for hours. The very smell of it cratered their bellies. But in pulling the kettle off the fire, the weakened cook slipped on a reindeer pelt and spilled everything, flooding the dirty cabin floor. The men stared at their feet for a second—then dropped to all fours and started scooping and slurping. Halfway through the "meal," one man joked about finding a hair in his soup, which in their delirium they found hilarious.

A few weeks later they'd even run out of humor. "What was the point of our suffering here in the mountains?" one soldier thought. "Was there any chance that [the sabotage mission] might succeed and that we might escape alive?" Confined to their cabin most hours, they were beyond stir-crazy by then, and the irritations of daily life—little arguments about whose turn it was to cook breakfast or take out the trash—metastasized into full-blown rows. The only fresh game for miles was reindeer, and on the rare occasions they caught one, they devoured it down to the marrow, learning to relish even the eyeballs and lips (which tasted uncannily like chestnuts). They also wolfed down the half-digested lichen in the reindeers' bellies, called *gørr*, their only source of vitamin C. All the while the bloody Brits kept telling them to sit tight. Operation Gunnerside was coming.

Gunnerside would follow the same basic plot as Freshman. Commandos would storm the Vemork power plant and destroy the heavy-water cells in the basement. But instead of thirty commandos, Gunnerside would employ a leaner, stealthier force of ten—the four Grouse men, plus six more who'd parachute in later. Their commander would be a tall, twenty-three-year-old explosives expert named Joachim Rønneberg, who looked thin to the point of being tubercular. And unlike the Freshman team, all ten soldiers would be Hardanger locals who knew the terrain and were expert skiers.

Apparently not superstitious, Rønneberg's crew of six trained on the same wooden mockup of Vemork as the doomed Freshman commandos had. Because the British remained skeptical of Norwegian troops, the six had to work doubly hard to prove themselves, memorizing every entrance to the basement—doors, windows, utility shafts—and drilling relentlessly on placing explosives in the dark.

They used half-pound plugs of Nobel 808, a burnt-orange compound that smelled like almonds; they got a feel for its strength by blowing practice holes in brick walls.

There was one risk they couldn't mitigate, however, no matter how hard they drilled—the risk of reprisals. The Nazis discouraged native resistance in Norway by going biblical on Allied collaborators, punishing even their innocent friends and neighbors. In one coastal village west of Vemork, for instance, the local German commander had caught several people with bags of British baking flour the previous April; he suspected (correctly) that they'd received them in exchange for helping to sabotage some nearby sites. His investigation began with a Gestapo agent infiltrating the village disguised as a Bible salesman. He found enough evidence to justify a raid a few days later, during which the Nazis discovered more sabotage equipment, as well as caches of weapons. Having nothing to lose now (so they thought), a few underground fighters opened fire and killed a Nazi soldier. In response, the German commander burned down every house in the village and executed eighteen men. His rage unslaked, he then sunk every boat in the harbor, killed every domesticated animal, and shipped every last surviving inhabitant—from toddlers to toothless grandmas—to a concentration camp. *Sic semper* collaborators: he essentially erased the town from the map.

In using local Norwegian soldiers, then, Operation Gunnerside was gambling with the lives of every plant worker and villager near Vemork. But once again, fear of an atomic Reich trampled every other consideration.

The Gunnerside raid was originally scheduled for Christmas Eve 1942—an unexpected and therefore ideal night to strike. Instead of using gliders, the six additional troops would parachute in a few days beforehand. Alas, foul weather foiled the drop, and the Grouse quartet had to wait until the next full moon period. Making the best of things, they snagged a reindeer for Christmas dinner and feasted on fried tongue and liver soup; for dessert they had brains seasoned with blood. To make things extra jolly, they erected a stunted juniper branch for a Christmas tree and listened to carols on their makeshift radio, placing the receiver inside a pie tin to amplify the sound.

The next full moon fell on January 21. But, déjà vu, the British pilot got lost the night of the drop and spent two hours circling the rendezvous point, unable to spot it. The Gunnerside crew in the plane begged for permission to jump anyway—we can "sniff our way to the dance floor," they assured him—but the pilot refused. And before the commandos could talk him around, sparks exploded beneath the wings. The plane was taking flak, and the pilot had to abort the drop and race back to Scotland. The British radioed an apology to the Grouse team, but it was cold comfort: they now faced another whole month on the plateau.

The Gunnerside six were no less frustrated. Rønneberg and company didn't really grasp what heavy water was, and the phrase *nuclear fission* meant nothing to them. But if so many good men had already died on the Freshman mission—and if their superiors still wanted to attack Vemork, despite the increased danger—well, that told them enough.

Thankfully, everything came together during the February 17 drop. The weather was clear, and for once a British pilot found the rendezvous point. A green light flashed in the cargo hold, and the Gunnerside commandos began tossing their equipment out before anyone could change his mind. Seconds later they took the plunge themselves.

Naturally, a blizzard kicked up as soon as they landed on the plateau, an epic five-day whiteout. The six men were barely able to find a cabin to break into, and they hunkered down in their special rabbit-fur underwear to wait out the storm. At one point the winds grew so fierce that they feared the cabin would lift off the ground, Dorothy Gale–like, and go tumbling through the air. When the storm ended they poked their heads out—and immediately realized that the pilot hadn't found the rendezvous point after all. They were supposed to land above the tree line, the elevation at which trees stop growing. But even through the snow, they could see a few gnarled trunks. Scouring maps in the cabin, they determined that they were twenty miles northeast of the meeting point with the Grouse team. Time to start walking.

But before they got going, they spotted movement in the distance—a skier. Was he a friend, a quisling, a Nazi? They had no idea. A German patrol seemed most likely, though, so they ducked back inside the cabin, hoping he would pass by. No such luck. He was coming straight for them. Which meant that even if they hid, he was bound to notice their footprints. They had no choice but confront the lone wolf.

They waited until he was just steps from the door. Then they pounced, surrounding him in seconds. He wasn't dressed like a Nazi, but that didn't mean anything. "What are you doing in the mountains?" Rønneberg demanded.

He turned out to be an old poacher, skiing from village to village to sell reindeer meat. He seemed harmless—until some sharp questions exposed him as a Nazi sympathizer. He then shocked the soldiers by insulting the exiled Norwegian king and his family. "They've never done me any good. They can just stay where they are." Given this defiance, the poacher seemed likely to squeal, and some of the commandos voted to shoot him, arguing that he'd betray the mission. Rønneberg overruled them. The poacher knew the landscape intimately, he argued, including places to hide. Far better to make the man our guide.

It turned out to be an inspired decision. The Gunnerside crew had hundreds of pounds of equipment with them, and they faced an awful week of drudgery now, fighting headwinds at every step as they dragged their gear across the plateau. Indeed, it would have been a Sisyphean nightmare if not for the poacher. He had a real gift for finding gentle counters and skirting rough patches; Rønneberg thought his skiing "beautiful to watch." Even with all their equipment, they were soon making good progress.

Until a crisis hit two days later. Without warning, two black dots appeared on the horizon. More skiers—probably Nazi patrols this time. Had the old man led them into a trap? Cursing him, the Gunnerside group ducked behind some cover.

The skiers slowly grew larger as they approached. They eventually mounted a nearby hillock and began scanning the terrain with binoculars. When it became clear they weren't moving on soon, Rønneberg sent a scout down to get a read on them.

Creeping up to the hill, the scout finally got close enough to make out the two men's faces. Or at least parts of their faces: they had heavy beards, and their skin looked oddly yellow. Disconcertingly, their movements seemed careful, guarded—they moved like soldiers, not weekend skiers on a lark.

Feeling he had to do something, the scout removed his pistol and, bracing himself, coughed to get their attention. The two men whipped around, their own guns leveled. All three stared at each other for a second—then whooped in delight. The "Nazis" were actually members of the Grouse team, out searching for Gunnerside. They'd grown beards to keep warm, and their skin had yellowed from malnutrition.

After a joyful Grouse-Gunnerside reunion, Rønneberg decided to let the old poacher go, albeit with a threat. Rat us out, he warned, and we'll tell the Nazis you served as our guide. That will get you executed, too. All eight soldiers then watched him ski off. He still moved beautifully, but they feared that Rønneberg's mercy would come back to bite them.

After a few days of rest—and of feasting on raisins and chocolate for the Grouse team—the saboteurs got to work. The assault was scheduled for February 27, a Saturday. The night before, Rønneberg sent a scout to Vemork for the latest intelligence, because he still had one crucial decision to make—how to approach the plant. Vemork sat on a rock ledge overlooking a gorge, which limited their options. One possible approach was the road the workers came in on, which led to a bridge that spanned the gorge. This was the easiest approach, and taking out the two guards on the bridge seemed straightforward. But if the guards managed to sound the alarm or shout for help, the mission would be over. Killing guards would also damn the local villagers to reprisals. A second approach involved descending a set of stairs hacked into the cliff behind the rock ledge; these ran alongside the sluices that fed water into the plant. The steps were icy, however, and rigged with booby traps and mines. For these reasons Rønneberg favored a third approach, the only one that was unguarded—in part because it was unguardable. It required climbing down the sheer rock face of the six-hundred-foot gorge in front of the plant, then scaling the other side. A few commandos had dismissed this route as impossible, but reconnaissance photos showed plants growing on the rock face. And if plants can grip the walls, one of the commandos argued, so can human hands. Sure enough, the scout that night confirmed that the gorge walls were dotted with juniper bushes and shrubs. For Rønneberg that decided it—they would climb in and out. Whatever doom his comrades might have foreseen, they were good soldiers and fell in line.

After checking over their gear on Saturday afternoon, nine of the ten Norwegian soldiers (the tenth stayed behind to guard the hut) set off for Vemork at 8 p.m. They were discovered almost immediately. Four youths from a local village—two men, two women—had snuck off to some nearby cabins to do you-know-what, and they bumbled into the Gunnerside crew. Embarrassingly, two of the commandos knew one of the men. He was no spy, but they seized him by the collar and threatened him with all sorts of hellfire if he breathed a word about their presence. They then herded the lovers into a single hut and told them not to leave until noon the next day. The wide-eyed youths obeyed.

The soldiers skied most of the twenty miles in, wearing white camouflage suits and rucksacks. Along the way they dropped below the tree line and began tramping through a patchy forest. They could hear the waterfall that powered Vemork now, and as they came into a clearing they suddenly spotted—off in the distance, in the moonlight—the building that had dominated their thoughts for months. It must have been a majestic sight. As one historian said, "If a medieval king had searched all of Norway for an impregnable place to build a castle, he could not have found anything better than the shelf of rock on which the Norsk Hydro plant stands." In peacetime, the eight-story granite building would have been glowing like a glitzy hotel, every window ablaze. During the war, with the windows blacked out at night, it looked like a giant sarcophagus. The Gunnerside crew could nevertheless hear, even above the waterfall, the hum of machinery. Vemork had produced ten pounds of reactor Juice that very day.

Ducking back into the forest, the commandos found a switchback road winding down toward the gorge. It was tedious going, so they eventually decided to cut straight down the gradient instead. This was basically a controlled plummet through the snowy undergrowth: their backs were flat against the ground and their skis were splayed in front of them as brakes. Every few feet they loosed a mini-avalanche.

Halfway down, some of the party reached a stretch of road and stopped a moment to let their companions catch up. Seconds later, their world got much brighter. They swiveled, and saw two pairs of headlights on the road—two buses bringing workers in. They leapt for cover behind a snowbank, but could only watch in horror as their companions continued skidding down the slope.

The men on the slope had seen the headlights, too, and began grabbing at tree roots and rocks, flailing for anything to slow their awful momentum. A few did manage to stop, but two of them continued tumbling and almost crash-landed onto the roof of one bus.

Somehow, though, none of the bus passengers noticed. It was late and dark and everyone was exhausted. They continued to stare dully forward, their minds so consumed with work or war that even the sight of something amazing—two soldiers falling out of the sky—couldn't penetrate their gloom. The bus rumbled on and disappeared around a bend. After taking a moment to catch their breath and shake their heads, the Gunnerside nine continued descending.

At the bottom of the slope they buried their skis in an impromptu igloo for the return trip—if there was one. Then they removed the guns and grenades from their rucksacks and stripped off their white snowsuits to reveal khaki British uniforms beneath. They'd decided to dress as British soldiers to minimize the chance of reprisals against the plant workers and local villagers. This cover would crumble if the Germans captured and interrogated them, of course, but that seemed unlikely. For one thing, most of the Gunnerside crew didn't expect to live through the assault: just before leaving Great Britain they'd all written goodbye letters to their loved ones. And in case of capture, each man took from his rucksack, in addition to the guns and grenades, a rubber capsule filled with cyanide.

They reached the lip of the gorge at 10 p.m. As expected, two sentries were patrolling the suspension bridge, with fifteen more guards in a hut just beyond that. A machine gun glowered from the rooftop, and they knew from intelligence reports that three more patrols were likely wandering the grounds. But as Rønneberg suspected, no one was guarding the gorge itself. Grinning at their luck, they started down.

At the bottom of the gorge they found a swift stream and scouted around for an ice bridge, testing various spots with their toes. The only suitable one felt as slick as oil and was submerged in rushing water. However precarious, it held their weight, and all nine saboteurs inched across.

Now for the real challenge, the six-hundred-foot ascent. Water was trickling down the sides of the gorge, making the rock face slick, and their fingers had already gone numb in the cold. If not for the gnarled junipers, the climb would have been impossible. The bushes also saved one commando's life. He was pulling himself up to a small ledge when his grip faltered. He kicked and scrambled and managed not to fall— but found himself dangling by one hand two hundred feet off the ground. Calming himself as best he could, he began patting the cliff face around him with his free hand, feeling for a new grip. Nothing— just bare, wet rock. With no other choice, he began swinging side to side, groping a little farther each time. His fingertips finally brushed the edge of a bush; but he couldn't reach it while still holding on with his other hand. The wind began to kick up now, threatening to pry him loose from the rock, and he realized he had to act soon.

Fighting every instinct in his body, he began swinging again, really kicking his legs out. After one final lurch, he let go. For a moment he was suspended in air, holding nothing. Then he snagged the bush like a Nordic Tarzan. It gave some, bending like a branch—but finally held. After a few deep breaths he resumed climbing. Four hundred feet to go.

At 11 p.m. all nine men heaved themselves over the lip of the gorge, then crept forward and ducked behind a transformer to rest. The guard was changing soon, so they took a moment to eat some chocolate and crackers and relieve themselves. As all young soldiers do, they also cracked jokes and talked of women. There was no need to go over the mission again. Everyone knew his role: five men would provide cover, with a demolition team of four penetrating the plant.

After the guard change, they waited another half hour for the new sentries to get good and bored before moving forward. Wherever possible, they followed routes through the snow that Nazi patrols had already trod, stepping in their footprints to conceal their tracks. At the end of a railroad spur on the perimeter of the plant, they found an unguarded gate and cut the lock. At this point five men peeled off and trained their tommy guns on the barracks and guard tower to provide cover. The demolition team, including Rønneberg, snipped a hole in a fence and made for the heavy-water plant. The team then further fissioned into two pairs of two and split up to try different entrances.

Spies inside the plant had assured the British that the Nazis were lax about security, but the commandos found every last door bolted shut. The windows were all locked, too. Frustratingly, through a chip in the blackout paint over one window, Rønneberg could see the heavy-water cells in the basement: they looked like oversized fire extinguishers, shiny gray cylinders four feet tall. His heart pounding, he was powerfully tempted to smash the window and take his chances—they were so close! He restrained himself and tramped on.

As a last resort Rønneberg and his partner tried a utility duct. They reached it via a short ladder, and after clawing aside the snow packed into the entrance, they began belly-crawling forward, wriggling like worms amid the cables and wires. Unfortunately, the trailing commando's pistol came loose during this effort. It dropped, and clanged on the metal beneath them when it fell. They froze, bracing for an alarm. But the throb of machines masked the noise, and after a moment of rest, they slithered on.

Having memorized the plant's layout, they snaked around to a point near the heavy water room and dropped fifteen feet to the floor through a hatch, landing in a parachute roll. After months of preparation, they were now facing the final door. A sign there read, in German and Norwegian, NO ADMITTANCE EXCEPT ON BUSINESS. They pulled their guns, nodded, and burst in on the guard.

By all rights the Germans should have posted a top soldier here—an alpha killer, their Ajax or Achilles. After all, the entire Nazi nuclear bomb project depended on this room. As it was, they'd posted a frumpy Norwegian pensioner inside, who took a good three seconds to turn around and just about messed himself when he did, finding a pistol buried in his solar plexus. Realizing he was no threat, the commandos disarmed him, pushed him into a corner, and told him to shut up.

Simple enough instructions, yet the guard proved incapable of following them. As the saboteurs unpacked the almond-scented Nobel 808 and started kneading it into half-pound sausages, this pompous old goat—name of Gustav—had the audacity to begin bossing them around. There were eighteen fuel cells in the room, nine on each side, and the commandos planned to wrap a sausage around the base of each and daisy-chain them together with detonator cords. But every time they got near one, Gustav had a fit. Those are electrolysis cells, he called out. Very expensive. Be careful about shocks. Oh, and mind the lye in the corner there! You're liable to burn yourself. It was a scene of Shakespearean low comedy: the know-it-all dogsbody prattling on and on, oblivious to the danger he's in. The gravity of the situation finally struck Gustav when, after several seconds of squinting, he realized that the commandos were handling fuses and detonators. Good god, he squealed, be careful. You're liable to blow up all this equipment.

"That's pretty much our intention," deadpanned Rønneberg's partner.

That shut Gustav up. And now that the commandos could get a few words in, they enacted a little farce of their own. Still worried about reprisals, they wanted Gustav to believe they were British. So they began reciting a few lines they'd memorized, making heavy-handed references to merrie olde England. Incongruously, they held this dialogue in Norwegian, but Gustav looked impressed.

All the while, the duo continued to wire up the explosives. They'd just about reached the halfway point when they heard a crash behind them—and wheeled to see a shattered windowpane. Someone had punched through from outside.

The Gunnerside duo froze, braced for guards, grenades, the tip of a gun. Had Gustav secretly signaled someone? Had the old poacher betrayed them at last? Neither, it turned out. A second after the pane shattered, a familiar gloved hand reached through. The other two commandos, tired of plodding about in the snow, had said to hell with stealth and smashed the basement window.

It was a bold move—bold and idiotic. Rønneberg had to stop working to help the two inside, and as he knocked aside the shards in the frame, he sliced his right hand open. This rendered him useless for laying explosives, which was a problem considering that he was the explosives expert. All he could do now was twist fuses together with his other hand. Smashing the blacked-out window also allowed light to pour into the yard—a veritable beacon to any patrols. So another commando had to stand outside and block the hole with his arse, which rendered him useless as well.

And if all this chaos weren't enough, the commandos soon heard an *ahem* from offstage. Gustav again.

What now? Rønneberg growled.

My glasses, he said. I've misplaced them somewhere in the room. I must have them.

You can't be serious.

Well, I can't see without them.

Tough.

Please help me look. With the war on, I won't be able to get another pair if they're destroyed. I'll lose my job.

Rønneberg couldn't believe it. Here he was, an elite commando leading the most dangerous mission of his career, with the fate of the Nazi atomic bomb in the balance—and now he had to drop everything and search for an old man's spectacles.

"Where the hell did you put them?" he snapped. Gustav motioned toward the desk. Rønneberg found his glasses case there, tossed it to Gustav in the corner, and returned to the fuses.

Only to hear, a moment later, another *ahem*. Rønneberg turned, incredulous, to find Gustav staring down into the empty shell of the case. "The glasses aren't in there."

"Where the hell are they, then?"

"They were there when you came in," Gustav whimpered, as if this was all Rønneberg's fault. Laying his work aside again, the commando began ransacking the desk, all but flinging things aside. He finally found the glasses folded between the pages of a ledger and handed them over. Now shut up.

Thankfully, the other commandos had nearly finished wiring the explosives by this point and could turn their attention to escaping. Having relieved Gustav of his keys, they left the heavy-water room for the hallway outside and unlocked a door that led into the yard; they also dragged Gustav into the hallway so he could run for safety once they were gone. About all that remained to do was light the fuse.

That's when they heard footsteps.

Everyone froze. They listened harder, and sure enough, they heard the click of shoes on a staircase adjacent to the hallway. The commandos swore and took cover, pistols drawn. A moment later another guard appeared, one far younger than Gustav. Luckily, he proved no more quick-witted, and he yelped in surprise as the commandos surrounded him. A few questions revealed him as a fellow Norwegian, someone with no desire to risk his life alerting the Nazis. So the commandos shoved him next to Gustav and returned their attention to the explosives.

At last, at 1:13 a.m., everything was ready. Rønneberg reached into his pocket, removed a few patches from a British parachute squadron, and tossed them on the ground—one last diversionary tactic to prevent reprisals. He then struck a match and lit the main fuse. It would burn for thirty seconds. Cutting things mighty close, Rønneberg then released Gustav and the young guard to run upstairs and lie down. Keep your mouths open until you hear the blast, he advised. Otherwise, the sudden pressure change will explode your eardrums. Only after the guards had scampered off did the commandos themselves turn and sprint.

From the outside, the noise of the explosion was disappointingly meek — "all bass and no treble," as one historian remarked. This left some of the Gunnerside team fretting, convinced that the explosives had failed. They needn't have worried. The thick walls of Vemork had simply muffled the pyrotechnics; inside the basement, the Nobel 808 had done its job beautifully, splintering the fuel cells and spraying heavy water everywhere. Shrapnel from the cells also pierced the pipes along the ceiling, flooding the room with regular water and washing the equivalent of 770 pounds of pure Juice down the drain.

Even the Nazis who showed up later to assess the damage had to admire the precision of the work. They'd initially taken a hundred hostages from the local population to exact revenge, but upon realizing that no mere amateurs could have pulled this off, all hundred were freed. The official German report on the incident stated that there was "not extensive damage, except at the place where the attempt was made, and there the devastation was total."

The nine commandos left the way they came. They hadn't fired a single shot, and they politely shut the open railroad gate behind them. Then they careened down the gorge, recrossed the ice bridge, and scrambled up the other side, hoping to make a quick escape. Unfortunately Rønneberg's hand, still bleeding, marked an easy trail of crimson for the Nazis to follow in the snow. Worse, Vemork's sirens began screaming not long after they ascended the gorge. The commandos looked back to see several search parties setting out, their flashlight beams swinging like scythes in the dark. After grabbing their skis from the igloo, the crew took off along a nearby road, ducking out of sight whenever a vehicle full of guards roared past.

This being Hardanger, snow began to fall as soon as they reached the top of another switchback road. But for once they welcomed a good old-fashioned blizzard. The snow covered their tracks, covered even the blood, and while it made for hard going, it bogged down the search parties far more than it did these veteran skiers. A few hours later they reached their cabin unmolested. They managed to stay awake just long enough to drink a toast of whiskey. Afterward they collapsed for eighteen hours of sleep, then rose the next day and scattered to the ever-present wind.

CHAPTER 27

Consolations of Philosophy

Werner Heisenberg had a frustrating winter in 1942–43. He was still building reactors, still doing fission research, but somehow the pathetic, striving Kurt Diebner had vaulted past him. Heisenberg had always favored a certain geometry for his reactors, with alternating layers of heavy water and uranium; sometimes the layers were circular shells, more often sandwichlike slabs. In either case, the arrangement wasn't ideal. After a uranium atom fissions, the neutrons that emerge can fly off in any direction, but not all directions were equally good. Imagine a neutron flying straight up or down from the surface of a flat slab. It would quickly encounter heavy water and slow down, thereby helping the chain reaction along. But if the neutron flew out *sideways,* it would stay within the uranium layer and encounter only other uranium atoms—meaning it would never slow down. These sideways neutrons were wasted.

Diebner sidestepped this problem with an innovative "lattice" design: instead of slabs or shells, he broke up the uranium into tiny cubes and suspended them in a tank of heavy water. This expanded the geometry, since the flying neutrons could now find material to slow them down in all three dimensions. (Enrico Fermi's reactor pile in Chicago, with small plugs of uranium embedded in graphite, used the same idea.) Setting up Diebner's experiments took weeks of

207

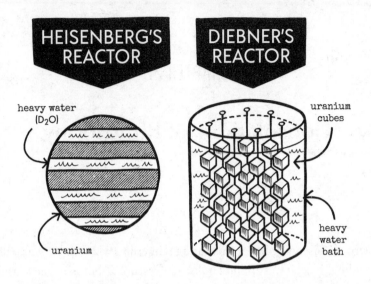

drudgery, and he had to coat every last uranium cube in lacquer to prevent it from mixing with heavy water and catching fire, but the huge spike in neutron production—110 percent, compared with Heisenberg's 13 percent—proved the superiority of his design. And the really vexing thing was that Diebner did all this with Heisenberg's table scraps. Heisenberg always got first dibs on heavy water and uranium, while Diebner made do with leftovers. The military hack nevertheless beat the pants off the Nobel laureate.

Adding to Heisenberg's frustration was the fact that Nazi leaders wouldn't leave him alone to work. As an elite scientist he often had to attend silly state functions, and on March 1, 1943, he and Otto Hahn and other scientists were summoned to a particularly asinine lecture in Berlin on the health effects of bombing raids.

Now, a naïve person might think that bombing raids had only negative health effects, but according to the lecturer that night, one

Dr. Schardin, there was a definite silver lining to the billowing clouds of burning smoke. Sure, you might be crushed under twenty tons of rubble during a raid or torn to bits by shrapnel, but if you put those complications aside and focused only on concussive forces, then bombs were actually a pleasant way to die, he said. They caused such a spike in air pressure that your brain essentially liquefied: all the blood vessels burst and a massive stroke ensued, killing you before you realized it—all in all, a painless, almost serene death. The take-away was that there's really nothing to fear from a bombing raid.

Excitingly, just as Schardin was wrapping up, the Berlin air-raid sirens began to wail. Here was a chance to put theory into practice! Heisenberg, Hahn, and the rest of the audience herded into the base-ment to wait the attack out. Few of them had experienced a raid before, and this was one hell of a baptism. Bombs poured down for what seemed like hours, each one detonating with a tectonic boom. The walls around them heaved and quaked, and every so often they heard a building collapse aboveground. Soon the power went out, and medics barged in with a bloodied woman moaning in pain. During the heaviest blitz, two bombs went off nearly simultaneously above them, and the double whammy of pressure left the audience dazed. Hahn and his wicked tongue piped up: "I bet Schardin doesn't believe in his own theories right now." For one moment they all laughed.

After the all-clear sounded, the scientists clambered out of the shelter and confronted the scene around them. The neighborhood had been gutted, a mess of twisted steel and shattered walls. Because Allied bombs often included incendiaries like phosphorus, pools of eerily flaming liquid had collected in the streets. Everyone present scattered, rushing off to see if their homes or labs were still standing.

Heisenberg had more reason than most to worry. He'd brought his twins, Wolfgang and Maria, now five years old, from Leipzig to Ber-lin that week to celebrate their maternal grandfather's birthday; he never imagined he'd be exposing them to bombs and incendiaries. And his initial hope upon leaving the shelter—that the destruction was confined to central Berlin—proved false: he could see fires lighting the streets for miles in every direction, burning deep into the suburbs. The city had suspended all buses and trains that night, so he and a colleague immediately took off walking.

Despite—or perhaps because of—his anxiety, Heisenberg spent most of the ninety-minute walk chattering about abstract topics. Like a modern Marcus Aurelius, he'd taken consolation in philosophy during the tumult of the war. Whenever he and his family could escape to their mountain cabin in southern Germany, the Eagle's Nest, he would pull out a treatise he'd been working on and add his latest revelations. He called it *The Order of Reality.* In the grand tradition of Teutonic philosophy, there was plenty of metaphysical bloviating about the nature of the cosmos, but it also included sections about ethics, especially the responsibility of intellectuals in unjust societies. He concluded that individuals are essentially pawns, buffeted by grand historical forces, and that beyond being decent to friends and loved ones, they can't affect society much. These unnamed individuals should therefore retreat into the higher realms of art and science, and should moreover be absolved of all moral responsibility for their intellectual work, even if that work ends up being exploited by unjust rulers. It doesn't take a Ph.D. in psychology to see that Heisenberg was wrestling with his own conscience here, most probably his research into uranium fission. Conveniently, he absolved himself of wrongdoing, and he considered the 161-page treatise such a profound contribution to human thought that he modestly gave copies to friends as Christmas gifts.

That night in Berlin, walking with his colleague, Heisenberg expanded on another favorite theme of his, the state of German science and its role in Germany's future. Today the conversation reads like a surreal Platonic dialogue, pitting the pessimistic colleague, who believed Germany doomed, against the chipper Heisenberg, who stressed his optimism about a scientifically inspired future. All the while they were weaving around bomb craters and piles of flaming rubble—the direct result of recent scientific and technological advances. At one point Heisenberg got so caught up in his rhapsody that he stepped into a pool of flaming phosphorus and had to hotfoot over to a puddle to douse his shoes.

Heisenberg also offered his thoughts on the character of the German race, whom he described as dreamers, people too inclined to believe in old myths about heroes (or Führers) rising up and leading them to greatness. But, he assured his colleague, such illusions have finally been shattered. Never again will the German *volk* indulge in silly fantasizing, never again will they neglect or ignore truth. They will learn to think scientifically, he insisted, and ground themselves in the reality confronting them.

Just as he was saying this Heisenberg splashed into another puddle of phosphorus, and his shoe caught fire a second time. This intrusion of reality didn't stop the discussion, though, as Heisenberg cheerfully doused his foot again. The colleague finally suggested, with some hesitation, that perhaps "it wouldn't be a bad thing if we simply bothered about the facts that stare us in the face." Heisenberg, his shoes no longer smoldering, couldn't have agreed more.

Shortly after this the two men wished each other luck and parted ways for separate suburbs. At which point reality once again intruded on Heisenberg's night. For as he approached the home of his wife's parents, where he and his children were staying, he realized that it was on fire. He ran up and found all the doors and shutters blown in—a bad sign. He kicked his way inside and sprinted up to the attic. There, he found his aged mother-in-law gamely trying to put out the fire; like a World War I trench gunner she was wearing a steel helmet to protect her noggin from falling debris. Heisenberg persuaded her to give up and hurried her outside. She assured him that his wife and children had escaped unharmed, albeit barely; they were staying at a neighbor's house near the local botanical gardens.

With his family safe, Heisenberg ran to the house next door, also on fire. The roof had caved in and the staircase inside had collapsed, so he began scaling the exterior walls like Spider-Mensch, working his way up to the attic. He found an old man there absentmindedly tossing handfuls of water onto the inferno. The water did no good, of course—it hissed and evaporated instantly—but the man was numb with shock and not in his right mind. He was likely to die, in fact; he kept backing into what Heisenberg called "an ever-diminishing circle" amid the flames. Only the appearance of a soot-smeared Heisenberg—who leapt through a "wall of flame" to reach him—startled the man back to reality. In a touchingly formal gesture, he bowed to his savior. Heisenberg then saved his life by helping him down the outside walls.

Overall, that night showed Heisenberg at his best and worst: obtuse and brave, a hero and a fool. He was so troubled by the moral implications of uranium research that he constructed a wholly new (and wholly bogus) philosophical system simply to absolve himself. And yet he never hesitated when confronted with one of the toughest moral dilemmas a human being can face, barging into a fire and risking his life to save a stranger. That night revealed why so many people admired Heisenberg, and why they found him so, so frustrating.

In the months that followed, the Allied bombing raids over Germany intensified, often lasting hours without relief. Berlin, Leipzig—nowhere seemed safe, especially in the north of the country. Fearing for their lives, Heisenberg permanently moved his family to the Eagle's Nest cabin in southern Germany in the spring of 1943, and good thing, too. The family protested the move—the cabin was cold and remote and uncomfortable—but by year's end their home in Leipzig had been leveled by bombs. Had Heisenberg not insisted, they probably all would have perished there.

Meanwhile, German scientific officials were also looking to evacuate. A few months after the Berlin raid, all Reich scientists received orders to begin scouting new locations for their labs. Heisenberg and the Uranium Club focused on a few sleepy villages in the Swabian Alps near the Black Forest in southern Germany. One town in particular, Haigerloch, seemed ideal for fission research. It was obscure and remote, and was situated near a cliff, which made it hard for bombers to attack. As a bonus, the cliff had a sturdy cave carved into its base. It was the perfect place for Heisenberg to fire up a new and much more powerful Uranium Machine—and show Kurt Diebner once and for all who the real genius was.

CHAPTER 28

"The Fun Will Start"

In March 1943, one of the most fateful conversations of the war took place between two imprisoned German generals, one a congenial cynic and the other a bitter optimist. The congenial cynic was Wilhelm von Thoma, who looked more like a prisoner of war than a general: gaunt and hungry-eyed, with ears sticking out from his head. In the fall of 1942 he'd been commanding a tank division in Egypt, and after a sound defeat at the hands of the British, he prepared to retreat. But his commanding officer, Erwin Rommel, received a telegram from Hitler ordering the tank corps to stay put. Either win an incredible victory, the Führer demanded, or die in glory. From a military perspective this was lunacy, and Rommel knew it. He nevertheless obeyed the order out of loyalty to the regime. In contrast von Thoma denounced the order as "unparalleled madness," and two days later, on November 4, he put on a clean uniform, climbed into his tank, and rolled toward enemy fire. After deliberately taking two hits, he crawled out the top and stood calmly amid the flaming husks of Panzers on the battlefield, waiting to be taken prisoner.

He ended up at Trent Park, a luxurious manor a few dozen miles north of London; it dated to the days of Henry IV. When the owner died on the eve of World War II, the British government converted it into a cushy POW camp for senior Nazi officers, which was probably

214

better than they deserved. The estate featured marble statues, huge cedars and oaks, and ponds with wild ducks, and although the lawns and courtyards were ringed with barbed wire, the prisoners had leave to stroll around them. Inside, the house had hot and cold running water and walls decorated with prints of German art. The common area included a radio console and a painting studio; one prisoner made such a nice copy of the famous Rembrandt-school painting *Man in the Golden Helmet* that the British hung it in the dining room. There was even a film room and a tobacco and beer shop. In short, these Nazi warmongers lived vastly more comfortable lives than most British folk at the time. The main complaint among the Germans, poor fellows, was that the food was "too abundant."

By war's end sixty-three generals were ensconced at Trent Park, but von Thoma was just the second to arrive. He therefore made friendly with the first general there—the bitter optimist, Ludwig Crüwell—even though the two men normally would have had nothing to do with each other. Crüwell looked like a stereotypical Prussian general: hard eyes, hard mouth, cruel stare, but he had a complicated relationship with the Nazis. During a military purge in June 1934 party goons had tried to murder him, and he'd despised the regime ever since. As a loyal soldier, however, he always did his duty, and he prosecuted Hitler's war without hesitation. (If nothing else, Crüwell had several young children, and feared that a German defeat would ruin their future prospects.) Despite his reservations, the war began brilliantly for him, as troops under his command captured Belgrade in 1941; observers compared him to Hannibal for his tactical brilliance. After a quick promotion, he was transferred to North Africa in August of that year.

His meteoric rise ended abruptly the following May. One afternoon he flew what should have been a routine reconnaissance mission to inspect the position of some Italian troops. Italian officers on the ground were supposed to light flares to indicate the location of the front line, so Crüwell's plane didn't overshoot it. In a typical Italian blunder—the Germans were constantly cursing their incompetence— the officer in charge started yakking on the telephone just before Crüwell arrived and forgot the flares. Crüwell's plane sailed right past him and was shot down a few minutes later. He was taken prisoner and ended up in Trent Park.

Crüwell did not adjust well to life there, losing twenty-five pounds in three months. He also abused himself in the hope of securing a medical discharge to Germany, taking cold showers to induce chills and clawing his legs with his fingernails until he had open sores. The British saw through these tricks and refused to release him. Crüwell sank into a sullen depression, and only the arrival of von Thoma, a fellow general, loosened him up.

The two spent many hours chatting together, despite markedly different temperaments. Crüwell was terse and unsmiling, and always maintained strict discipline; von Thoma rambled for hours and disregarded military regulations as frivolous in prison. The two men did share a contempt for Nazi leaders: Goebbels's "beer-house tirades" disgusted them, and they deplored Hermann Göring—with his palaces full of looted art and his gold belt buckles "as big as an octavo volume"—as a greedy fop. They also swapped rumors about Hitler's fits of madness (he supposedly fell to the ground sometimes and snapped like a dog) and mocked his conduct of the war. But while both agreed that things were dire at the moment, they differed in their long-term forecasts. Before his capture Crüwell had experienced only German victories, and he remained optimistic. Indeed, he refused to even entertain the thought of defeat, lest he go mad with anguish. Von Thoma, meanwhile, cheerily declared the war lost already and delighted in tormenting Crüwell about it.

Hearing top German officers disparage Hitler was certainly heartening for the British. But the fifty-one-year-old generals also provided vital military intelligence in a conversation on March 22, 1943. During a drawn-out debate about the future of Germany, von Thoma mentioned his surprise that London was still standing; bombs should have flattened it by now. Crüwell asked what he meant, and von Thoma relayed some gossip he'd heard before his capture about revolutionary new rockets being tested at Peenemünde, on the Baltic coast—the V-weapons. "They've got huge things which they've brought up there," he explained. "They go fifteen kilometers [nine miles] into the stratosphere" and rain down "frightful" terror. Something had obviously delayed their deployment, but London no doubt had a nasty surprise in store. He then quoted a senior official at Peenemünde: "Wait until next year and the fun will start!"

Unbeknownst to von Thoma and Crüwell, the British were recording this conversation. As soon as the government took over Trent Park in 1939, intelligence agents had walled off a secret room in the house—dubbed the "M room," for mic'd—and outfitted it with recording equipment that captured conversations in seven-minute stretches on gramophone discs. They then snaked microphones into the light fixtures in the common area and other places where the Germans would likely gather. Recording conversations between prisoners without their consent violated the Geneva conventions, but the British brushed aside such concerns.

Ironically, most German officers at Trent Park realized that the British were probably trying to eavesdrop on them: in private they admonished each other over and over about the danger of loose lips and careless chatter. *You never know what the enemy is up to.* But in the very next breath they'd start bragging about all the things they'd withheld from the British during official interrogations, as well as a dozen other things the Limeys would probably love to know. Indeed, to help new prisoners adjust, von Thoma often held group therapy sessions in which he encouraged them to open up about their war experience; as one historian commented, "No stool pigeon could have done it better." British agents had a riot listening to all this, and sometimes provided newspapers with fake stories to goad the prisoners into speaking more.

Trent Park was ruinously expensive to run, but when the generals were fat and happy, they let their guard down, and the information the British gleaned was invaluable. An impromptu discussion between Crüwell and another officer on submarine tactics, for instance, provided a vital edge for the Allies on D-Day and ultimately consigned hundreds of U-boats to Davy Jones's locker. But the most important intelligence concerned V-weapons.

British officials had been debating for months about how seriously to take Peenemünde. Some declared the threat overblown—a farrago of rumor and circumstantial evidence. Yet a few lucky breaks argued otherwise. In late 1942 a Danish chemist dining at a restaurant in Berlin happened to be seated next to two engineers from Peenemünde, who got roaring drunk and began blabbing about the latest rocket tests; the Dane swiftly reported this to Allied intelligence. Then, a photographer on a reconnaissance mission found himself with a half-used roll of film as his plane passed over the screaming head of Peenemünde. He started snapping, and happened to capture a cloud of exhaust at the moment of launch. Follow-up work revealed a power station nearby, as well as railroad spurs. All this was still ambiguous, but von Thoma's chilling report about the imminent "fun" clinched the case.

Incidentally, von Thoma was correct in guessing that something had delayed the production of V-weapons. Several things, actually. The components used to launch and steer the rockets required extensive testing, and their construction involved something like eighty different chemical elements and alloys, many of them rare and hard to procure. Then a series of fires during some test launches in February 1943 hampered progress further. Given von Thoma's cynicism, he also would have chuckled over a third reason. In early March of that year, Hitler had a vision while sleeping one night, and the gods apparently revealed unto him that the V-2 missile program would fail; a few days later he suspended the entire project. This left the head of rocket production at Peenemünde shaking with fury: not only do we have to battle wartime shortages and technical challenges, he fumed, but also "the dreams of our Supreme War Lord."

Albert Speer, the one sane Nazi in Hitler's circle, talked some sense into the Führer over the next few weeks, and just after von Thoma and Crüwell's conversation, work at Peenemünde resumed. Slowly but steadily, then, the V-weapons were getting close to deployment. And unslowly and unsteadily, the Allies began to panic.

CHAPTER 29

Seeing Red

After running spy missions in Baja Mexico, newly promoted Lieutenant Colonel Boris Pash took over West Coast security for the Manhattan Project in late 1942, and he spent the next several months investigating security breaches throughout the western United States. Little did he realize that the biggest threat the project faced was right under his nose, in Northern California.

Although the United States and Soviet Union had formed an alliance to fight Germany in 1941, the Soviets had no compunctions about spying on American laboratories, especially the famed Radiation Lab at the University of California at Berkeley. The Rad Lab was a vital component of the Manhattan Project because it had the world's most powerful cyclotrons, which were essential for enriching uranium. So when the Federal Bureau of Investigation turned up evidence that Soviet spies had infiltrated the Rad Lab, officials were horrified. Especially because every last spy seemed connected with Robert Oppenheimer, who'd worked at the Rad Lab before taking over at Los Alamos.

The FBI first drew a bead on Oppenheimer after (illegally) wiretapping the home of a prominent communist in the Bay Area named Steve Nelson. During the most serious incident, which took place at 1:30 a.m. in late March 1943, a student of Oppenheimer's revealed to Nelson the results of several secret fission experiments. From him and other contacts Nelson eventually pieced together detailed knowledge about research at Oak Ridge, Tennessee, and Los Alamos. Worse, Nelson was then observed passing information to Soviet agents. During the conversation with the student, Oppenheimer's name came up several times. To be sure, Nelson explicitly denied Oppenheimer's involvement in espionage; he in fact complained about the physicist's resistance to spying and called him "jittery." But the context also made it clear that Nelson knew Oppenheimer personally, which in the slippery world of FBI intelligence cast further doubt on Oppenheimer's loyalty. The bureau began trawling through Oppenheimer's phone records and posting agents in trench coats and fedoras outside the Berkeley Faculty Club to monitor him (very subtle). It also prepared transcripts of Nelson's most damning conversations and forwarded them to Boris Pash.

To say that Pash was alarmed is an understatement. "Even by the standards of military security officers," one historian noted, "[Pash] was passionately and belligerently anticommunist," and this was the worst possible news he could imagine. The Reds had penetrated the most secret project of the war—and in his backyard, on his watch. Within twenty-four hours of receiving the transcripts, Pash flew to Washington to brief General Leslie Groves in person. Groves then threw his considerable bulk around and forced the FBI to turn over its files on Oppenheimer and channel all intelligence to Pash in the future.

Rushing back to the Bay Area, Pash rented a house in Oakland as a command center, just south of the Berkeley campus, and opened a three-pronged attack on communism at the Rad Lab. First, his team installed $6,000 worth of spy equipment in the house, including a telephone switchboard to monitor all the wiretaps they established at people's homes, especially Oppenheimer's. Second, they flooded Berkeley and Los Alamos with security agents—aptly nicknamed "creeps"—to follow suspects around and monitor their comings and goings. Finally, they placed spies of their own at the Rad Lab. These included two "bodyguards" for Oppenheimer, to drive him around and protect him from supposed assassins. Their real job was to report everything Oppenheimer did to Pash.

As these tactics suggest, Pash's office wasn't afraid to trample legal niceties. Pash himself had once proposed "shanghaiing" the most flagrant of Oppenheimer's student-spies and taking them out to international waters, where pesky U.S. laws didn't apply. There, he vowed to interrogate them "in the Russian manner" to extract their secrets. It's not clear if he was joking, but the threat proved unnecessary. Between the wiretaps and the tails, Pash and the FBI caught several Rad Lab students (as well as scientists at other sites) meeting with Soviet agents. As for punishing them, government officials had to get creative. Because the wiretaps were illegal, they couldn't prosecute the scientists in court; they resorted to drafting them and shipping them off to remote military sites instead. One scientist was banished to a base above the Arctic Circle in Alaska, the American gulag, where he spent the balance of the war folding long underwear.

Because he'd insisted on hiring Oppenheimer, Groves took special interest in the Rad Lab case—a fact that caused security problems of its own. While visiting the Oakland spy house with Pash once, Groves arrived in full uniform. Problem was, this was a nice, quiet residential area, and people would surely wonder why a general had showed up. Realizing his error, the tall, pudgy Groves grabbed the bantamweight Pash's raincoat from the back seat of the car and stuffed himself inside. He then sprinted for the house with all the grace of a buffalo. Pash hooted over the incident for years, but he wasn't much smoother in the field. One day while reconnoitering the Berkeley campus, Pash bumped into a former student of his from Hollywood High. "Coach, what are you doing here?" the boy asked. After a few seconds of sputtering, Pash turned and ran.

Although he busted several of Oppenheimer's students, Pash never caught Oppenheimer himself doing anything suspicious—in part because Oppenheimer realized he was being watched and took countermeasures. With his driver-bodyguards, for instance, he would lower his voice and roll down the windows in the back seat whenever he wanted to talk with someone, creating a protective bubble of noise. Still, the stress of being under constant surveillance got to Oppenheimer. Already a waif, he dropped from 130 pounds to a skeletal 110; friends remember, bizarrely, that he could slip his frame inside a toddler's high chair and sit down in it. As a precaution he also stopped speaking with his brother, Frank, a dues-paying member of the Communist Party. Things got so bad that, just months after arriving at Los Alamos, he confessed to a friend that he wanted to quit the Manhattan Project altogether.

To relieve the pressure on himself, Oppenheimer decided to submit to an interview with one of Pash's colleagues in August 1943. His thinking seems to have been this: *They know I ran with a dubious crowd in my past. So to prove my loyalty, I'll submit to an interview and explain everything—maybe even give up the names of some subversives. That will win their trust, and they'll stop harassing me.* The plan backfired spectacularly, with one moment during the interview proving especially damning. A British engineer living in Berkeley had been sniffing around for technical secrets to hand over to the Soviets. The engineer thought Oppenheimer would play ball, so he asked a mutual friend of theirs—a professor of French at UC-Berkeley—to sound Oppenheimer out at a dinner party. This had happened six months earlier, and Oppenheimer assured Pash's assistant that he'd refused to help.

To Oppenheimer, the point of the story was the British engineer—he wanted Pash's people to turn the glare of their scrutiny onto him. He apparently didn't grasp that, in relating this tale, he was also implicating his friend, the professor of French, in treason. When asked for the professor's name, Oppenheimer deflected the question, but when Pash heard about the incident later that afternoon, he demanded his own meeting with Oppenheimer. Oppenheimer reluctantly agreed, and the two antagonists met face-to-face on the Berkeley campus the next day, in a room that Pash had hastily wired with hidden microphones.

Even Oppenheimer's defenders would concede that he had an arrogant streak, and he considered Pash his intellectual inferior, which was no doubt true. That didn't mean Pash was a dunce. He didn't interrogate the physicist in the Russian manner, but he did ask several pointed questions. He also deftly turned the focus of the interview away from the British engineer (whom Pash knew about anyway) and onto Oppenheimer himself. Why had he waited so many months to come forward with this story? And if he was so loyal and so concerned with security, why wouldn't he name the go-between, the professor who'd approached him?

Cornered, Oppenheimer tried to talk his way out of the jam and only entangled himself further. At one point he admitted that he would "feel friendly" about sharing atomic secrets with Russia, just not through backdoor channels. Pash was aghast. Oppenheimer then started speculating—completely unprompted—about what the British engineer would do with any top-secret documents he obtained. Probably shrink them onto microfilm, Oppenheimer guessed, and deliver them to the local consulate. Which left Pash wondering how on earth a supposedly innocent scientist would know the Soviet protocol for delivering documents to Moscow.

If Oppenheimer had hoped to reassure Pash, he miscalculated badly. He later called his testimony that day a "tissue of lies" and a "cock-and-bull story," and in refusing to answer the thorniest questions, he looked shifty and evasive. Always a little histrionic, Oppenheimer ended the interview by saying, "I would be perfectly willing to be shot if I had done anything wrong." Pash was unmoved. Earlier in the interview he'd compared himself to a bloodhound, and he let Oppenheimer know that he didn't intend to give up the scent now that he had it.

Pash spent the next few months keeping this promise. "We never let Oppenheimer out of our sight," he said. "We knew his every step. Every letter was read, every telephone call was overheard, every contact checked and studied." Once, when Oppenheimer left a suitcase in his car while eating at a restaurant, two creeps broke into the sedan and wrenched the bag open. They found a fifth of gin, a bottle of twenty-seven-year-old brandy, underwear, batteries, and diarrhea medication—but alas, no spy documents. In another solid legal move, agents took sworn testimony against Oppenheimer from a twelve-year-old boy.

For his part, Oppenheimer continued trying to prove his loyalty. In a third interview, he named several more supposed communists for security officials—students of his, a secretary, a close friend and his wife. He also suggested that the Manhattan Project plant undercover agents in a local scientific union to further spy on his colleagues. All this might come as a disappointment to those who, based on the harassment he suffered in the 1950s, consider Oppenheimer a martyr to McCarthyism. But in these early days, when he was desperate to make his reputation, he comes off as a real fink sometimes, willing to betray almost anyone.

Pash's fanaticism served him no better. In tracking Oppenheimer and other suspects, he often adopted extreme measures, like stopping a passenger train to hunt for subversives, which infuriated railroad officials. He also had a bad habit of stealing classified documents from the offices of his superiors, which he'd then return the next day with a smirk, to teach them a lesson about security. They were not amused. Indeed, Pash pissed off pretty much every senior officer he worked for that year, and by late 1943 several of them were looking for a way to get him out of their hair—overseas if possible, and preferably in the line of fire. It just so happened that Groves had an assignment that met those qualifications.

CHAPTER 30

Beautiful Peenemünde

In the second half of 1943 the British began seeing signs that, as the German generals had promised, the "fun" with the V-weapons was about to start. Common sense held that the Nazis would launch the rockets from northern France, the closest point of attack to London. And in a sickening noncoincidence, reconnaissance planes caught the Germans hard at work building massive infrastructure there. In some cases they erected what looked like huge skis—launch rails to guide rockets to their targets. In other cases they were excavating trenches the size of canyons, to build gigantic bunkers. Even scarier, an escaped French prisoner who turned up at the American embassy in Switzerland in June 1943 claimed to have escorted a cask of heavy water to Peenemünde recently. He didn't know what it was for, but given that heavy water had no conceivable use in rocketry, only nuclear research, connecting the dots wasn't hard. The Nazis were clearly developing atomic rockets at Peenemünde, then building the apparatus in northern France to hurl them at London.

Still, not everyone reached such dire conclusions. A powerful faction within British intelligence thought the whole rocket danger overblown, and they dismissed Peenemünde as a factory for making airplane bombs, or even harmless sledge pumps. Between Nazi tanks and fighter planes, they argued, we have enough to worry about without inventing new threats. Pretty soon a civil war broke out within British intelligence, and things got so heated that Winston Churchill finally had to arbitrate during a late-night meeting on June 29. It took place in a subterranean chamber deep beneath the government offices at Whitehall. There, behind a thick green door with an observation slit, stood the Cabinet War Room, a dark windowless space with a U-shaped table covered in blue cloth.

The antagonists that night were two of Churchill's confidants, Duncan Sandys and Frederick Lindemann, Lord Cherwell. The German-born Cherwell was an accomplished Oxford physicist with a nasty temperament: he reportedly despised Jews, women, and Africans, and instead channeled all his affection toward Churchill, whom he practically fell in love with. He served as the prime minister's personal scientific advisor, and in that capacity had been downplaying the threat of Peenemünde. Opposed to him was Sandys, who was both an intelligence officer and Churchill's son-in-law, having married Diana Churchill eight years earlier. Sandys had nearly lost a foot early in the war after his driver fell asleep at the wheel one night and crashed. He endured several grueling months of rehab, and Churchill took pity on him and appointed him head of a committee investigating Peenemünde. This sign of favor gave Cherwell fits, and from that moment forward he despised his young rival.

Sandys presented his case first that night, recapping all the evidence his team had gathered: prisoner testimony, spy reports, the conversation between von Thoma and Crüwell, reconnaissance photos of rockets at Peenemünde, pictures of the concrete bunkers being built in northern France. He capped his argument with a chilling statistic. Given their apparent size, the V-rockets could kill or wound an estimated four thousand people per strike; at just one rocket per hour, two million British citizens could suffer casualties in the first month alone.

After the prosecution rested, Cherwell began the defense. He wanted to humiliate Sandys, but had to be cagey about undermining

Churchill's son-in-law. So that night he claimed he merely wanted to play the "avocatus diaboli," the devil's advocate, and raise a few points about Sandys's evidence. First of all, prisoners and spies were notoriously unreliable, he said. We shouldn't trust them too far. As for the reconnaissance photos, they were taken from such great heights that the details were ambiguous at best. Where Sandys saw "rockets," he saw only fuzzy white blots—they might be anything. Plus, if the Germans really were testing huge, explosive rockets on the Baltic coast, then British agents in Sweden—just ninety miles away—would surely have heard something by now. Finally, in a fit of pride, he insisted that the Germans couldn't have developed such sophisticated rockets yet because, well, British engineers hadn't developed anything similar, and the Krauts surely couldn't outpace our boys. Cherwell concluded that the tube-shaped blots in the photos were probably not rockets at all but "aerial torpedoes" to launch from airplanes, quite conventional. Or perhaps the whole site was a hoax, intended to distract the Allies from the real danger elsewhere.

Churchill mulled all this over. He understood precious little science, and he had just watched his two closest technical advisors make diametric cases. Luckily, Churchill had a third vote to help him decide, courtesy of R. V. Jones. Originally a protégée of Cherwell, the thirty-one-year-old Jones now worked in intelligence with Sandys and could therefore provide a balanced view. Jones had already impressed Churchill with his acumen on other matters, and at this point the prime minister turned to the young man and shook his finger. "Now, Dr. Jones," he boomed, "may we hear the truth!"

Jones gathered himself—this was the biggest presentation of his career. He was especially nervous because he was about to eviscerate his mentor Cherwell's case. He started by pointing out that, however unreliable *individual* spies or prisoners might be, in this case they'd provided too much corroborating detail to dismiss everything as hearsay. And if the construction sites or rocket-testing grounds were a hoax, they seemed a singularly stupid one. Why pour millions of

Reichsmarks and literal tons of precious concrete into dummy bunkers and fake launch pads? The sheer scale of the sites argued for their authenticity. He also rebutted the argument about the tubes in the photographs being aerial torpedoes: given their apparent size and mass, no known plane could have lifted them.

At this point Churchill yelled "Stop!" and turned to Cherwell with a malicious grin. "Hear that?" he said. "That's a *weighty* point against you." Churchill also delighted in the irony that Cherwell had discovered Jones in the first place: "Remember," he chortled, "it was you who introduced him to me."

Cherwell tried to rally, but to no avail. Churchill was convinced, and from that day forward the German rocket program at Peenemünde became an official threat to the Allies. What they learned over the next six weeks only heightened the danger.

By the middle of the war, Paul Rosbaud—the publisher in Berlin who'd taken the nom de guerre of the Griffin—had perfected a sort of petty sabotage against the Nazis. He plastered letters with excessive stamps to waste paper, and exacerbated wartime shortages of copper by hoarding small change and removing copper fixtures from the bathrooms of trains and flinging them out the window. But aside from such pranks, the Griffin did his most valuable work gathering intelligence. Although the British had ignored his 1941 report about V-weapons, he had never stopped pursuing leads about Peenemünde, and his persistence paid off in August 1943.

The Griffin's usual method for extracting secrets was to get people drunk over dinner, and this time his mark was the nuclear physicist Pascual Jordan. Jordan had once been considered the equal of Werner Heisenberg and probably would have won a Nobel Prize had he not been such a committed Nazi. The great tragedy of his life was that he stuttered, and not mildly. He stuttered constantly, in every conversation, sometimes butchering words so badly that people averted their eyes in mortification—which only made him stutter more. Rosbaud sensed, in fact, a connection between Jordan's stuttering and his politics. The Nazis glorified good looks and physical perfection, and Rosbaud theorized that scores of "hunchbacks, people with limping legs, and stammerers became members of the Party" to compensate for feelings of inadequacy.

When the war started, Jordan began working at Peenemünde, ostensibly as a meteorologist and later as a rocket engineer. This story didn't sit right with Rosbaud. Jordan was a nuclear scientist, and a good one. Could he be testing atomic payloads at Peenemünde? The Griffin decided to investigate, so he and some friends in the resistance invited Jordan to dinner one night in August 1943 and began plying him with drinks. This loosened Jordan's tongue in both senses of the word: for whatever reason, he stuttered less when drunk and therefore tended to talk more at such times. The booze lowered his inhibitions as well, and with one of Rosbaud's buxom lady friends making eyes at him, he was soon regaling the table with tales about research at Peenemünde. He even let slip the schedule for launching the first rockets. It was an uproarious evening, and Jordan had a grand time. He never noticed the Griffin's friend in the corner taking notes.

Jordan stumbled home and no doubt had a hell of a katzenjammer the next day; who knows how much he even remembered of the night. The Griffin's friend, meanwhile, dashed off an immediate report to London.

Given all the warning signs, the British decided to bomb Peenemünde as soon as possible, a mission they called Operation Hydra.

Inevitably, though, the warring factions within Churchill's government got into another row over Hydra, this time about what exactly to bomb. One faction wanted to restrict the attack to the power plant and rocket factories. Ultimately, this side lost. To kill the hydra of myth, Hercules had to cut off and burn every last head, and the British decided on a similarly thorough approach with Peenemünde. This meant targeting not just infrastructure but people—it meant deliberately killing scientists and engineers by bombing the nearby barracks.

The attack began on August 17 when a Royal Air Force squadron roared past the northern coast of Germany on a heading for Berlin. German radar picked up the invaders, and the Luftwaffe dispatched every Nazi fighter in the vicinity, 158 total, to protect the heart of the Reich. The feint worked brilliantly. While the Germans were distracted with Berlin, a second wave of nearly six hundred bombers took off from England and swept into Peenemünde at 11:25 p.m. They dropped three million pounds of explosives and obliterated most everything in sight, scoring several direct hits on the barracks. They also spied a housing development for scientists in a nearby forest and flattened a hundred more homes there.

The most vivid account of the raid comes from the diary of a German secretary. She recalled running through a boulevard of flames on the Peenemünde campus as buildings collapsed all around her. At one point she almost splashed into a pool of blood, and gasped to see a severed leg floating in it. Another eyewitness spotted the military commander of the grounds—the same man who'd boasted to von Thoma that "the fun will start soon"—sobbing in despair as his life's work burned around him: "Peenemünde," he wailed, "my beautiful Peenemünde!"

Most engineers and scientists fled the scene, intent on saving their skins, but the wunderkind Wernher von Braun refused to abandon his beautiful Peenemünde. He grabbed the aforementioned secretary by the hand and shouted, a bit melodramatically, "We must rescue the secret documents!" With a few other brave souls they plunged into von Braun's burning office building and felt their way along a smoke-choked corridor to a staircase. Upon reaching the second story, they found the entire center of the floor collapsed, and had to edge around the gaping hole to a safe in the corner. Von Braun spun it open, and the secretary spent the next hour dashing up and down the stairs with armfuls of paper and tossing them into a second safe outside. She finally collapsed in exhaustion, while the oven of the fires continued to cook the air around her.

At first blush Operation Hydra seemed like a disaster for Peenemünde. The very next morning the chief of staff of the Luftwaffe (whose planes had circled uselessly over Berlin, leaving Peenemünde exposed) shot himself in the head in disgrace; his secretary found his body in his office. But von Braun had managed to rescue the most important documents, saving this hard-won knowledge. Even worse for the Allies, the raid managed to kill very few scientists. Thanks to nearby sirens, most of them had fled their homes before the bombers arrived, and of the 120 Germans who died, only two were considered top researchers. Sadly, the barracks that suffered the direct hits turned out to be quarters for slaves, who couldn't run. Six hundred of them died that night, including several Luxembourgers who'd provided key intelligence for the Allies—intelligence that had allowed them to plan the raid in the first place. The British also lost forty planes and 215 troops to antiaircraft fire. Overall, Operation Hydra had left several important heads unburned.

Hydra also had another, subtler effect. Although a poor strategist, Hitler had a lively military imagination that sometimes served him well. Instead of clinging to World War I–era weapons and tactics (like the French Maginot Line), he championed tanks and fighter planes and the innovative Blitzkrieg. On the other hand, Hitler's imagination sometimes led him to gamble on grandiose projects, and right after the gutting of Peenemünde, he decided to double down on his most fantastical scheme yet: the so-called high-pressure gun, or Hochdruckpumpe.

The Hochdruckpumpe consisted of a 416-foot-long gun barrel capable of launching 600 nine-foot rockets every hour. Hitler dubbed the pump the V-3, his third and greatest vengeance weapon, and he intended to aim it at the heart of London. In preparation, he ordered his military engineers to erect even more concrete bunkers in northern France. Specifically, the V-3 launch site would sit near Mimoyecques — the tiny village that Joe Kennedy and his sister had blithely driven by in their Chrysler convertible on his tour of the Continent a few years prior.

CHAPTER 31

PT-109

For the time being, Joe Kennedy remained unaware of developments in Mimoyecques. He had a more pressing threat to monitor anyway—his kid brother.

Bored with naval intelligence, Jack had begun a fling with a Danish newspaper columnist named Inga Arvad in early 1943. (He called her "Inga-Binga" in private.) Unfortunately, Arvad was a little too cozy with Hitler, having accompanied him to the 1936 Olympics in Berlin. She'd also attended Göring's wedding and chummed around with Goebbels. Given her charms, the FBI worried that she was wheedling information out of Jack—or had turned him into a spy—so agents wiretapped his telephone and began following the couple around. They didn't turn up much, but the navy wanted to dump Jack anyhow—and would have, except for his father. So in February 1943 the navy transferred Jack to a mobile torpedo unit training in Rhode Island. It was considered a win-win. Jack got an exciting mission, and the navy eliminated a security risk by separating him from Arvad. For her part, Inga-Binga thought the transfer absurd. Jack already suffered from back problems ("He looks like a limping monkey from behind," she told a friend), and sending him off to combat seemed reckless. Nevertheless, in April 1943 the navy deployed Jack to the Solomon Islands, east of Papua New Guinea, the site of some of the most ferocious combat of the war.

The boat Jack commanded, PT-109, was essentially a nautical wasp. PTs had light wooden hulls that made them sleek and fast and easy to maneuver. Under cover of night, they'd buzz up to larger, slower enemy ships and sting them with 2,600-pound torpedoes. They did have machine guns for defense, but given that flimsy hull, they couldn't take much fire and relied on speed and cunning instead. In short, working a PT was one of the most dangerous jobs in the navy, and after hearing about Jack's assignment, Joe grew alarmed at the potential hazard—to himself. The South Pacific was the theater where the navy shined, and his baby brother was virtually guaranteed to win a medal there. Meanwhile, Joe had recently been transferred from Puerto Rico to Norfolk, Virginia—even farther removed from combat than before.

Joe's shot at redemption arrived in July 1943, when navy officials called a meeting in Norfolk to recruit pilots for a "very dangerous" mission. Joe volunteered immediately, practically standing up on his seat and waving both hands. The mission involved tracking Nazi subs in the Bay of Biscay, off western France. Germany had several submarine pens in the area, and U-boat harassment of Allied shipping was hampering the war effort.

Joe's mission would be to hunt down and occasionally attack the subs. The hitch was that, because the skies over the bay were crawling with Luftwaffe fighters, he would have to learn to fly a sleeker, more nimble plane. And unfortunately, this plane had one of the most complex cockpits ever devised, with some 50 dials and 150 switches. Most pilots found the details overwhelming—they'd joined the navy to fly, not memorize checklists and punch buttons. The complexity played to Joe's strengths, however, and he managed to master the controls in just six days. Still, he needed more experience flying the plane before he saw combat, and he got stuck ferrying newly constructed planes from a warehouse in San Diego to Norfolk, the same dull route week after week after week. It's like piloting for Pan Am, he complained. He could only imagine the sort of glorious danger Jack was getting into.

He didn't need to imagine for long. The incident is now as much a part of American presidential lore as George Washington's cherry tree or Abraham Lincoln's log cabin. On the night of August 2, 1943, Jack's boat was buzzing around the Solomon Islands and taking pot-shots at Japanese ships when the destroyer *Amagiri* (*Heavenly Mist*) rammed PT-109's starboard side and splintered her hull. Two of Kennedy's crew died on impact, and the rest tumbled out into the water. The other PTs in the vicinity scattered, certain that no one could have survived such a wreck.

In fact, eleven had survived, and they found themselves clinging to the scraps of the overturned hull. As the commanding officer, Jack addressed them as they floated in the dark. He first mentioned the possibility of surrendering, saying he wouldn't judge anyone for doing so: "A lot of you men have families and some of you have children." As for him, he would try to escape, or fight if it came to that. "I have nothing to lose," he said. Stirred by his speech, his men voted to join him and set out for the nearest island.

Some swam, and those who had injuries grabbed planks and kicked. One crewman, Patrick McMahon, was so severely burned that he couldn't do even that, so Jack—despite having suffered injuries himself—took the strap of McMahon's life vest between his teeth and towed him through the water, straining with his bad back. After four hours battling the waves, the eleven men collapsed onto a spit of sand with the carefree name of Plum Pudding Island. It was dry but offered no food or fresh water; the Japanese patrolled the area anyway, so they couldn't hide there long. After resting for four hours, Jack—the "limping monkey," the man who'd gotten into the navy only because his father arranged for a sham medical exam—swam on. He scouted several more islands, then returned to Plum Pudding and rallied his men to make a break for Olasana Island, which at least had coconuts. Another long swim followed, but once again all eleven reached shore.

Unlike the American navy—which had already held a memorial service for PT-109—the Australian navy dispatched two native islanders in canoes to look for the lost crew, just in case. After a series of near misses on various islands, Jack made contact with them, and at the natives' suggestion, he scratched an SOS message onto a coconut. They paddled off with it, and eight more islanders returned to get Jack himself the next day, hiding him under palm fronds in the belly of their canoe. At last, on the sixth day after the accident, Jack returned with a rescue crew, whooping and hollering as they approached Olasana—"his relief and exhilaration enhanced," one historian noted, "by a couple of doses of medicinal brandy." Except for those who'd perished on impact, every last one of his men was saved.

It didn't take newspapers long to start trumpeting the story of Jack, especially his towing McMahon with his teeth. He became one of the brightest heroes of the war—the ambassador's son who'd deliberately sought out a dangerous assignment and had acted with incredible courage. He became so famous that, later in the war, Plum Pudding Island was renamed Kennedy Island in his honor.

The only person not impressed was Joe. He refused to ask the family for news of Jack after his rescue, a coldness that infuriated Kennedy Senior. And there was more turmoil to come. A month after the PT-109 wreck, in early September, Kennedy Senior celebrated his fifty-fifth birthday in Hyannis Port, Massachusetts. Jack was there, having flown home to recover from his injuries (although he never fully would). Joe attended as well, on leave from Norfolk. There was cake, along with presents and drinks, but the climax of the evening came when a close friend of Kennedy Senior, a local judge—the same one who'd tutored Joe at Harvard—toasted the ambassador for his service to his country. He then toasted him again for being "the father of our hero, our own hero, Lieutenant John F. Kennedy." A roar went up from everyone there.

Almost everyone. A little later another family friend was walking through the halls of the Kennedy home when he passed a closed bedroom door and stopped short. He thought he could hear someone inside, and suddenly realized who it was—Joe, weeping bitterly.

CHAPTER 32

Blabbermouth

With the diplomat Ernest von Weizsäcker off scheming in the Vatican, Denmark had no champion in Berlin anymore, and in 1943 the political situation there deteriorated quickly. Emboldened by a Soviet victory at Stalingrad in February, the Danish underground stepped up a campaign of sabotage over the next few months. The Nazis retaliated by shooting political prisoners and seizing the Royal Palace. This in turn led to a widespread labor strike in August, at which point Hitler lost patience. He'd always considered the relative freedom of Jews in Denmark "loathsome," and he demanded corrective action. So in mid-September Gestapo agents broke into the Jewish Community Center in Copenhagen and took one and only one item—a list of names and addresses. A massive raid was in the works.

Unlike virtually every other Nazi raid, however, this one had a happy ending. In one of the most remarkable feats of the war, the Danish underground helped eight thousand Jews—95 percent of the Jewish population of Denmark—disappear overnight on the first of October. Most fled via the sea, piling onto skiffs, barges, and pontoons and floating across the frigid Kattegat Strait to Sweden. When Gestapo squadrons swept into Copenhagen the next morning, they were stunned to find virtually every target on their list missing, including Niels Bohr.

Bohr had actually been smuggled out two nights earlier; a family friend who worked as a clerk for the Gestapo had come across a warrant for his imminent arrest and tipped him off. Given the strict curfews in Denmark then—anyone caught outside after dark was shot—he and his wife had left their home during the afternoon, carrying just one bag between them. After a few blocks they passed a man on a street corner who nodded—a signal that the plan was a go. They walked to a field on the edge of town and proceeded to the coast, where they huddled in a small shack on the beach until nightfall. When a fishing boat finally appeared, they crawled across the sand to meet it, then were spirited across the water to Sweden, where they slept the last few hours of the night in an empty prison cell. Bohr's sons and their families followed in the mass exodus the next night. A granddaughter of Bohr's was concealed in a shopping basket carried by a Swedish official.

Unfortunately, Bohr was no safer in Stockholm than in Copenhagen. Furious about being outfoxed, the Gestapo ordered him seized at all costs. Bohr's Danish bodyguards therefore had to shuttle him from place to place within the city, devising elaborate ruses to shake off spies. At one point a taxi took him to a safe house run by Swedish intelligence, where he dashed up to the attic and climbed outside. After crawling across the roofs of several adjacent homes, he arrived at another safe house and slipped inside the attic window, then descended to the street where a different cab was waiting to whisk him away.

It didn't help matters that Bohr was impossible to keep hidden. He had a gigantic, conspicuous head, and his distinctive mug made him recognizable everywhere in Scandinavia. Worse, Bohr refused to lie low. He kept running off to meetings with Swedish ministers to plead for further aid for Denmark. And although his bodyguards lectured him repeatedly on the need for stealth, Bohr was an incorrigible talker and kept forgetting that he was in hiding. Whenever the phone rang, he'd lunge for it and announce, "This is Bohr." Increasingly frazzled, his Danish guards turned to officials in Sweden for help. The Swedes, however, pooh-poohed the danger. "This is Stockholm, not Chicago," one said with a laugh. To which a Danish official answered that he'd trust an American gangster over a Gestapo officer any day of the week.

Finding no help in Sweden, the Danes decided to smuggle Bohr to Great Britain in an airplane—a risky venture. To respect Sweden's neutrality, the Royal Air Force had to strip the plane, a Mosquito, of all weapons. It would have to fly dangerously high as a result, to dodge the German antiaircraft batteries in Norway; but this seemed safer than keeping Bohr in Stockholm. So on October 5, his handlers snuck him into a car (his family stayed behind) and rushed him to the airport. They bundled him into the Mosquito at 10 p.m. and left the scene, not wanting to draw attention to the plane. Relieved and exhausted, his main Danish guard went home, poured himself a generous glass of champagne, and collapsed.

An hour later he heard a knock on the door. No doubt fearing the worst—had the plane crashed? had the Nazis shot it down?—he rushed to open it. There stood Bohr. Apparently the plane had had engine trouble, an oil leak, and had been grounded. Bohr had blithely hailed a cab and puttered back to Stockholm by himself, oblivious to the danger. The guard yanked him inside and spent all night outside Bohr's door with a revolver.

The underground arranged for another plane the next night, and if anything, the second run proved an even bigger fiasco. Just before the flight, the RAF crew handed Bohr a parachute, a Mae West life jacket, and some flares, in case they got shot down over the sea. Bohr also received a lesson on how to use an oxygen mask. He'd be sitting on a mattress in the unpressurized bomb bay of the Mosquito during the flight, and because they had to fly so high over the antiaircraft guns, everyone aboard would need supplemental oxygen. Alas, blabbermouth Bohr talked so much during the preflight lesson that he absorbed none of the instructions. Worse, he realized after takeoff that the standard-issue flight helmet didn't fit over his gargantuan noggin. As a result, when the plane reached a dangerous height and the pilot announced through the helmet's headset that Bohr should strap on his oxygen mask, the physicist didn't hear him. A few minutes after they'd cleared twenty thousand feet, the pilot radioed back to check in on Bohr—and heard dead silence. He tried again, and got nothing, over and over.

After clearing the German lines, the pilot shot back down to a breathable altitude, but there was still no sign of life from Bohr. It must have been a sickening feeling: the Allies had snatched one of the world's great scientists from the clutches of the Nazis, only to see him die in the cargo hold of an airplane. When it landed, the ground crew threw open the doors and rushed in to offer medical care—and found Bohr sitting upright, alert and chipper. I've just had the most wonderful nap, he said, and proceeded to talk and talk and tell them all about it.

Because Bohr had been cut off from developments in nuclear physics, a few colleagues and government officials soon met him at the Savoy Hotel in London to update him on their atomic research. "Never in history," one historian noted, "had so much secret information been discussed by so few." According to some accounts of the meeting, one of them also mentioned Bohr's clever "Maud Ray Kent" / "radyum taken" telegram from earlier in the war. When asked about it, Bohr wrinkled his brow in bafflement. There was no anagram, he said. He'd simply been reaching out to an English nanny who'd cared for his children before the war and had grown close to them. Her name was Maud Ray, and she lived in Kent.

That wasn't the only surprise Bohr had in store—he also had news about Heisenberg. Unbeknownst to the Allies, Heisenberg had visited Bohr in Copenhagen in September 1941 to discuss nuclear fission. The fight that broke out that night has since passed into the annals of scientific legend, and historians have never quite parsed exactly what happened, in part because Bohr and Heisenberg remembered the conversation so differently. One problem was that openly discussing fission research with Bohr would have been treason, so Heisenberg tried to be subtle and elliptical, counting on Bohr to read between the lines and infer his true meaning. Unfortunately, for all his love of conversation, Bohr was one of history's worst listeners and absorbed nothing. His sudden anger at Heisenberg—who was working energetically for the Reich, after all—probably distorted what he heard anyway. Regardless, we do know a few things: that they discussed uranium fission, however obliquely; that Heisenberg asked whether it was "morally permissible" to engage in such research during wartime; and that Heisenberg sketched a crude design of what Bohr thought was an atomic bomb. They ended up quarreling, and when Heisenberg finally left that evening, Bohr despaired. The two men had been close for over a decade, less friends than father-son; but those few hours destroyed their relationship. And no matter what Heisenberg said (or meant to say), Bohr walked away convinced that German physicists were working on atomic weapons.

Bohr relayed this news to the group at the Savoy. His only consolation was the fact that nuclear weapons were impossible to build, since no one could ever enrich enough uranium. That's when the meeting turned awkward. Actually, the scientists told him, not only were bombs possible, but the Americans were building one. For once, Bohr was speechless. As the scientists laid out their calculations and explained the vast scale of the Manhattan Project, Bohr saw the truth at last: atomic bombs would soon be a reality.

The Vemork power plant in Norway produced heavy water for the Nazi atomic bomb project. It sat next to a mighty waterfall and overlooked a steep gorge. *(Photo courtesy National Library of Norway)*

Charles Henry George Howard, the 20th of Earl of Suffolk, better known as "Mad Jack." *(John Oxley Library, author unknown, PD-1996)*

Joe Kennedy Jr. as a U.S. Navy pilot. (Photo courtesy John F. Kennedy Presidential Library & Museum)

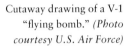

Cutaway drawing of a V-1 "flying bomb." (Photo courtesy U.S. Air Force)

Crater left by a V-1 strike in 1944 in Belgium. (Photo courtesy Library and Archives Canada)

The heavy-water cells in the basement of the electrolysis building at the Vemork power plant. *(Photo courtesy Norsk Hydro/NIA)*

The first self-sustained chain reaction in history was achieved in an unused squash court at the University of Chicago. *(Photo courtesy Chicago History Museum, ICHi-033305; Gary Sheahan, artist)*

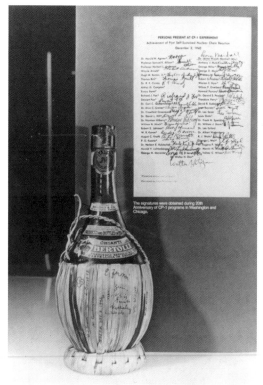

A Chianti bottle signed by the physicists present at the first self-sustained chain reaction in Chicago. *(Photo courtesy Argonne National Laboratory)*

The Gunnerside commando team that raided the heavy-water plant in Norway, including Joachim Rønneberg (*front, far right*). (*Photo courtesy Norway's Resistance Museum*)

Heavy-water cells damaged with explosives in the Gunnerside commando raid. (*Norsk Industriarbeidermuseum*)

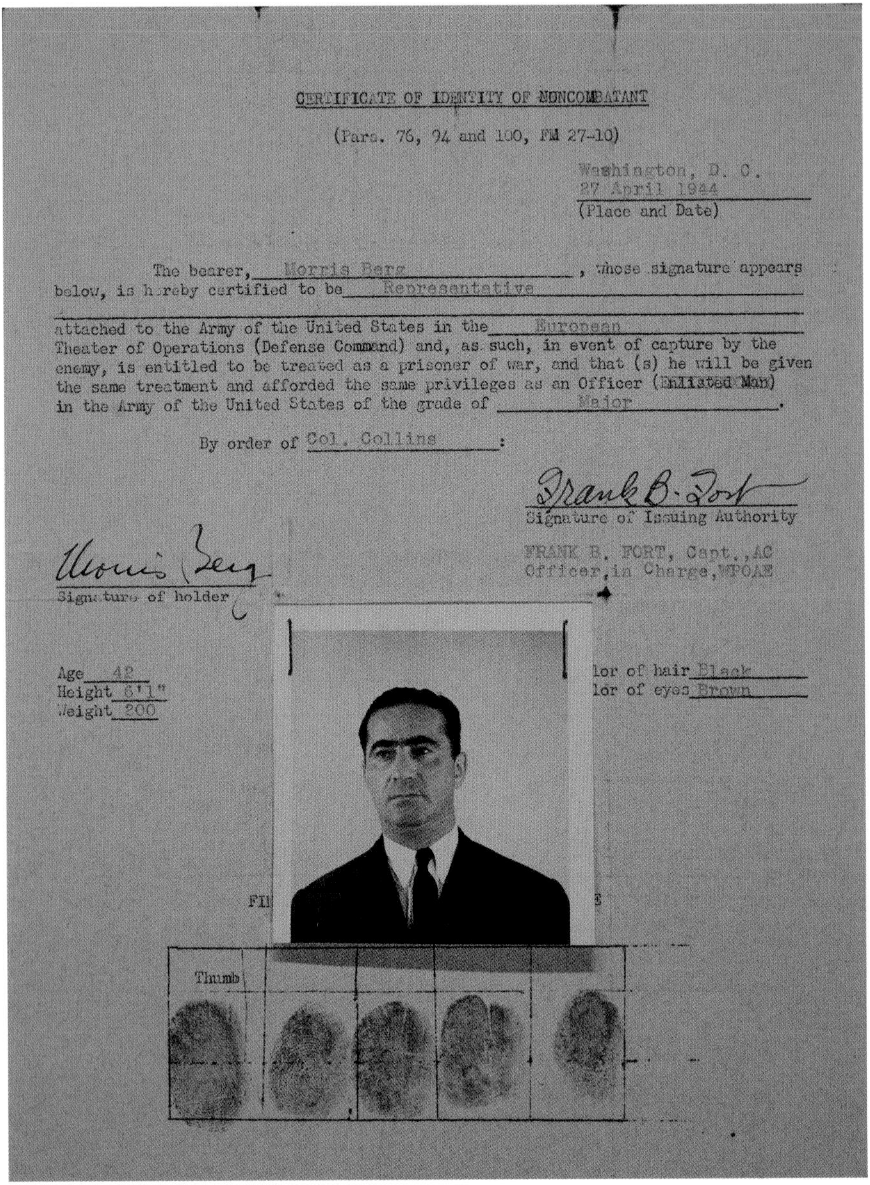

CERTIFICATE OF IDENTITY OF NONCOMBATANT

(Para. 76, 94 and 100, FM 27-10)

Washington, D. C.
27 April 1944
(Place and Date)

The bearer, Morris Berg , whose signature appears below, is hereby certified to be Representative

attached to the Army of the United States in the European Theater of Operations (Defense Command) and, as such, in event of capture by the enemy, is entitled to be treated as a prisoner of war, and that (s) he will be given the same treatment and afforded the same privileges as an Officer (Enlisted Man) in the Army of the United States of the grade of Major .

By order of Col. Collins :

Frank B. Fort
Signature of Issuing Authority

FRANK B. FORT, Capt.,AC
Officer in Charge, MPOAE

Morris Berg
Signature of holder

Age 42
Height 6'1"
Weight 200

lor of hair Black
lor of eyes Brown

Thumb

Moe Berg's application for the Office of Strategic Services (OSS), forerunner of the CIA. *(Department of Rare Books and Special Collections, Princeton University Library, Moe Berg collection)*

Reconnaissance photo of Wernher von Braun's rocket testing grounds at Peenemünde. Disagreements over tiny, ambiguous details, like the features labeled here, provoked plenty of heated arguments within British intelligence. *(NASA, author No. 540 Squadron RAF Flight Sergeant E. P. H. Peek, PD-1996)*

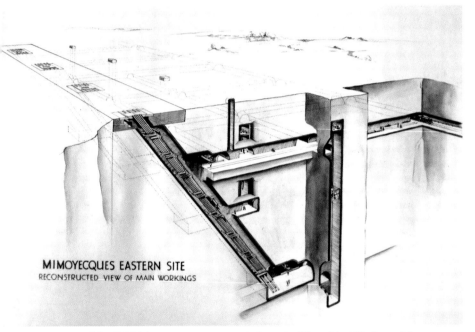

MIMOYECQUES EASTERN SITE
RECONSTRUCTED VIEW OF MAIN WORKINGS

Drawing of the V-3 high-pressure pump gun (a.k.a., the "Busy Lizzie") planned for Mimoyecques. *(United Kingdom Government, author T. R. B. Sanders)*

The landscape around Mimoyecques, heavily scarred by bomb craters. The ominous concrete bunker is the fan-shaped object in the upper middle. *(Photo credit Imperial War Museum)*

Moe Berg in Italy. A note on the back says, "Days after the fall of Rome. <u>Only</u> pictures I have seen of Moe in any kind of uniform." *(Department of Rare Books and Special Collections, Princeton University Library, Moe Berg collection)*

CHAPTER 33

Heavy Water under Fire

In August 1943, catcher Moe Berg officially joined the misfits of the Office of Strategic Services and shipped out to a former Girl Scout camp in rural Maryland for training. True to OSS style, it was demanding, innovative, and kooky all at once.

No record exists of Berg's course of study there, but the Sorbonne this wasn't. An agent typically learned to pick locks, crack secret messages, bug telephones, gouge people's eyes out with his fingers, and kill enemy patrols without making a sound. In the morning he might construct hidden cameras out of matchboxes, and in the afternoon learn the finer points of blowing up stone versus metal bridges — a lesson taught by a munitions expert who was missing several fingers. The most memorable OSS exercise was the "funhouse." After being shaken out of their bunks at midnight, recruits had to break into this ramshackle home and, gun in hand, navigate its twisting passages in the dark. The floors dropped unevenly underfoot in places, and OSS deliberately tried to disorient them by piping in German voices through hidden speakers. To complete the "mission," recruits had to disarm booby traps and shoot papier-mâché Nazis that popped up like monsters at a carnival house of horrors.

On a more serious note, agents like Berg got briefed on how to use L-pills — cyanide capsules coated in rubber. (L stood for lethal.)

Thanks to the rubber coating, you could hold the pills safely in your mouth for hours if necessary; you could even swallow them and they'd pass through you undigested. But if you bit down with your teeth, the cyanide would squirt out and you'd seize up and die within seconds. Agents were instructed to slip the pills under their tongues if taken prisoner by the Nazis. When the inevitable torture started— teeth smashed in, fingernails pulled out, ears ripped off—they could commit suicide honorably rather than give away secrets.

After a few weeks of training, OSS instructors turned the recruits loose in the real world. One exercise involved sneaking onto a bridge or dam, clobbering some poor night watchman, and setting dummy charges at vulnerable points to simulate a demolition. Several trainees ended up in jail after such stunts, which presumably counted against their grades. As a final exam the recruits had to infiltrate an American defense plant and filch something classified, to prove they could per-form under pressure. One man forged the signature of the U.S. com-merce secretary, talked his way into a munitions plant, and walked out with a top-secret bombsight—an A+ job. Berg tried something similar, forging a note on White House stationery and weaseling his way into an airplane factory. Alas, an astute worker confronted him, and the legendary Berg charm failed to mollify him. Berg finally con-fessed that he was working undercover, and the OSS suffered a black eye over the incident.

Despite flubbing his final, Berg passed the training course and was assigned to a few high-profile missions in the fall of 1943. One paired him up with famed astronomer Edwin Hubble. The two men, along with five other agents, were supposed to parachute behind enemy lines in Europe to carry out unspecified undercover work. No one knows exactly why Wild Bill Donovan thought it a good idea to send a fifty-something telescope jockey and a forty-something ex-catcher career-ing into hostile territory on their first mission, but that was Donovan for you. He never let his judgment cloud his enthusiasm. Sadly for history buffs, nothing ever came of the Hubble-Berg mission. Berg's next assignment, though, to the Vemork heavy-water plant in Norway, had real and deadly consequences.

Word of the brilliant Gunnerside commando raid reached General Leslie Groves in April 1943, five weeks after it happened. Being kept out of the loop on Vemork miffed him, as did having to piece the story together from news reports in the media and announcements by the Nazi government—two groups he hated with roughly equal ardor. Once Groves confronted the British, however, they happily explained the raid to him, bragging that they'd knocked the plant offline for two years.

The very next day the British issued a correction. Maybe they hadn't knocked it out for two years, but it was down for one year at least, you could count on that. This hasty revision alarmed Groves. Why had they cut their estimate in half? The British told him not to worry. They had it all taken care of, yessiree, and there was no need—none—to ask any follow-up questions. Groves pressed for more information and got stonewalled, which irritated him further.

That July the British honored the Gunnerside commandos—all of whom had escaped the desolate Hardanger plateau after the raid— with a lavish dinner in London. It took place at the Ritz, and the saboteurs were delighted to see that the main course was grouse. Each man earned a medal of some sort, and they raised toast after toast to their success. It was one of the grandest nights of their lives.

It was also a sham. The British had been hearing rumors for months that Vemork was producing heavy water again, and one week before the Ritz gala a report from a resistance fighter confirmed their fears. It turned out that the pathetic, striving Kurt Diebner had sent his top assistant to the plant shortly after the raid to hasten the clearing of wreckage and the installation of new equipment. Plant engineers had worked "with a knife at our throats," one remembered, and Diebner's man got the heavy-water cells functioning again in just six weeks. In fact, the Nazis took advantage of the downtime to expand from eighteen high-concentration cells in the basement to twenty-six. Production at Vemork increased accordingly, from eleven pounds of heavy water per day before the assault to almost fifteen by mid-June. The British kept mum about all this during the dinner for the Norwegian commandos. But the truth was, their brave mission had gone for naught.

In the end, the only thing the commando raid did destroy was the remaining goodwill between American and British officials working in atomic intelligence. Early in the war the British and Americans had exchanged information freely, but the security-obsessed Groves had put a stop to that, for both sound and petty reasons. Oddly, Groves's father, although born in 1856, had never quite gotten over the British monarchy's treachery toward the American colonies, and his son had inherited his old man's hatred for the Union Jack. More rationally, Groves knew the Americans were doing all the heavy atomic lifting now, and if the British didn't have the decency to inform him about things like the Vemork raid, then to hell with them. The United States would build a nuke alone.

Most frustrating of all, the British hadn't even crippled the plant properly. Production had barely slowed, and Groves grew determined to wipe out Vemork for good. No longer trusting British intelligence, he decided he needed his own man on the ground in Norway to assess the situation. That's how OSS became involved in atomic espionage.

The agency was going through a tough period at the time. Given its freewheelin' ways, most officials in Washington viewed it with suspicion if not disgust. (*We're turning* these *yahoos loose in foreign countries?*) Donovan desperately needed credibility, so when Groves dropped by his office one day in October 1943 — the first senior military official to deign to meet with him — Wild Bill did all he could to impress Groves. After they talked, he ordered his personal chauffer to drive Groves home, then ran out to the parking lot to hold the door open for the general, like a clumsy high-school sophomore on a date. After the visit Donovan could refuse Groves nothing, and when Groves inquired about sending someone to Norway to scout Vemork, Donovan suggested one of his most promising agents, Moe Berg.

A scant few weeks later, Berg boarded a plane in England and parachuted out over Norway. Luckily, no blizzard hampered his landing, and Norwegian freedom fighters whisked him off to Oslo. There he interrogated various scientists associated with Vemork, and they confirmed that the Germans had resumed producing heavy water. After being smuggled back out of Norway, Berg alerted Donovan, who relayed the intel to Groves.

Groves didn't hesitate: it was time to destroy Vemork. No cutesy glider mission, no commando raid. Like an archangel, he would rain down vengeance from above and bomb the hell out of the place. As historians have noted, he'd spent his career building things in the Army Corps of Engineers, so the decision to reduce Vemork to rubble was a curious departure. But Groves had always longed to lead troops in battle, and this was as close as he'd probably ever get. So on November 16, he ordered one of the army's elite bombing groups, the Bloody Hundredth, to attack.

To minimize casualties, the attack was scheduled for shortly after 11:30 a.m., when most plant workers left for lunch. But the planes of the Bloody Hundredth—*Bigassbird, Hang the Expense, Raunchy Wolf,* and 142 others—encountered so little resistance on the flight over that they arrived at the Norwegian coast twenty minutes early and had to dawdle in the air. Bombs finally started dropping at 11:43 a.m., with a barrage of 700 thousand-pounders. Twenty minutes later, thirty-nine additional planes, which had been scheduled to bomb Oslo but found the conditions there too cloudy, also swung through. Mistaking a nearby nitrogen plant for Vemork, they dropped hundreds more five-hundred-pounders on the village of Rjukan.

But if Groves set out to destroy Vemork, he failed. Only twelve bombs hit the plant buildings, and while the raid did knock out several generators and the suspension bridge over the gorge, the mountainous terrain and the plant's layout had rendered it fairly immune to bombing, just as the Norwegians had predicted long ago. What's more, the bungling secondary raid on Rjukan killed twenty-two villagers, with at least one bomb hitting an air-raid shelter. (Witnesses reported finding stray limbs and a decapitated head outside.) The exiled Norwegian government in London was furious. And with the filtration cells in the basement still bubbling away, heavy-water production at Vemork continued unabated.

Vemork proved an exercise in impotent rage for the Allies. The only benefit was that it forced Leslie Groves to confront some hard truths about his own operation: that he knew virtually nothing about the German atomic bomb project, and that the little he did know came filtered through the bloody British. Both situations were intolerable, and he decided to rectify this deficit in two ways. The first involved setting up an eclectic team of scientists and soldiers to gather intelligence on the front lines in Europe. Groves also needed a sort of spy-at-large to slink into the shadows of Europe and ferret out rumors. He again consulted Donovan, who offered Moe Berg to the Manhattan Project on a long-term loan.

Groves hesitated at the suggestion. However admirably Berg had performed in Norway, he was one of the most famous athletes in the country—his picture had appeared in newspapers hundreds if not thousands of times, making undercover work tricky. Still, Berg had upsides, too. He spoke several languages, and he could charm both military officers, who were awed by his intelligence and advanced degrees, and scientists, who fawned over his athletic prowess. It's possible, in fact, that Groves himself was dazzled by Berg's wit and accomplishments. (Berg would later claim that Groves had secretly wanted to be a pro ballplayer—a catcher, even—but couldn't throw.)

So like many a Major League ball club in the 1930s, the general overlooked the catcher's faults and added him to his roster, making Berg the first dedicated atomic spy in history.

Given that Berg stood a strong chance of being captured and interrogated, Groves refused to brief him on work going on at Los Alamos. An atomic spy clearly needed to know some technical details, however, and Berg spent the next several months cramming to learn the strangest language he'd ever encountered, that of quantum mechanics. To help Berg out, OSS arranged for physics tutors, and as with Joe Kennedy, Berg got only the best, including several Ivy League professors. (According to a dubious legend, these tutors included Albert Einstein. The wild-haired one supposedly even promised that if Berg would teach him "the theory of baseball," then he'd teach Berg theoretical physics. After a moment's further thought, Einstein said never mind: "I am sure you will learn relativity much faster than I will learn baseball.")

Berg eventually learned enough science to start reading papers by the colossi of twentieth-century physics—Chadwick and Fermi, Meitner and Hahn, Irène and Frédéric Joliot-Curie. He grew especially enamored with Werner Heisenberg's uncertainty principle and its disturbing philosophical implications. In short, the uncertainty principle puts a limit on how well we can know both the position and the speed of any particle at the same time. And in drawing such limits, it makes these basic, fundamental quantities seem unreliable, even slippery. By extension (or so some people claim) the principle also undermines our certainty about the world in general, because if even bedrock physics is uncertain at its core, can we really be sure of anything? Could any knowledge have a true foundation now? Such questions fascinated Berg, and this fascination would serve him well during the next year, as Heisenberg himself would become the number one target of American atomic espionage.

CHAPTER 34

Alsos

After fleeing Denmark, Niels Bohr sailed to the United States to advise on the Manhattan Project, and he proved just as much of a security nightmare in America as he had in Europe. He first visited New York, a dangerous city for an oblivious jaywalker like him; more than once he almost got creamed crossing the street. Then, on a cross-country train ride to New Mexico, he kept blowing his cover by forgetting his code name (Nicholas Baker), and an armed guard had to sleep outside his room at night to keep him from wandering off. Worst of all, he blabbed about fission research to anyone who'd listen. Things got so bad that General Leslie Groves had to drop everything he was doing to join Bohr for the last leg of the trip, which he spent lecturing the physicist—"for twelve straight hours," Groves recalled—about the need for discretion. Bohr instantly saw the wisdom in this and promised not to say another unauthorized word to anyone. He even succeeded in keeping his vow for a good five minutes after arriving in Los Alamos. But as soon as Bohr saw his old colleagues, at a reception in his honor, he started babbling again—spilling every secret Groves had just warned him to keep mum about. The man was simply incapable of keeping his trap shut.

Although the Great Dane proved a valuable mentor at Los Alamos—at fifty-nine, he was the eldest scientist at the lab, thirty years older than the average—the immediate result of his trip was to heighten the paranoia about the Nazi atomic bomb, especially for Groves. Groves was not a hysterical man by nature, but as one of his staff noted, "He worried the hell out of the German bomb project during the war." Bohr's account of his conversation with Heisenberg in 1941 only deepened the general's unease. He also rehashed the Heisenberg *Sturm und Drang* for Oppenheimer and other top officials at Los Alamos; this included pulling out the sketch Heisenberg had made, which caused quite a stir. To be sure, everyone concluded that it looked more like a nuclear reactor than a bomb, but the sketch was two years old at that point; Germany had no doubt made great progress since. And if nothing else, you could use reactors to breed plutonium.

Or worse. In addition to plutonium, running a reactor created all sorts of nasty by-products that were ideal for so-called dirty bombs. Although dirty bombs also require radioactive material, they differ from fission bombs in important ways. Fission bombs kill by releasing gobs of energy all at once; they vaporize you. Dirty bombs kill by releasing deadly isotopes that wriggle inside your body; they poison you. And while fission bombs require a nuclear explosion, dirty bombs don't. You simply have to spread the dirty radioactive material around, which you can do with regular explosives; you can even mix the material with smoke or powder and use crop dusters to spray it on troops or cities.

As of 1943 there was no hard evidence that Germany was making dirty bombs, but the very idea of them contaminated the minds of Manhattan Project scientists, filling them with lurid visions. In the summer of 1943, project officials installed secret nuclear-defense systems in Boston, Chicago, New York, San Francisco, and Washington, with Geiger counters wired to air-raid sirens in case of attack with radioactive species. There was talk of preemptive strikes as well. Enrico Fermi pulled Robert Oppenheimer aside one day and suggested manufacturing deadly strontium-90 to poison food and water supplies in Germany. Oppenheimer met this horrifying suggestion with enthusiasm, and abandoned it only after determining that they probably couldn't kill enough people to make it worthwhile. He wanted at least half a million dead Germans, or why bother?

The paranoia reached its peak, or nadir, in late 1943. Based on projections about the rate of German research, several scientists convinced themselves that the Nazis probably had enough radioactive material by then to make several dirty bombs. Nazi propaganda minister Joseph Goebbels then ratcheted up the tension by declaring that Germany would soon unleash a revolutionary "uranium torpedo" on the Allies. The only question was when, and for various reasons American fears began to coalesce around dates in December. For one thing, security always slackened during the holiday season. For another, Hitler clearly loved stagecraft and grand gestures — witness the Berlin Olympics and his storm troopers goose-stepping through Paris. He would certainly plan the attack to maximize its emotional impact. And what day could be more devastating than Christmas or New Year's Eve? Their imaginations now at full gallop, a few American nuclear scientists actually sent their families to hideouts in the countryside in late December to protect them. They then endured a grim holiday week alone by the phone, their stomachs churning with acid, awaiting news of the atomic apocalypse.

Nothing like that happened, obviously, but the hysteria once again highlighted the fact that Manhattan Project leaders had no idea what German scientists were really up to. At the time, the United States had pathetic intelligence capabilities, and there were no espionage units with scientific expertise, which meant they were probably overlooking vital clues. (The vast majority of people, for example, still considered uranium a useless metal.) And the problem would only get worse over the next year. By late 1943 the Allies had gained footholds in Italy and were planning to attack occupied France in 1944. Every newly conquered city meant new chances to gain precious atomic intelligence — or alternatively, to let it slip away.

To fix this problem, one of Groves's deputies came up with a plan.

Rather than rely on third-hand rumors from abroad, he decided the Manhattan Project should build its own intelligence unit to scour Europe. It would consist of both scientists and soldiers, and they would spend their days infiltrating labs, deciphering secret documents, and interrogating captured scientists. This was something new in the history of warfare: no one had ever turned scientists loose like this on an espionage spree. The team would report directly to Groves and would operate in strict secrecy, allowing no one in the field to know what they were searching for. The closer they could get to the front lines, the better.

The program became known as the Alsos mission, a name based on a multilingual pun: ἄλσος means "grove" in Greek. But when the object of the pun, Groves himself, discovered this Easter egg, he was furious. He didn't find it cute, and furthermore considered it a security hazard, since anyone who knew his role on the Manhattan Project could then infer what this scientific outfit was doing in Europe. (There's evidence, in fact, that a few British agents did deduce the purpose of Alsos based solely on the name.) No one in Groves's office ever confessed to this etymological crime, and by the time Groves found out about it, the name had already started circulating within the Pentagon. Changing it would only draw more attention, so he grudgingly let it stand.

Groves eventually widened the scope of Alsos beyond nuclear science, figuring that as long as people were ransacking German labs, they might as well learn all they could about radar, rockets, jet engines, and biological weapons, too. But in large part, these other topics served as a beard, a way to obscure the mission's real goal: hunting down secrets about the Nazi nuclear bomb. Alsos also had the authority to seize atomic matériel like uranium and heavy water, and even scientists themselves. As the mission evolved, in fact, manhunts became its top priority. All the uranium on earth wouldn't do much good without a Hahn or a Heisenberg to sculpt it into weapons.

Given their hostility to Groves, the British hated Alsos. The existing intelligence apparatus worked just fine, they insisted; there was no need to deploy a bunch of skittish scientists to the front, especially when they ran a high risk of being captured and interrogated, no doubt "in the Russian manner." Groves cared not a whit for what the British thought, but he shared their concern on this last point. Therefore none of the scientists who worked on the Manhattan Project were eligible for Alsos—they simply knew too many secrets.

Ruling out everyone on the Manhattan Project, however, left very few candidates for the role of chief scientific officer on Alsos. Ideally, Groves wanted a nuclear physicist with some experience studying neutrons and cyclotrons. He couldn't be a theoretical egghead, though: the job would be as much detective work as anything. Given the dangers involved, he would have to be pretty eager to do something for the war effort, and knowledge of Europe and its languages would be a nice bonus. But where on earth would they find somebody like that?

Luckily, the selection of chief military officer went more smoothly. Groves already had a candidate in mind, in fact, a former science teacher with experience in irregular warfare—someone just as obsessed as he was with security and someone who needed to get transferred out of his current post before he enraged every general in the army. Boris Theodore Pash.

PART V

1944

CHAPTER 35

Busy Lizzie

Despite the raid on Peenemünde, the British continued to tremble over the German V-rockets: they remained a "crossbow," as one official put it, "aimed at England's heart." The situation looked menacing enough by February 1944 that Winston Churchill felt compelled to warn the House of Commons about the huge concrete bunkers in northern France. The Reich had forty thousand workers toiling away there by that point, and the sites were expanding at an incredible clip: some contained more concrete than Hoover Dam. Pouring so much money and matériel into unproven weapons was a big risk, especially given how strapped the German economy was, but as one historian noted, "The Führer always brightened when mention was made of some wild and grandiose scheme. If it involved the use of hundreds of thousands of tons of reinforced concrete, he was ecstatic." By that criterion, no scheme in the Reich could have made him happier than the ominous bunkers.

Especially because one of the bunkers, at Mimoyecques, would house Hitler's beloved V-3—officially named the Hochdruckpumpe (high-pressure pump) but informally dubbed the Busy Lizzie. According to the first sketchy reports the Allies received, the V-3 would stretch twice as long as the V-2, measuring an incredible 92 feet, with a 118-foot wingspan. It reportedly weighed 40,000 pounds, half of that explosives, and could reach speeds of 435 miles per hour. The most fantastical spec was its range, more than 6,000 miles. Which meant, forget London—Busy Lizzie could smite New York and Washington.

(Attacking the United States might sound absurdly ambitious for Germany, but it was one of Hitler's fondest dreams during the war, and his henchmen developed several schemes to do so. The loopiest was probably Projekt Huckepack, the Piggyback Project. It called for filling a bomber to the snout with fuel and pushing it as far over the Atlantic Ocean as it could go and still make it back to Germany. Right before it turned around, a slender, pencil-shaped plane would detach from its belly and continue on. If everything went right, this piggyback plane could just reach Manhattan and unload its bombs. Afterward it would splash down at a rendezvous point in the ocean, where a submarine would surface and pick up the crew. Now, the pencil bomber would have limited firepower, perhaps only a few bombs, which hardly makes the effort seem worthwhile—unless those bombs were unimaginably powerful. As powerful as, say, nuclear bombs. And there are hints that the Nazis considered such a tactic, turning Germany's Manhattan Project against Manhattan itself. Even if it carried conventional bombs, one successful piggy-back strike would force the United States to divert soldiers and sailors to its coastline rather than shipping them to Europe. Moreover, as al-Qaeda proved decades later, striking New York would have provided the Reich with an immense psychological boost.)

Perhaps not surprisingly, the rumors about the incredible size of the V-3 turned out to be just that—not credible. In truth the V-3 rockets measured less than a tenth of their reported 92 feet, and their range fell far short of 6,000 miles. V-3s weren't even real rockets, since they couldn't launch on their own. But that didn't mean the V-3 wasn't scary—there's a reason Hitler relished it. The Busy Lizzie was essentially a 416-foot rifle that shot nine-foot bullets. The difference was that, instead of using one explosion to propel a bullet, the way a normal rifle does, Lizzie used several precisely timed explosions along the length of the barrel to accelerate the bullet in stages, pumping it to near-supersonic speeds. And while a single nine-foot bullet obviously wouldn't do as much damage as a V-1 or V-2, Lizzie made up for her lack of punch with an incredible rate of fire: the Nazis planned to operate twenty-five pump guns near Mimoyecques, which would allow them to shoot a round every six seconds. That would mean up to 14,400 V-3s crashing down on London every day. As Goebbels said, "Twice as many inhabitants are crammed into London as Berlin. For three and a half years they have had no sirens. Imagine the terrific awakening that's coming!"

Now, a 416-foot gun sitting out in the open would of course have been an irresistible target for bombers; it wouldn't have lasted even an afternoon. So to protect Lizzie, some five thousand German engineers started digging. They opened huge tunnels in the limestone hills out-side of Mimoyecques, and anchored the rifle barrels in the bedrock thirty-five stories beneath the surface. (Anchoring the barrels like that also helped absorb the unfathomable recoil.) Because the guns loaded from the breech (the back end), the engineers also had to dig access tunnels for the nine-foot bullets. For additional protection, the open-ings of the muzzles at the surface were surrounded by eighteen-foot-thick domes of concrete and covered (when not in use) with eight-inch-thick steel plates, which were camouflaged with haystacks.

To be frank, the Busy Lizzie was a long shot. Accelerating a bullet that size with a series of explosives would have required ridiculously precise timing, probably beyond the reach of electronics back then. In fact, at the V-3 proving grounds west of Berlin, test after test failed in the spring of 1944, with the bullets either exploding inside the barrels or tumbling out of control during flight. Given all this, sev-eral historians dispute that the V-3 ever could have worked. That said, Albert Speer—by far the most sober and realistic Nazi in Hit-ler's circle—swore that end-stage tests were on the right track. And modified V-3s were successfully fired against troops in Belgium and Luxembourg later in the war.

Regardless of the V-3's actual potential, what mattered at the time was that Nazi leaders believed it would work—as did the Allies. The danger seemed especially acute given the existence of a German nuclear weapon program. No firm evidence ever linked Mimoyecques with Vemork or the Uranium Club, but there were always allusions, whispers, and once people started associating the two in their minds, the idea continued to fester. After all, what other weapon could justify so much money, so much labor, so much concrete? It had to be nukes. One Manhattan Project physicist was so scared of atomic rockets that he started listening to the BBC broadcast twice every day—not for the news, but to reassure himself that London still existed. Even General Dwight D. Eisenhower, the supreme Allied commander in Europe, could only shake his head when briefed on the sites and mutter, "You scare the hell out of me."

Clearly, the Allies needed to neutralize this threat—and quickly, before the Germans reinforced the bunkers beyond the point that bombing them would do any good. Eisenhower therefore decided that strikes against rocket sites, including the bunkers, would receive the highest priority of any mission in Europe aside from the D-Day invasion. The U.S. army and navy accordingly got to work, and developed a scheme so outlandish it would have made the kooks in the OSS smile. The navy arm was called Operation Anvil, and Joe Kennedy was one of its first volunteers.

CHAPTER 36

Groves's Second Assault

In early 1944, General Leslie Groves went on the offensive again, albeit in an unorthodox way. Rather than bomb only military and industrial sites, he decided that the time had come to take out scientific targets. His primary goal was to evict researchers from their "comfortable" labs and delay their work, but he certainly wouldn't oppose more permanent measures. As one report put it, "the killing of scientific personnel...would be particularly advantageous."

After fielding suggestions from Manhattan Project leaders, Groves focused the strikes on the cherry tree–lined streets of Dahlem, a suburb of Berlin where Otto Hahn and Werner Heisenberg worked. Because no one had ever bombed a laboratory before, the area was virtually defenseless on the evening of February 15, 1944, when a squadron of planes swooped in and unloaded. Both labs suffered heavy damage, especially Hahn's nuclear chemistry institute; one bomb landed more or less in his desk chair and blew out the entire southern wall of his office. The building's rafters soon caught fire, and the surviving scientists tried to salvage books and equipment by passing them hand to hand in a chain. They then stepped back and watched the roof blaze red against the nighttime sky—a "terrible-beautiful sight," one recalled.

By lucky coincidence, Hahn was absent from Berlin at the time. He was in fact scouting around for new lab space in the Black Forest region of southern Germany, to put himself beyond the reach of raids like this. The trip no doubt saved his life. Still, the sixty-four-year-old chemist lost most of his life's work to the bombs; he especially bemoaned the destruction of personal letters from scientists like Ernest Rutherford.

The Dahlem strike had secondary effects as well. On the German side, it hastened the evacuation of Uranium Club members to the Black Forest. On the Allied side, it made German scientists seem like legitimate targets for attack, and in doing so helped revive another unorthodox plot. Way back in 1942, two of Samuel Goudsmit's friends had drunkenly proposed kidnapping Werner Heisenberg in Zurich. They'd passed the idea to Robert Oppenheimer, who in turn alerted his superiors, but as far as Goudsmit knew, the higher-ups had dismissed the idea. In reality, Oppenheimer was still chewing it over, and had even expanded its scope: because if snatching one German scientist was good, snatching several German scientists was better. To help military folks identify them, Oppenheimer began compiling dossiers on seven German nuclear scientists. And as the idea was bandied about at meetings in Los Alamos, someone there (Groves never revealed who) finally cut to the chase and said, If you're so worried about enemy scientists, why not just rub them out?

Most military officers would have recoiled at this. Generals have no compunctions about ordering assaults in which thousands, even tens of thousands, die, but in that era they rarely talked about killing individual people; that was considered unseemly, even slimy, the province of assassins and spies. In his typically tactless way, however, Groves didn't care. To his thinking, the scientists designing weapons for the Nazis were every bit as dangerous as the storm troopers wielding them, probably more. Killing them first seemed fair.

Still, given the lingering stigma of assassination, Groves passed the idea up the chain of command to see what his superiors thought. The answer came back in February 1944: "Tell Groves to take care of his own dirty work." If this was intended as a rebuke, it failed: all Groves absorbed was that no one had told him no. He wasn't quite sold on assassination yet, but at least the option was on the table.

CHAPTER 37

The Ferry

Although the air strike on the Vemork heavy-water plant failed to knock out the cells in the basement, the assault proved a success in the long run because it convinced the Germans that the Allies would keep attacking the plant over and over. Given that, why not produce heavy water somewhere safer? Nazi officials had already been sketching out plans for a D_2O plant deep inside Germany, and the Vemork raid accelerated its development.

In the meantime, Vemork still had fourteen tons of Juice on hand in various stages of concentration, Juice the Uranium Club needed for its research. So in early 1944 Vemork arranged to ship it all to Germany. Security was of course a concern, and after considering several routes, officials decided on a convoluted plan: The heavy water would leave Vemork via railroad, traveling until it reached a nearby fjord lake. It would then cross Lake Tinnsjö on a ferry, and connect with another railroad on its way to the North Sea. However complex logistically, it seemed the safest course.

Inevitably, though, word of the plan leaked out to publisher Paul Rosbaud in Berlin, and the Griffin alerted British intelligence. The Allies obviously couldn't allow so much heavy water to reach the Reich, but stopping it en route wouldn't be easy. The two railroad legs of the journey would be heavily defended, as would the transit across the North Sea into Germany. The only vulnerable point seemed to be the trip across Lake Tinnsjö. The ferry there, the *Hydro*, was a commercial vessel and largely defenseless. A bomb placed belowdecks could easily sink it, and given the impressive depth of Tinnsjö, 1,300 feet, the sunken cargo would be irretrievable.

There was just one problem. As a commercial ferry the *Hydro* would be carrying passengers, mostly local Norwegians, who would die in any attack. As soon as the Norwegian underground heard about the plan, they radioed London to protest. Was killing more innocent people necessary? London responded the next day, tellingly using the passive voice: "The matter has been considered, and it is decided that it is very important to destroy the Juice." The Norwegians hardened their hearts; more of their countrymen would have to die.

The mission began with a clandestine scouting trip. In mid-February 1944, a man in a dark blue suit boarded the *Hydro* just before its 10 a.m. launch. He stood on the upper deck cradling a violin case and smoking a pipe to keep warm. He proved rather undexterous for a musician, however: halfway through the trip he dropped his pipe down a grate and into the engine room in the bowels of the ferry. Acting sheepish, he crept down to retrieve it—and managed to reconnoiter the engines while he was there. On the way back up, he bumped into the ship's engineer and struck up a conversation. The engineer was happy to chat—it was a sleepy job, with nothing much to do—and when the violinist confessed his fascination with ferries and gave the engineer a pinch of tobacco for his trouble, the engineer offered to give him a full tour, stem to stern. Delightful.

Back on the observation deck, the violinist grabbed his instrument case, feeling for the heft of the gun inside. He spent the remainder of the trip mentally going over what he'd seen below. He was a veteran of the Gunnerside mission, and compared to the obstacles at Vemork, the *Hydro* seemed laughably easy to sabotage. Upon disembarking he reported the ship's layout to the Norwegian underground, adding that the ferry passed over the deepest stretch of water thirty minutes after departing. Given the ship's schedule and the vagaries of the boarding process, he estimated that blowing the *Hydro* at 10:45 a.m. would be ideal.

As for the bomb, the underground planned to use Nobel 808 again, the same almond-scented explosive the Gunnerside team had used. This time, though, the commandos couldn't just light a fuse and run; they would have to plant the bomb ahead of time and rig up a timer. The design they settled on consisted of detonators wired to an alarm clock with its bells removed. The saboteurs would sneak onboard the ferry the night before, wind the clock, and set the alarm. When it went off in the morning, the metal clapper would fly up and, with the bells missing, strike a brass contact instead. This would complete a circuit between the alarm clock and the detonators, sending a pulse of electricity into the Nobel 808. *Boom.* The only remaining question was the size of the hole. A large hole would sink the ship quickly but would doom everyone belowdecks to a watery grave. A small hole would save lives but might give the guards aboard a chance to salvage the heavy water. Balancing mercy and murder, the saboteurs decided to sink the ship in five minutes, which according to their calculations meant a hole twelve feet in circumference.

A few days before the heavy water shipped out, workers at Vemork began decanting it into four dozen 400-liter drums labeled POTASH LYE. The manager overseeing the operation was in on the plot and knew that his involvement would likely cost him his life: after the sabotage, the Nazis would surely arrest and interrogate him. He did his job anyway, and managed to drag things out long enough to ensure that the barrels would board the *Hydro* on February 20, a Sunday, the morning when the fewest local people rode the ferry.

From the saboteurs' point of view, the weeks of preparation could not have gone more smoothly. But they paid for their early good luck with a series of catastrophes on the eve of the mission. The saboteur who'd scouted the *Hydro* had dressed as a violinist because a symphony orchestra was touring the area, and a stranger with a violin case wouldn't attract much attention. But at a party on Saturday, February 19, one of the co-conspirators overheard a musician—a world-class violinist who'd accompanied the orchestra—mention his plans to ship out on the ferry the next morning. After choking on his drink, the conspirator tried persuading him to stick around and, um, see the sights. (*The plateau isn't* that *bleak—try the skiing!*) No luck. The violinist had a concert in Oslo the next night and couldn't stay. The conspirator, feeling wretched, could only nod. That same night, another conspirator's mother announced her own plans to take the Sunday ferry, and he had just as little luck dissuading her. The son was tempted to tell her the truth, and wrestled hard with his conscience. But the mission was too important, and he stayed silent.

A few hours before midnight, Nazi guards began loading the barrels of heavy water onto railroad cars at Vemork. A dozen soldiers with machine guns then climbed aboard, straddling the drums as an extra precaution. After reaching the lake, the guards uncoupled the cars at the ferry dock and stationed several klieg lights around them, along with more troops. They would load the cars onto the *Hydro* in the morning.

In the meantime, the team of saboteurs was dealing with another crisis. Reaching the ferry docks at night wouldn't be easy—they had to cross nine miles of craggy terrain. They would need a car to do so, and because cars were scarce during wartime, they'd made arrangements to "steal" the vehicle of a sympathetic local doctor. Unfortunately, the man was a better physician than a mechanic, and when the saboteurs broke into his garage just before midnight and inserted the key, all they heard was *rur-ruur-ruuuuur.* The engine wouldn't turn over. They tried again, and heard the same dull dirge. Two of them jumped out at this point and popped open the hood. They tested the battery, the fuel pumps, the gas line, the spark plugs. Nothing: *rur-ruur-ruuuuur.*

Aware of the minutes slipping away, they became increasingly frustrated; one kicked the car and spat, "You dumb brute!" Finally, after an hour, they cracked open the carburetor and found enough sludge to choke a horse. After a hasty cleaning, which left them looking like chimneysweeps, they tried the key again—and the engine caught. They slammed the hood down, wrenched the car into gear, and raced down the icy roads to the docks.

Upon arriving, they noticed the phalanx of troops surrounding the heavy-water drums. Incredibly, though, the Nazis had neglected to guard the ferry itself. Feeling their luck turning, three of the five saboteurs peeled off to approach the ship. Although the crunchy ice underfoot made them wince, boarding the vessel proved unnervingly easy: they simply tossed a rope over the side and pulled themselves up. Their luck continued inside, as the *Hydro*'s crew was celebrating Saturday night with a loud and well-lubricated game of poker. No one noticed the saboteurs as they slipped down into third-class and followed the route the violinist-scout had described. They'd just about reached the hatches that led to the lowest deck when another disaster befell them, and a guard stepped out of the shadows.

The trio dove for cover, but too late. The guard had seen them. They braced for an alarm.

"Is that you, Knut?"

The saboteurs peeked out, baffled. It turned out that the guard knew one of them; they went to the same athletic club.

Scrambling for a cover story, Knut stood up and told the guard they were fleeing the Gestapo. "We're expecting a raid, and we have something to hide," he said. Could they store it below?

"Why didn't you say so," the guard said, suddenly chipper. He pointed out the hatch. "This won't be the first time something's been hidden below."

While Knut stayed and distracted the guard, the other two saboteurs, still half stunned, lowered themselves into the bilge. They found a foot of cold, foul water down there and began splashing their way toward the bow. Planting the bomb there would cause the *Hydro* to pitch forward as she sank, pulling the screw and rudder out of the water and preventing her from steering for shore. They pulled the explosives from their rucksack, laid out a twelve-foot ring underwater, and carefully wired up the detonating caps, a ticklish job. After taping the clock to the steel ribs of the hull, they set the alarm and rejoined Knut and the guard, who were still gabbing. They finally slipped off the ship at 4 a.m.

The day dawned cool and clear a few hours later, and a few hours after that, the *Hydro* pushed off at 10 a.m. The captain that day came from a nautical family; he had a brother who piloted ships across the Atlantic. The brother had been torpedoed twice already, and each time it happened the ferry captain thanked the stars above for his cushy job.

At 10:45 he felt the *Hydro* lurch beneath him. At first he thought they'd run aground, but given the depth of Lake Tinnsjö, that was impossible. Perhaps they'd been torpedoed, he thought. But no, U-boats didn't patrol fjords. Before he could puzzle the matter out, the ferry pitched forward precipitously. A minute later chaos erupted on deck as the railroad cars holding the heavy water snapped their restraints and crashed into the ship's guardrails.

Down below, the passengers in third class proved quicker on the uptake than the captain. "We've been bombed!" somebody cried. Water began gushing into the windowless hold, and a pipe soon burst, hissing steam. People scrambled over each other for the exit, until the lights on the ceiling flickered and cut out, making it impossible to see.

The ferry sank perfectly on schedule, within five minutes. Only one lifeboat got launched, and of the fifty-three souls aboard, twenty-six drowned, including fourteen Norwegians. This brought the total body count for the Vemork missions—including the Freshman disaster and the aerial strike—to roughly ninety people over sixteen months.

Thankfully, the saboteur's mother was not among the victims. After hearing her plans and deciding that he couldn't betray the mission, the son had dumped a whole bottle of laxative into her dinner. After a wretched night of "food poisoning," she'd decided not to travel after all. The famous concert violinist did board the *Hydro* that morning, keen to get to Oslo. But he saved himself by leaping aboard the life raft. He even rescued his prized violin, which he found bobbing in the water.

As for the manager at Vemork who'd overseen the transfer of the heavy water—and who would now come under heavy suspicion—he escaped unscathed as well. The Nazis made inquiries, to be sure, but found that he'd checked himself into a hospital the night before the bombing, clutching his side and moaning. He'd gone under the knife for an appendectomy, and the Nazis concluded he'd been innocent. What they didn't know was that his sister, a nurse, also worked at the hospital and had arranged the whole thing. The surgery was a sham.

Of the twenty-seven survivors floating in the lake that morning, a few were found clinging, ironically enough, to barrels of "potash lye." (The barrels hadn't been filled completely, and the air pockets inside kept them afloat.) The Nazis managed to salvage four of these drums, which contained the equivalent of thirty-two gallons of pure Juice. It was the last heavy water the Uranium Club would receive from Norway. The rest of it—paid for nearly ounce for ounce with human blood—sank to the bottom of the fjord.

CHAPTER 38

Sharks

If there was one thing that drove Boris Pash crazy it was sitting around doing nothing, and for the first few months of Operation Shark, he did nothing but nothing. After being yanked off the Oppenheimer case, Pash was summoned to Washington in the fall of 1943 and put in charge of the Alsos mission. As one of his first official acts, he grew a trim line of whiskers across his upper lip; perhaps this *mustache de guerre* reminded him of his Russian Revolution days. The first Alsos assignment would take place in Italy, and preparation for it ruined Pash's Thanksgiving holiday. But two years to the day after Pearl Harbor, the forty-two-year-old coach kissed his wife Lydia goodbye (she reportedly hated the mustache) and flew off to fight in his third war.

Direct flights to Europe didn't exist then, so Pash and his crew of thirteen military personnel and six scientists had to hop the Trampoline to Victory across the ocean: Miami, Puerto Rico, British Guiana, Brazil, French West Africa, Marrakesh, and so on. It took eight long days, but Pash's crew arrived in Naples on December 15 in fine fettle. "The morale of the members of the mission is excellent," he boasted to Washington. It helped that the crew had adopted a snazzy logo: a Greek α on a blue background, with a red lightning bolt

271

streaking across; the α represented both *Alsos* and *atom*, while the lightning represented fission.

α

Priority number one in Italy was tracking down Edoardo Amaldi, the former assistant of Enrico Fermi who used to race him down the hallway with irradiated samples. Given Amaldi's scientific credentials and the close ties between Italy and Germany, it seemed likely that the Uranium Club had been in contact with him in Rome. Pash needed to find out what Amaldi knew. Perhaps he'd even worked jointly with the Germans on a fascist atomic bomb.

Pash was eager to start hunting scientists, but the Alsos mission soon got caught on several snags. One was their own fault. As a bastard unit, unattached to any larger group, Alsos needed to borrow vehicles and supplies from other divisions to reach Rome. Some units might have been willing to help, but Pash pointedly declined to explain why he needed the gear, citing the secrecy of his mission. He resorted to bluster instead, which worked fine on privates but not on officers, who wanted to know what the hell he was up to. A standoff ensued, and the general in charge of the area, Pash said, "flatly stated that he would not do anything for us unless we told him the whole story." Pash refused.

It soon became clear that the road to Rome was barred anyway. The Italian campaign had started off promising enough for the Allies. They'd conquered Sicily in July 1943, and shortly afterward the Italian government had deposed and imprisoned Benito Mussolini. (Nazi paratroopers later rescued him and smuggled him to Germany.) Allied forces then invaded the southern Italian mainland, from which point they hoped to shinny up the peninsula into Europe proper and threaten the underbelly of Germany. Unfortunately, Hitler could see this gambit as clearly as the Allies, and after Mussolini's arrest, tens of thousands of Nazi troops—each one cursing the name of Italy—poured into the Boot to fortify the place. The ascending Allied and descending Axis armies eventually crashed together eighty miles south of Rome, with much of the fighting centered around the ancient mountaintop monastery of Monte Cassino, where 55,000 Allied troops would eventually sacrifice their lives.

This quagmire left the Alsos crew stranded in Naples, a gritty town in the best of times and especially depressing during the winter of 1943–44. The food was rotten, most buildings lacked power and running water, and the only benefit of the constant drizzle was that it suppressed the dry, choking dust that otherwise coated everything in a white film. An outbreak of typhus required spraying the region with DDT. An outbreak of syphilis then decimated the troops, causing staff shortages and forcing army health officers to nail Burma Shave–like PSAs to trees along the roads. ("Girls who take boarders / bring social disorders. / Take a pro[phylactic]." And "A happy guy / a fool was he. / Forgot his pro, / now has VD.") As a result of all this, New Year's Day passed in a malaise of inaction for Alsos, as did the first weeks of January. Trying to be productive, Pash and his scientists— some of whom reportedly posed as correspondents for *National Geographic*—began hunting down physicists around Naples, hoping they'd heard rumors of what Amaldi was up to. They knew nothing, and interviewing them proved futile. The mission's "excellent" morale began to plummet, especially when—having nothing else to do—technical personnel were assigned menial chores like KP. Here they were, elite scientists on a top-secret mission, one that could save the world from an atomic Hitler—and how did they spend their days? Peeling potatoes.

Worse, each day's delay only increased the chances that the Nazis would bundle Amaldi up and whisk him away to Berlin, putting him beyond the reach of Alsos. That's why an increasingly desperate Pash drew up plans to grab the scientist first, a mission he dubbed Operation Shark. It called for sending an American agent or two into Rome to persuade Amaldi to defect. They'd then sneak him out of the city and rendezvous with a submarine twenty miles away, on the Mediterranean coast. From there the submarine would smuggle Amaldi back to Naples for interrogation. Similar missions along the Adriatic Sea, on Italy's eastern coast, would target other physicists. And if the physicists resisted? Well, the agents would bludgeon them on the head and drag them along by their hair. Easy-peasy.

At first Pash envisioned leading these raids into Rome himself, but Washington quickly put the kibosh on that—he knew too much about the Manhattan Project to risk capture. So Pash threw his energy into organizing the raids, which turned out to be labor enough. For one thing, contacting Amaldi and other targets proved tougher than expected. The Nazis, ruthless in a way the Italians weren't, had wiped out the underground resistance in Rome, making it difficult to reach anyone outside official channels. That meant sending in undercover agents, which meant involving OSS. Wild Bill Donovan's boys didn't balk at the idea: Operation Shark was just the sort of caper they relished, and OSS was already working with Mafia dons in southern Italy to open doors there, so the agency had a good chance of succeeding. But dealing with OSS added more layers of bureaucracy and caused more delays.

Pash finally got word that OSS agents had contacted Amaldi in Rome, as well as another physicist in Turin several weeks later; arrangements to slip them out were under way. All OSS needed was a submarine. That detail, however, was proving a headache. The British technically controlled the seas around Italy, so Pash had to secure their permission to slip a sub in. But the British didn't trust the bastard brigade any more than the Americans did, because Pash once again refused to explain his mission. German U-boats still haunted the seas anyway, making a submarine run risky in the extreme. Pash ended up wasting whole weeks flying back and forth to London and Washington and Naples trying to arrange things.

It didn't help that OSS kept changing the mission for reasons that Pash couldn't fathom. They suddenly wanted to swap out the submarine for PT boats, the kind Jack Kennedy had captained. However frustrated, Pash bit his tongue and hustled up some PTs. Even worse, OSS twice scheduled extractions and then scrubbed them without explanation—once without bothering to tell Pash. He had stayed up all night to hear how the mission went, fretful and anxious, and come dawn was furious to learn that nothing had happened.

Pash had just about blown his fuse when, in late spring, after months of preparation, the entire operation collapsed. On May 30, U.S. military officers arrested an Italian agent code-named Morris. OSS had hired Morris to help infiltrate Rome and make contact with Amaldi and others. Unfortunately, they hadn't done their homework, and Morris turned out to be a double agent; he'd been tipping off the Nazis for months about operations in Italy and North Africa. When the military raided his apartment in Naples, they found classified documents related to Shark, including the names of people involved and a timeline for the operation.

The army appeared to have stopped Morris before he warned the Germans about Shark. But no one knew for sure, and Pash couldn't risk an agent walking into a trap with Amaldi. Operation Shark was dead.

Most of the fault for this mess rested with OSS, but Pash—a supposed security expert—also took some blame. Even worse for his reputation, his entire scientific team soon quit, convinced Alsos was a waste of time. The bastard brigade's chief bastard had already angered several officers in Naples, and now Washington was looking askance as well. Most frustrating of all, after months of work, Pash had learned exactly zero about the Nazi atomic bomb.

CHAPTER 39

Biscay Blues

It was a long, cold, depressing winter for Joe Kennedy in the Bay of Biscay—dangerous from a military point of view and dispiriting from a personal one.

At the war's outset a German admiral named Karl Dönitz had mocked Allied efforts to hunt submarines with planes, saying, "An aircraft can no more kill a U-boat than a crow can kill a mole." At the time Dönitz was right, but new technologies soon drove the moles to near extinction. Allied scientists managed to shrink radar equipment to fit inside planes, which the Germans had considered impossible. They also shrunk down their sonar to fit inside buoys, and made the gear robust enough to drop the buoys from planes into the sea, where they pinged any subs that passed by. German engineers did fight back with defense measures, including the so-called *pillenwerfer*, which was basically a gigantic Alka-Seltzer tablet. When released into the water surrounding the sub, it produced trillions of bubbles that absorbed some of the incoming waves instead of reflecting them back. But Allied sonar and radar just kept getting better, and plane crews also learned to spot subs visually—either by the oil slicks they left behind or their sharklike shadows from above. They could then use phosphorus flares to expose them when they surfaced for air. Once located, submarines became vulnerable to attack by depth charges, and in tandem these tactics all but annihilated the U-boat fleets off France. For every twenty German subs that set out in the winter of 1943–44, the Allies destroyed nineteen.

That's not to say the Allies didn't suffer, too. German fighters often escorted U-boats, and because they were much nimbler than the sub-hunting planes, they sent dozens of flaming wrecks into the sea. On one excursion in November, Kennedy's crew had to fight off two German hawks simultaneously and barely survived; a few months later they had to make an emergency landing when an engine got knocked out. Other planes proved less lucky. By March 1944, Joe's squadron had lost three dozen of its 106 men, including several friends of his.

What really upset Joe was that, despite all the danger, hunting subs offered no chance for personal glory. Crews succeeded or failed en masse, and Joe's patrol planes seldom killed subs themselves, only marked them for others. Unlike his kid brother, Joe never had a chance to distinguish himself as a hero, and he considered ditching Europe altogether and joining a unit in the Pacific.

Little did he know, his salvation was already in the works. In the spring of 1944 General Eisenhower made destroying the huge concrete bunkers in northern France a priority, and over the next few months the Allies dropped fifteen million pounds of explosives on them. Several attacks featured the new 12,000-pound "earthquake bombs," which caused tectonic booms when they landed. The mysterious V-3 bunker at Mimoyecques was pummeled especially hard, enduring a dozen raids involving 2,200 planes; the surrounding landscape looked stricken with smallpox afterward, pitted and scarred with overlapping craters. Overall, these raids cost the Allies 450 planes and 2,900 men—and did zero good. Reconnaissance photos showed that activity at the bunkers never slowed a whit. Soldiers and supplies kept pouring in, and the bunkers kept expanding. The most devastating bombs in the Allied arsenal were impotent.

Allied leaders began scouting around for new strategies, and they soon hit upon a doozy. In March 1944 a navy captain suggested taking old, "war weary" Liberator bombers and, instead of selling them for scrap, turning the planes themselves into bombs. That is, he suggested packing them with explosives and slamming them into targets. Given the expense involved, not to mention the danger, the project made sense only for the most impregnable sites, and the bunkers in northern France shot to the top of the list.

The army and navy worked independently on these flying bombs—the navy called its scheme Project Anvil, the army Project Aphrodite—but the general idea was the same. Each mission required two planes, a mother ship and a baby ship. The babies were the war-weary bombers, and to get the projects rolling, the military turned loose a crew of mechanics to strip every last component from them, inside and out: ammo bins, bomb racks, benches, tables, turrets, guns; they even removed load-bearing structures, and had to shore up the bombers' interiors with timbers. In all, the mechanics (who had a hell of a good time with this) removed four tons of metal, lightening the bombers by almost one-quarter. As soon as the mechanics finished, other crews came along and filled the babies back up with explosives. They used either napalm (gelled gasoline) or nitrostarch (acids mixed with cornstarch, a common ingredient in hand grenades of that era). The explosives came packed in crates, and each bomber held roughly ten tons of crates. Thus loaded, the bomber became a bomb.

Getting one of these flying bombs off the ground was harrowing. The one component of the original plane left intact was the pilot's seat, and if you've ever overloaded a moving van, you can imagine the feeling of rumbling down the runway here: furniture piled to the ceiling all around you, looming over your shoulder, shifting ominously with every bump. Except in this case, it's not lamps and bookshelves knocking about, it's napalm. And really, the pilot was the lucky one. While the plane had only one seat, the Anvil/Aphrodite missions required two people to fly, a pilot and an engineer. That meant the engineer had to squat down during takeoff, wrap his arms and legs around the pilot's seat, and hope to god for a smooth ascent.

Once airborne, the pilot would settle into a cruising altitude of 1,800 feet. The engineer would peel himself off the floor and set up the plane's autopilot—a series of gyroscopes that controlled pitch, roll, and yaw. Each gyro had to be calibrated independently, and they would do most of the work of getting the flying bomb to France. But not all of it. Actually steering the plane into the bunker would require finer control than the autopilot could provide, and here's where the mother ship came in. The mother had remote-control capabilities, allowing it to nudge the baby left or right and send it into a dive. The only problem was, given the crude state of electronics at the time, you couldn't just buy a remote-control kit and install it. Several different custom-made circuits had to be wired up and calibrated, and this could only be done midflight. All told, this equipment took the engineer roughly half an hour to set up.

When this step was finished, the engineer would radio a code word to the mother ship (e.g., "flyball" or "change pictures") and the mother ship would take over. His mission complete, the engineer would descend into the belly of the plane via trapdoor and wriggle through a two-foot-wide crawl space toward an emergency escape hatch. There, he'd strap on a parachute and—trying to ignore the propeller whirring a few feet from his head—fling himself out into the wild blue yonder.

Meanwhile, the pilot would go through a series of checks to make sure the remote-control circuits worked. With this confirmed, he'd point the baby toward France and send it into a gentle dive at 175 miles an hour. Then he'd lower himself through the trapdoor and retrace the engineer's path toward the parachute hatch. He'd have to hurry here, because the plane was already sinking, and parachuting out at any height below a thousand feet was tantamount to suicide. Last thing before jumping, he'd arm the explosives. This involved locating several cables on a large bank of electronics and jerking them out—essentially cocking the world's largest revolver. Then, Geronimo! Assuming the plane was still over England, he'd land in a field or a pasture. If not, he'd take a dip in the Channel.

From here on out, the mother ship—flying high above the baby—would steer it over to France. (The baby was painted white or yellow on top to make it easier to see.) Once the baby drew within range of the bunkers, someone in the mother ship would hit the "dump switch" to snap the elevators on the baby's wings into the down position. This would send it careening into the bunkers and rid the world of Hitler's atomic rockets in a huge explosion.

The most dangerous part of the mission involved the so-called jump team, the pilot and engineer in the baby ship. Because while this wasn't quite a Japanese kamikaze mission—in theory, they both survived—it wasn't much safer. You might as well try to surf a torpedo up to an aircraft carrier. The number of ways the mission could go awry is too long to list, but a few things stand out. Explosives are inherently unstable, and electronics back then were pretty dodgy, so a rocky takeoff or some midair turbulence could detonate the payload. Equally dangerous was the act of exiting the plane. It's easy to forget in our age of skydiving grandmothers, but parachuting—especially at such low altitudes—was once considered a desperate maneuver: military manuals often advised pilots to crash-land if at all possible, even if their planes were full of fuel, rather than risk jumping out. And after all this peril, Anvil/Aphrodite had a slim chance of succeeding anyway. As large planes, bombers could hold an impressive amount of explosives, but they were actually too well engineered for this type of mission: they were so stable in flight that it was hard to make them plummet on command. Worst of all, while the concrete bunkers in France spanned several football fields, the most vulnerable points, especially the V-3 gun muzzles at Mimoyecques, were only a few yards wide.

But once again, fear of Hitler's "wonder weapons" overrode all other considerations, and in the spring of 1944 the U.S. Navy began plans to convert war-weary PB4Y-1 Liberators into flying bombs. Fortunately for Joe Kennedy, the Liberator happened to be the plane with the super-complicated cockpit that he'd already mastered—and Project Anvil was just the sort of recklessly heroic mission he'd been dying to fly.

CHAPTER 40

The Fat Captain

After months of idling, Alsos finally saw some action in Italy in the summer of 1944. The Allies toppled the mountaintop monastery of Monte Cassino in mid-May, and the rest of the exhausted country fell in short order, including Rome on June 4. Boris Pash packed the remainder of his crew into jeeps and raced the 120 miles north from Naples. They entered the city on June 5 at 8 a.m. and immediately set off to find Edoardo Amaldi, hoping against hope that the Germans hadn't absconded with him.

They hadn't. The moment Pash and company stopped their jeeps near Amaldi's house, a mob of children surrounded them, awed to see real Americans. The soldiers smiled and patted their heads but didn't linger. Operation Shark documents had described Amaldi as a slim five foot five, with "blue eyes, chestnut hair, [and] tortoiseshell glasses," and Pash was immensely relieved to find someone of that description at the address indicated. After six futile months Alsos had snagged its first target.

Pash explained to Amaldi that Alsos would have to detain and debrief him, which Amaldi accepted with good humor. Indeed, he was flattered that the mighty Americans took an interest in him; they even called him *doctor*. When the Germans had taken over Rome, he said, they'd cared only about rounding up military officials, ignoring scientists completely.

However relieved he was to see Amaldi, Pash didn't remember their conversation quite as fondly. He was actually cringing inside the whole time, waiting for Amaldi to rebuke him for the botched Shark mission. Finally Pash couldn't stand the tension any longer and broached the subject himself: "We are certainly sorry, Dr. Amaldi, to have caused you difficulty and danger."

The physicist asked what he meant.

"I'm referring to the time our agent contacted you, with that plan to cross the lines."

Amaldi looked bemused. "There must be some mistake, Colonel Pash. I had not heard of such a plan."

Now it was Pash's turn to look perplexed. "Do you mean to tell me, Doctor, that you were not alerted to be at a beach area in late February, to be picked up by a PT boat?"

Amaldi blinked. He had no idea what Pash was talking about.

The truth hit Pash all at once. The delays with Shark, the shifting plans, the unexplained cancellations: OSS had been bullshitting him the whole time. Perhaps OSS really had intended to help him and had merely bungled things. Or perhaps they'd never intended to help and had simply been stringing Pash along. Perhaps it was all part of some larger scheme he couldn't discern. Regardless, they'd played Pash for a fool. He managed to maintain his composure in front of Amaldi, saying that he must have misunderstood something back at his base. "But the wheels in my head," he recalled, "were grinding out sparks."

When they finished talking, Pash gave Amaldi strict orders to stay in Rome, then left to hunt down another physicist. After that, Pash retired to his room at a local hotel, along with the rest of Alsos, planning to debrief Amaldi and others the next day. But just as Pash was settling down to a postdinner glass of Chianti, Amaldi knocked on his door, looking shaken.

There's an American captain in the lobby below, Amaldi said. He has orders to take me away.

"By what authority?" Pash asked.

"Orders from the president of the United States," Amaldi said. American attention might be flattering, but this seemed absurd.

Confused, Pash went to investigate. In the lobby he found a "big hunk" of a man with heavy eyebrows lounging in a chair. The captain looked Pash up and down and smirked. "Colonel," he said, "looks like you and I are going to have to reach an understanding." Amaldi, he said, was going with him.

All the frustration of the past six months erupted inside Pash. "Attention!" he barked. The captain looked startled. "Attention!" Pash barked again, and the man jumped to his feet. He stood a good half-foot taller than Pash, but the colonel got right in his face. "What is this all about?"

The captain protested that Pash was interfering with a vital mission regarding Amaldi. "This is secret, Colonel. But since you got yourself into this—well, I'll tell you. I am to deliver him to the Alsos mission in Naples."

As an old football coach, Pash knew how to rip someone a new asshole. "You're looking at the Alsos mission!" he screamed. "No doubt you're from OSS. Your job was to get Dr. Amaldi out of Rome several months ago. You failed. You didn't even go after him."

There was plenty more invective where that came from, and Pash eventually dismissed the man: "You have no business in Rome. If I run across you again, I'll bring charges, and I can think of plenty. Now get out."

Pash spent the rest of the night smoldering. He'd endured six months of excuses and lies, and now this. Who the hell did OSS think they were? He had half a mind to pursue the matter further, make an official complaint.

But with the arrival of morning, he had more important things to worry about than some OSS lunk. It was June 6, 1944. The invasion of France had begun. Which meant Alsos now had much bigger targets to track down, including Frédéric and Irène Joliot-Curie.

CHAPTER 41

Augers & Peppermint

You've probably never spent much time thinking about the science of D-Day, but both geology and nuclear physics played a surprising role in this most important invasion of the war.

As for geology, the Allies planned to invade France with several thousand heavy-duty vehicles, which meant they had to choose the landing site carefully: trucks and tanks don't do much good if they're sunk to their axles in muck. Accordingly, General Eisenhower's office sponsored several geological raids on the European mainland to determine which beaches were best. The commandos on these raids would approach the coast at night in inflatable rafts or midget submarines, often swimming the last few miles to shore. Under cover of darkness they'd survey the beach and take soil samples for analysis back in the lab. They also plumbed the water depth at various points and mapped the speed and strength of currents. (Frustratingly, whenever a storm redistributed the mud and silt and sand, they'd have to return another night to remap everything.) Then, after a full night's work, they'd plunge into the surf again and swim several more miles back to the raft or sub.

Overall these troops did incredible work, combining the danger of undercover ops with the rigor of scientific research. But they did commit one memorable gaffe, leaving a soil auger behind one night where the Nazis seemed likely to find it. Because its presence on the beach could have betrayed where the invasion would take place, the Allies seriously debated rounding up every auger in the free world, loading them onto planes, and carpet-bombing the beaches of France, to throw the Nazis off. Sadly, the auger bombardment never took place, in part because of a wartime shortage of such tools.

In addition to gathering data firsthand, geologists studied reconnaissance photos. Just as nuclear physicists investigated the structure of atoms by pinging them with neutrons, geologists realized that bombs hurled down from airplanes could reveal a lot of information about the landscapes of Normandy. They were especially interested in the shapes of bomb craters. Strong, cohesive soils—which you could drive tanks on—produced U-shaped craters when bombed, with steep sides and large, solid chunks of material in the surrounding aureole. Weak or mucky soil, in contrast, produced V-shaped craters with mud splatters or modest chunks in the aureole. Similarly, the Nazi habit of stealing hay carts from French peasants and using them to haul heavy equipment around ended up biting the Germans in the derrière, since the depth of the ruts that resulted provided clues about the stability of the underlying earth.

Based on such work, geologists convinced Eisenhower and his staff to abandon their preferred D-Day invasion site and focus instead on Omaha Beach, several miles to the east, which offered better traction. Thanks to their efforts, the Allies successfully landed 8,851 heavy vehicles on the first day, and over 150,000 in the first seven weeks. In addition, at least one geologist landed with the troops on D-Day to offer on-the-ground (so to speak) expertise; small teams of geologists then helped lay down twenty crucial airstrips over the next two months. The soldiers who landed at Normandy deserve the lion's share of credit for the invasion, of course, but the geologists who told them where to set their feet played a vital role.

As for the role of nuclear physics in D-Day, the issue was, once again, dirty bombs. Even though the Nazis hadn't rung in the New Year with them, fear of dirty bombs continued to fester in the minds of American officials in the run-up to June 6. As noted, dirty bombs kill by spewing radioactive isotopes, and they're well suited to resisting invasions. Even a few hours of exposure can kill a person, and because radioactivity is spookily invisible to all our senses, soldiers might not realize they were being exposed until too late. Furthermore, rumors about radioactivity would probably sow widespread panic within the ranks. (Nuclear *bombs* were top-secret in 1944, but the public knew plenty about the dangers of radioactivity in general, thanks to lurid newspaper stories of people's jaws falling off and other horrible deaths after exposure to radium paint and uranium health tonics.) Finally, because radioactive atoms blend with the soil and can't be scrubbed clean, poisoning a parcel of land with the right isotopes would prevent an invading force from occupying it for months, even years.

Given the rate at which reactors work, Manhattan Project scientists estimated that the Germans could contaminate four square miles per week. That might not sound like much—the Nazis obviously couldn't barricade off all of northern France that way—but the Reich didn't need to poison the entire landscape. They needed to poison only certain areas: ports, beaches, railyards, airports, highways. Or they could spike reservoirs and crops, making it impossible to find clean food or water. By concentrating on key infrastructure, in other words, the Nazis could neuter the entire D-Day operation with a few ounces of isotopes.

The Manhattan Project first warned Allied commanders about the dangers of radioactivity in April 1944. As usual, the British rolled their eyes at the American worry over this: on first hearing about dirty bombs, Churchill furrowed his brow and said, "This all seems very fruity." Eisenhower, in contrast, was spooked, and soon initiated Operation Peppermint, history's first attempt to grapple with radioactive warfare.

In the first phase of Peppermint, planes with Geiger counters swept the northern coast of France, looking for hotspots of radioactivity. (Some officials feared that the concrete bunkers there actually housed atomic reactors, in which case they might be leaking radioactive species into the air.) In addition, several foot soldiers received training with Geiger counters, which they planned to carry into battle on D-Day. These lunchbox-sized counters were housed in watertight cases with gray enamel finish, and they emitted a tone whenever they detected something. The initial plans for Peppermint called for eight Geiger-counter teams of four men each spread among the invading armies. The number actually deployed remains unknown, although some did reportedly join the forces at Omaha Beach.

Medical personnel also played a role in Peppermint, albeit unknowingly. A few days before June 6 they received orders to keep an eye out for clusters of certain symptoms, including fatigue, nausea, rashes, and low white-blood-cell counts, all signs of radioactivity poisoning. To avoid spreading fear among the troops, these medical teams never learned the reason for their vigilance. Their superiors made up a story instead about an outbreak of exotic germs that needed tracking.

The final arm of Peppermint involved dentistry, of all things. Given the unreliability of mobile Geiger counters back then—their batteries and vacuum tubes were especially spotty—troops on the ground needed something more trustworthy to detect contamination. Someone suggested dental film, which is quite sensitive to radioactivity. (Radioactivity was in fact first discovered with photographic film in 1896, since exposure to radioactivity will stain film with foggy patches and/or black spots.) So several units among the D-Day forces were handed rolls of film and told to stop every so often in the field to develop them. Again, these troops weren't told why they were doing this; the cover story was that a few crates of expensive film had been ruined recently, and the higher-ups wanted to nail down the cause. It must have seemed like sheer raving lunacy, amid the chaos of D-Day, to stop and develop blank dental film while Nazis were shooting at you. But the soldiers no doubt chalked it up to the *Catch-22*–like madness of the army and did as they were bid. However crazy it seemed, their efforts removed a weighty fear from the minds of Allied leaders and allowed the D-Day operation to proceed smoothly.

CHAPTER 42

Remus

Moe Berg dang near went broke waiting to join the war. He'd finally landed the assignment he coveted—going undercover in Europe to ferret out atomic secrets—and OSS minced no words in declaring how important the work was. Like Shakespeare wrapping up a crucial scene, Berg's briefing officer informed him, in a rhyming couplet, that the outcome of the entire conflict might rest on his shoulders: if the Germans get the bomb, the officer warned, "We can lose [the war] at midnight less one. Find out what they're doing, and we've got it won."

Unfortunately, Berg's first assignment was in Rome, and with the American army stalled south of there, he couldn't just waltz into the Eternal City and start poking around. OSS toyed with various schemes to deposit him on Italian soil, including via submarine, but the agency needed special paperwork for that and couldn't get the right signatures. So Berg sat around the Mayflower Hotel in Washington for months, boning up on physics and racking up huge bills. For once a life of leisure frustrated him, and he took out his vexations in petty ways, picking fights with OSS staff and singling out colleagues, seemingly at random, as enemies. He'd then go to ridiculous lengths to avoid running into these devils, including hiding behind office furniture. Berg soon earned a reputation as a "brilliant but temperamental" diva.

Finally, in early May, Berg got orders to ship out. Because he'd be infiltrating Rome, OSS code-named him Remus, after one of the city's mythical, wolf-suckling founders. For equipment, he was issued a pistol and a cyanide-filled rubber L-capsule. He also put in a request for a dozen pairs of nylon stockings, which were scarce in Europe and which he probably planned to use to wheedle information out of women—or seduce them.

Berg hardly proved a smooth operator, though. On the flight out of Washington he was the only man not in uniform, wearing a dark blue suit and gray fedora. He had $2,000 cash in one suitcoat pocket, the pistol in another. And the first time he leaned over on the plane, the gun slipped out and clanged onto the floor—a total rookie move. "They gave me this as I came onboard," he muttered in apology. An officer more used to traveling with firearms suggested tucking it into his belt, but Berg's soft belly pushed the gun out twice more during the flight.

Like Boris Pash, Berg took a circuitous route to Italy, hopping from London to Portugal to Casablanca to Algeria to Sicily to Naples. As soon as he landed, his cover was nearly blown, for exactly the reason Groves had feared. Actor Humphrey Bogart, baseball star Lefty Gomez, and former heavyweight champ Jack Sharkey were all visiting Italy to boost the morale of troops, and the last two happened to bump into Berg in Naples. Gomez knew Berg well, having accompanied him on the all-star tour of Japan in 1934, and he called out Moe's name. Mortified, the catcher put a finger to his lips—*shhh*. He then turned and melted into the crowd on the street, leaving Gomez baffled.

Berg apparently thought it prudent to lie low for the next few weeks. In fact, he disappeared so thoroughly that even his bosses lost track of him. As a result, with Rome on the brink of collapse in early June, OSS was reduced to sending out cables to track down its own agent. "In the event that Berg has not taken action in Italy," one read, "he should leave immediately." London, Cairo, Algeria, Istanbul—they checked everywhere. Berg never responded.

He hadn't shirked his duty, though. He first heard about the liberation of Rome while dining with an American general in Bari, a city on the heel of Italy. After Berg explained his situation and laid on a little charm, the general loaned the catcher a private plane and pilot, who whisked him back to a town near Naples. Because Berg couldn't drive, he cadged a ride to Rome; friends called him "World War II's most prolific hitchhiker." He arrived four hours later and set out to find his top target there, nuclear scientist Edoardo Amaldi.

Back in Washington, Berg had made a risky decision not to take a commission in the army and wear a uniform. Civvies allowed him to blend in more easily as he moved around; it also helped him coax information from strangers, since people tend not to confide in soldiers. But the rules of warfare grant soldiers certain immunities that civilians lack; in particular, civilian spies can be shot without repercussions. Berg knew this, and decided to chance it.

But for some reason—perhaps it facilitated movement in a city under martial law; perhaps he thought it would give him extra authority; perhaps he was just taking another crazy risk—Berg donned a captain's uniform before dropping by Amaldi's home. He arrived there to find the physicist covered in grease, repairing his beloved bicycle, which he'd hidden from the Nazis during the occupation. The two men set to talking, and Berg discovered that the military had already contacted Amaldi; in fact, an officer had ordered him to stay in Rome. This irked Berg, who confided that *he* had orders from the president of the United States to take Amaldi away. Amaldi protested that he should at least inform the other officer, which Berg permitted. They traveled to the hotel where the officer was staying, and while Amaldi went upstairs, Berg took a load off in the lounge, mighty pleased with himself.

He expected Amaldi to return in a few minutes. Instead, a bantamweight officer came stomping down the stairs, absolutely fuming. Berg said coolly, "Colonel, looks like you and I are going to have to reach an understanding." The colonel, in no mood to negotiate, barked "Attention!" Berg just blinked. "Attention!" the colonel repeated, and Berg—apparently only now remembering he was in uniform—jumped up and saluted. Boris Pash then got in Berg's face: "What is this all about?"

Yes, the lazy "hunk" of a captain that Pash chewed out that night was none other than Moe Berg. "I am to deliver him to the Alsos mission," Berg sputtered. "You're looking at the Alsos mission!" Pash roared. At that moment Berg embodied every OSS frustration that Pash had endured over the past six months, and he gave full vent to his wrath. Given their mutual love of baseball, Coach Pash and the ex-catcher should have gotten along swimmingly. Instead, Pash ripped Berg six ways from Sunday, and he never tired of badmouthing the big lug afterward.

But if Pash thought he could intimidate Berg, he was wrong. However sheepish he felt, Berg dropped by Amaldi's home again the next day when Pash was absent and smoothed things over, in part by treating the half-starved physicist to a meal in a fancy restaurant with fresh meat and gold-plated cutlery. Berg also brought sweets for Amaldi's children, courtesy of Enrico Fermi. As a result of these kindnesses, Amaldi agreed to let Berg interview him about nuclear fission. Among other things, Berg learned that Amaldi, after being drafted into the Italian army, had served a tour in North Africa, then returned to his lab in Rome. But far from collaborating with the fascist government, as the Allies feared, Amaldi had abandoned all fission research. In September 1943 he'd even gone into hiding for five months, fearing for his life during the German occupation of Italy.

As for the state of fission research elsewhere, Amaldi admitted that he'd talked to several German scientists during the war. These included Heisenberg and Hahn, and he suspected that Heisenberg in particular was working on a bomb. But he hadn't been in touch with them since 1942, and could offer no details. For the skittish nuclear scientists in America, this was the worst possible news: it reinforced their fears about German intentions without offering any hard facts to constrain their imaginations.

Besides Amaldi, Berg interviewed several other Italian physicists over the next few weeks. To be sure, Alsos had already located most of them, even questioned a few. But Berg's interactions with the scientists were warmer, friendlier, and they highlight a key difference between him and Boris Pash. Pash was a heckuva soldier, resourceful and daring, but he worked best in the chaos near the front lines. After securing targets in Rome, in fact, he scurried off to participate in the D-Day invasions and left Italy behind. Berg proved more patient, and while he certainly blundered sometimes—James Bond never would have dropped a gun in public—he also excelled at the real work of intelligence agents: winning people's confidence and drawing out their secrets, skills the blustering Pash never learned. One Italian physicist, for instance, refused to talk at first, so Berg casually pulled a book of Petrarch's sonnets off the man's shelf and began reciting a few passages—in the original Italian, natch. Startled, the scientist asked Berg how he knew Petrarch. Berg had studied him at Princeton, it turned out, and they fell into a literary discussion. They were soon toasting the poet with wine, and almost before the scientist realized it, he was spilling everything he knew about atomic research.

And when charm didn't work, Berg was not above subterfuge. One top scientific target in Rome—Gian-Carlo Wick, a broad-shouldered physicist who spoke in slow, measured phrases—was decidedly unimpressed with Berg upon meeting him. While brilliant, Berg had an encyclopedic mind. He horded facts and took little pleasure in things like art or music for their own sake; he could recite page after page of biographical details about Mozart, but the music itself didn't move him. Wick in contrast was deeply cultured and deeply read, and he spoke Romance languages even more fluently than the catcher. He was a prototypical European intellectual, in other words, and he dismissed Berg as a shallow American.

Berg nevertheless got the last laugh. Despite his disdain, Wick granted Berg an interview, thinking that this poseur could do no harm. Wick had studied under Heisenberg in Leipzig a decade earlier (in addition to physics, they shared a passion for Ping-Pong), and he told Berg that they'd kept in touch during the war. Heisenberg had even sent Wick a postcard in January 1944 updating him on life in Germany.

Eventually Wick tired of the interview and asked Berg to leave. But on his way out the door, the catcher managed to filch the postcard from Wick's desk. That night he translated it from German to English and photographed it for intelligence officials. He then dropped by the physicist's office the next day and slipped it back into place. Wick never suspected a thing. He'd let his guard down, assuming that a lunkhead athlete could never put one over on a brilliant scholar like him. Sometimes you have to play a fool to catch a fool.

Heisenberg's postcard contained one line of real emotion—"The time in which one could think calmly on physics is so far away that it seems as if ages had passed." Otherwise it simply related what he'd been doing for the past few months, which was mostly dodging disasters. He mentioned the fire at his in-laws' home in Berlin and the bombing of his own home in Leipzig months later. As a result he'd moved his family to a cabin in the Bavarian Alps, and visited them as often as he could. This news probably seemed mundane to Heisenberg while he was writing it, but thanks to these unguarded words, OSS now knew of one place to look for him.

The postcard also mentioned Heisenberg's research. His lab in Leipzig had been leveled by bombs, he reported, while "the Berlin institute still stands." But recent intelligence reports had hinted that Heisenberg was actually shifting his lab south of Berlin, beyond the reach of American bombing raids. Berg questioned Wick on this point, and Wick conceded that Heisenberg was moving to a wooded region. He declined to say more, but this tidbit narrowed the search area substantially.

When Berg relayed the results of this interview, including the postcard, OSS was ecstatic; Wild Bill Donovan personally congratulated him. Depending on whom you asked, the catcher was either a lunkhead or a diva—but he'd delivered when it mattered.

CHAPTER 43

Aphrodite vs. Anvil

Because the Busy Lizzie V-3 gun was not yet ready by D-Day, the defense of the Reich fell to another weapon, the V-1 rocket. Hours after the invasion in Normandy began, the German high command in Berlin cabled a code phrase—"junk room"—to several units of soldiers stationed at sites across northern France. Nazi engineers toiled for days without sleep to prep the launch equipment, and the first V-1 slammed into Kent, near London, at 4:18 a.m. on June 13, 1944. Hundreds more rockets—each one long and finned, like an airborne shark—quickly followed.

The barrage had two main effects. First, it hindered the Allied military. At one point 40 percent of all bombing raids originating from Great Britain were devoted to attacking V-1 launch sites. Reconnaissance efforts were equally intense, with the Allies snapping more than a million photographs of them. Both these activities drained resources from other operations. Rocket attacks also sowed chaos in southern England by taking out factories, power plants, water mains, roads, and transport centers. Supply chains were seriously disrupted, and the Allied armies struggled to make headway against the Germans on the mainland as a consequence. A full month and a half after D-Day, they'd advanced just twenty-five miles inland from Omaha Beach—at a cost of 100,000 casualties.

Second, the missiles terrorized civilians. Because the Germans couldn't really aim V-1s, no one knew where they'd land. That fact, coupled with their speed (close to 400 miles per hour), made them virtually impossible to defend against. The British tried floating barrage balloons in their paths, which sometimes triggered them to explode early, and daredevil pilots learned to knock them off course by edging up to them—sometimes within six inches—and dipping their wings up and down to create turbulence. But these measures couldn't stop more than a handful of rockets, and millions of people went to bed every night fearing that the next one would have their name on it. The barrage got so bad—2,700 people died within the first two weeks—that a drunken Winston Churchill ordered the army to launch poison gas against German civilians in retaliation. Churchill's cabinet demurred, and the British government suppressed the incident for years. Instead, it began encouraging Londoners to move to the countryside. More than 1.5 million eventually fled.

Overall, there was a real fear that the German rocket attacks would reverse the tide of the war and drive the Allies off the continent for good. And the really scary thing was that worse weapons were on the way. So far, Germany had launched only V-1s, which were relatively small. Meanwhile, the giant concrete bunkers in northern France were still under construction, and whatever weapons they portended would surely be worse—especially if (rumors continued to swirl) the new rockets were nuclear. Because the bunkers were located in Calais, a good hundred miles northeast of where the Allied armies were pinned down, there was no hope of attacking them on the ground, either.

All this merely ratcheted up the pressure on the Allies to knock out the bunkers before they became operational. With conventional bombs proving impotent, the one real hope lay with the American military's plan to load war-weary planes with explosives and slam them into the sites using remote control.

Farsighted aviators had been tinkering with remote-control planes—drones, in effect—since the days of the Wright brothers, but electronic systems in the early 1900s couldn't send and receive signals over the necessary distances. During the war, the army in particular struggled with these limitations. Their Project Aphrodite drones could obey only three commands (left, right, dive), and those only haltingly.

By mid-1944, however, the navy's Project Anvil had achieved a breakthrough, thanks in part to an irascible thirty-five-year-old engineer in southeast Pennsylvania named Wilfred "Bud" Willy. A new radio-based system that he helped design could handle up to ten commands, allowing the mother plane to steer the baby with far greater precision. The mother could vary the baby's speed, for instance, and detonate the explosives remotely; it could even turn the baby's heater on and off. No one had ever seen such sophisticated drones.

The only hitch in Willy's design was the remote-detonation circuits. For some odd reason they kept popping on during test flights; had these been real missions, with real explosives, the babies would have blown prematurely. After weeks of scratching their heads, Willy's team finally chalked the problem up to, essentially, boogie-woogie. Because the mother ship had to fly three vertical miles above the baby, the baby needed pretty sensitive radio detectors. Sensitive enough that they appeared to be picking up stray signals from pop music and news radio stations in nearby Philadelphia, which were flipping the circuits on willy-nilly. Willy's team never quite resolved the problem, but they consoled themselves that stray radio signals would be less of an issue over the English Channel.

Meanwhile, both the navy and army were recruiting pilots in England for the mission, including Joe Kennedy. Although Joe and his crewmates had long ago flown enough missions for an honorable discharge, he talked his men into staying on through D-Day, hoping for another shot at glory. After D-Day, however, pilots hunting U-boats in the Bay of Biscay were reduced to a mere support role, and his crewmates said adieu. The only silver lining to being abandoned was that it freed up Joe for other work, and when officers at his base began recruiting pilots for another blatantly unsafe mission, he volunteered. For security reasons the recruiters couldn't tell him what the mission actually involved, but Joe didn't care. The word "dangerous" was enough for him.

The pilots of the new Special Attack Unit No. 1 were briefed on the mission in late June — *brief* being the operative word. A tight-lipped officer told them they'd be taking off in planes loaded with napalm, pointing them at France, then jumping out before they exploded. Any questions? Uh, a few. The officer refused to say more, though, and quickly dismissed the men. Still, the flyboys didn't let the danger dampen their spirits. Whenever a new pilot reported for duty with the army group, the others would call out, "Hey, fellows, another nut's arrived," and they'd all laugh. The navy group was giddy, too, especially Joe — he couldn't wait to read the headlines about himself back in Boston.

As training got under way, however, that enthusiasm gave way to frustration. The British military was understandably jumpy about having planes loaded with explosives sitting around at air bases, so base commanders kept tossing these hot potatoes around, forcing the Americans to transfer to other sites. Obviously the crews couldn't fly planes full of napalm around, so each time a transfer order came, they had to unload all the crates of explosives, fly the planes empty, and ship the crates ahead via trucks — a gigantic pain in the neck. Settling into the new base was never fun, either, since they invariably got dumped in some shithole. Probably the most miserable site was Woodbridge, in extreme southern England; it served as the emergency landing strip for any planes that suffered damage in France. If you'd lost an engine and could barely steer; if you were hemorrhaging gasoline by the gallon; if the pilot and copilot had both been shot and some poor radioman was behind the stick now, flying for his life—well, you said a Hail Mary and steered for Woodbridge, where spectacular fireballs were a regular feature of life.

After several transfers, the Aphrodite and Anvil crews ended up at an unused base in rural England where a farmer actually raised beets and turnips inside the triangle of runways. The place was also infested with rats, which nipped the pilots as they slept in their barracks. They tried shooting them at first, but the bullet holes they left in the wall only let in more cold and damp, so they gave up. Most navy pilots preferred the company of rats over army pilots anyway, and vice versa. Being in different branches of the service, the two groups had a natural antipathy for each other, and their isolation on the base only exacerbated the rivalry.

The pilots' biggest frustration, though, was the weather. The crews for the baby ships needed practice setting up the remote-control system in midair, and the crews for the mother ship needed practice flying the drones. But to hold practice flights, they also needed CAVU conditions—ceiling and visibility unlimited, with zero cloud cover up to 20,000 feet. Otherwise the mothers couldn't see the babies from so far above, much less steer them. In other words, they needed just the sort of gorgeous, sunny days that southern England almost never has. Whole weeks would pass with no action as gray clouds squatted above them. And oftentimes, even when the forecast did look promising, Kennedy and the other pilots would go through all the work and stress of prepping their planes only to have the flight scrubbed at the last second. Any measly bank of clouds rolling through spelled doom.

Officers tried to keep morale high by putting the pilots on triple rations, with extra candy and cigarettes. As a special treat, Winston Churchill also graced them with a visit. Churchill even tried, in his mischievous way, to squeeze his spherical bulk through one of the tiny emergency hatches in a baby plane to get a look at the explosives. Oddly, he was wearing golf shoes with metal spikes that day, and an army major suddenly realized that a stray footstep might puncture a cable and short-circuit the electronics inside, with fatal results. So the major ran up and grabbed Churchill by the most obvious target that presented itself—his generous bum—and yelled, "Mr. Prime Minister, you can't get into that airplane with those spikes." Churchill turned to see who'd goosed him and said, "Oh I can't, can't I?" Everyone tensed. But the major's courage amused Churchill, who conceded, "Very well, then. Just show me where to put my foot and I'll climb out."

Candy and visits from Churchill could soothe the pilots only so much, though. The weather continued to torment them, and they eventually began taking out their frustrations in boneheaded ways. One pastime involved playing bloody knuckles with their foreheads, butting each other like goats until someone cried uncle. They also went for joyrides in jeeps and deliberately crashed them into trees, bailing out at the last second. Ha ha.

Kennedy didn't participate in these hijinks, but if anything he was mired in an even deeper funk than the other pilots. For one thing, there were dire tidings from France. After weeks of little to no progress in Normandy, General George S. Patton's army suddenly engineered a breakout in early August. His troops began steamrolling the Germans there and liberating huge swaths of land. Most of the world cheered, but Joe despaired over the news. If Patton liberated France too quickly, how would he win any glory for himself?

Even worse, the navy mission ran into unexpected delays. To get their fancy remote-control planes to England, Bud Willy's team in Pennsylvania had to hopscotch across the Atlantic Ocean. Unfortunately, bad weather and a lack of deicing equipment had delayed them in Iceland for three weeks; Joe and other navy pilots had fallen behind on their training as a result. So when the weather in southern England finally cleared for a few days on August 4 and mission commanders began scheduling the first real attack runs, the army's Aphrodite crews—despite inferior technology—got first crack at the bunkers in northern France. Kennedy was furious. Those army chumps were going to blow up all the targets and leave him with nothing.

He needn't have worried. Project Aphrodite was an unmitigated disaster. Of the six army bombers that participated in the mission—including *Taint a Bird, The Careful Virgin,* and *Quarterback*—three nearly blew up right after takeoff: one ran into a field of barrage balloons, one took friendly fire from local antiaircraft units, and one stumbled into an outgoing raid and almost collided midair with another plane. Things only got worse from there. The mother ships lost track of one baby midflight, and another one veered off course toward London. Two babies eventually plunged into the sea and another exploded prematurely over English soil, killing eighty cows in a pasture. Three planes did make it across the Channel, only to have their remote-control switches fail, which sent them sailing over their targets. In the end only one plane-bomb hit something worth damaging, albeit not a bunker. After this baby missed its target, the mother ship decided to swing it around toward a German antiaircraft unit, whose gunners were no doubt licking their chops at this fat, easy mark. They opened fire, and ended up killing themselves in a gigantic greenish-yellow fireball.

Missing the bunkers was bad enough, but the Aphrodite crews also suffered an appalling number of casualties. One pilot died after jumping out and finding that his parachute wouldn't open. Several others nearly died because of similar malfunctions, including one man who had to tear his pack open mid-plummet and cast his parachute out by hand. Even when the parachutes worked, the men suffered sprained ankles, smashed teeth, and lacerations; one had his arm torn off at the shoulder. Not surprisingly, a furious army general grounded the Aphrodite mission on August 6. "This whole project," he fumed, "is put together with baling wire, chicken guts, and ignorance."

Joe Kennedy was delighted. The army's failure would only enhance his magnificence in comparison, and when Project Anvil got the go-ahead to fly a few days later, he jockeyed to be the first pilot up. Indeed, some historians suspect that Joe put the screws to his superiors, threatening to destroy their careers through his powerful father if he didn't get the initial mission. *You thought Hitler had a temper?* He got the nod.

For their part, the surviving army pilots were glad to be done with the fiasco, and after being grounded, they proceeded to get stinking drunk. During their bender, one of them saw Kennedy strut into the bar on base, and slurred, "If my old man was an ambassador, I'd get my ass transferred out of this outfit." Joe just laughed.

CHAPTER 44

Valkyrie

Werner Heisenberg had a rocky summer in 1944. He and Kurt Diebner continued to spat over access to uranium and heavy water. He could barely find time to work anyway, since the process of moving his lab south to the Black Forest region was taking longer than expected. The war was going terribly for Germany, too, and what he heard behind closed doors was even more troubling. One of Göring's top officers visited him in Berlin in July to ask about rumors that the Allies planned to drop atomic bombs on Dresden in six weeks. Heisenberg thought this unlikely but worried nonetheless.

On a personal level, Heisenberg was also feeling lonely and isolated in Berlin. He'd stashed his family in their dilapidated cabin in the Bavarian Alps and rarely saw them. He'd lost touch with the international scientific community as well. Hardly any letters got through anymore, and although he'd recently visited Copenhagen and Krakow and Budapest to give "cultural lectures," the scientists there remained unaccountably hostile when he showed up with his Nazi handlers.

About the only thing Heisenberg had to look forward to that summer was the Wednesday Club. This collection of twenty or so German aristocrats (diplomats, finance ministers, prominent professors) had been meeting every other Wednesday in Berlin since 1863 for dinner and an informal lecture; afterward they got drunk and sang songs from their schooldays. In short, it was an old boys' club, just what the laddish Heisenberg needed sometimes to revive his spirits.

To his and everyone else's surprise, however, the club had matured during the war and become something nobler—an outlet for people's frustrations with the regime. They could unburden themselves with impunity there, and by 1944 meetings had shaded into subversive territory, with mocking denunciations of that "schimpanski" Hitler and his minions. Some of the bolder members even spoke of deposing Hitler, and a few of them sounded Heisenberg out on the idea. He didn't take them seriously, of course, but he found such talk heartening. These were reasonable people, people who loved Germany and despised Hitler just like he did—people who understood that simultaneously wanting Germany to win the war and yet the Nazis to somehow lose the war wasn't muddleheaded nonsense, but the only sensible way to think.

Between the move south and his research on Uranium Machines, Heisenberg didn't have much time to attend the Wednesday Club during the first six months of 1944. So he was very much looking forward to the July 12 meeting, which he was hosting at his institute. He prepared a lecture on nuclear reactions inside stars, and picked fresh raspberries from the institute's garden for that evening's dessert. And for once during the war, everything went as well as he'd hoped. The lecture was a hit, and it inspired an interesting discussion afterward about the military and political consequences of nuclear energy. Then he and his fraternal brothers drank some lovely wine, and everyone had a grand time singing and shouting and popping off about the schimpanskis in Berlin.

A week later, on the afternoon of the nineteenth, Heisenberg was still coasting on a good mood. He finished up the official minutes for the meeting and dropped them off with a fellow member, then headed down to the *Bahnhof* to catch an overnight train south, to see his family. It must have been a pleasant journey—the lush countryside ahead of him, the stress of his research fading away behind. Arriving at his destination, he still had a two-hour walk through the mountains to the family cabin, so he grabbed his bags and set off.

A mile down the road he overtook a young soldier pulling a cart. Heisenberg tossed his luggage on top and offered to help. They probably exchanged a few pleasantries, until the soldier relayed some stunning news: someone had tried to assassinate Adolf Hitler earlier that day.

It later emerged that the assassination plot, code-named Valkyrie, had come achingly close to succeeding. A thirty-six-year-old German colonel had snuck a suitcase bomb into a meeting with Hitler and placed it next to him. But another officer unwittingly moved the suitcase to make room for his feet, pushing it beneath a large wooden table that absorbed most of the blast. So while the explosion knocked Hitler unconscious and shredded his trousers, he suffered no long-term injuries. Indeed, the attempt on his life invigorated him. He actually met with Mussolini later that day, and triumphantly showed Il Duce the smoldering ruins of the meeting room.

The news staggered Heisenberg. He thought immediately of all those drunken boasts at the Wednesday Club—had his friends been serious? Was he himself now in danger? Cautiously, he asked the soldier what he thought of the news. It was a fraught moment for both of them: under the Third Reich, you never unburdened your feelings to strangers, not about politics. The soldier finally said, "It's time something was done." That was all they dared say, but they understood each other well enough.

Over the next few days at the cabin Heisenberg listened to the news on the radio, and it was every bit as bad as he'd feared. Several members of the Wednesday Club—men he'd dined and caroused with, men he'd picked raspberries for a week earlier—were named as prime conspirators, and all of them were executed. A general purge of the German professional class followed, as five thousand people were eventually rounded up and shot, including a son of the famed quantum physicist Max Planck.

Heisenberg's emotions continued to roil over the next few days. If the Gestapo tied him to the Wednesday Club, even obliquely, he was doomed. And damn his luck, he'd just written up the minutes for the most recent meeting. Yet no midnight knock on the door ever came. No one even stopped by to ask him questions. Perhaps the conspirators had refused to name Heisenberg. Perhaps his family friend Himmler had intervened. Perhaps Heisenberg's status in the Uranium Club protected him. (Hitler himself had awarded Heisenberg a 1st Class Cross for War Services the previous October.) Whatever the reason, Heisenberg survived the purge. And to the dismay of Allied scientists, he proceeded from the family cabin to his new lab in southwest Germany. It was located in a cave in the village of Haigerloch, and it would house his most powerful Uranium Machine yet.

CHAPTER 45

Escape and Resistance

In the spring of 1944, the Gestapo in Paris arrested Frédéric Joliot again, and this time they meant business. In addition to the usual threats and roughing up, they got him fired from his teaching post, an ominous sign: dismissals like that usually foreshadowed a trip to a concentration camp. After discussing the situation with Irène, they both agreed that he needed to go into hiding, and soon. It was disappear or be disappeared.

First, they had to get their children to safety, and they decided that Irène should be the one to smuggle them out of France. She was feeling much healthier by that point in the war, fitter and stronger, thanks to several long stints in sanatoriums. (During her recovery she'd sent Joliot a flirty photograph to prove that she was regaining the weight she'd lost due to illness. "I am going to become a little elephant," she claimed. Joliot cheered every kilo: "I authorize you to gain some more. I love plump women.") Irène's recovery was also speeded by some newfangled drugs called antibiotics.

In early May she and the children escaped Paris for a village near the Swiss border. But rather than flee to Switzerland immediately, Irène made a rash decision. Her daughter, Hélène, age seventeen, needed to take her all-important baccalaureate exam in physics so she could qualify to work as a scientist someday. (These grueling, two-day tests were mandatory for French science students, even during wartime.) So Irène delayed their flight to freedom for four full weeks until the test date, then let Hélène travel alone to a nearby village to take her exam. Irène knew how easily female scientists could get pushed aside and—war or no war—she refused to let that happen to her daughter.

The girl aced the test, of course, finishing so quickly that she had time to go back over several problems and derive the answer a second way. And in the end, the four-week delay benefitted the family enormously. On the day they left for good, their guide into Switzerland led them to a secluded trail well shielded by pine trees. For extra safety, he also deployed several scouts—his own children—to sneak ahead and look for German patrols, signaling when the coast was clear. But they still might have been caught if not for the fact, the dumb-lucky fact, that Irène had chosen to flee on June 6, D-Day, when the German military had more important things to worry about than a few refugees. Had the family tried to escape one day earlier they might well have been arrested; one day later and they surely would have been, since the Germans clamped the border shut. As it was, Irène and the children, like scientific von Trapps, simply waltzed to freedom.

In Porrentruy, Switzerland, they were unceremoniously deloused and shunted into a detention center in a local manor. Characteristically, Irène paid little attention to her surroundings, and plopped down on a straw mat in their room (there was no furniture) to read a book on logarithms she'd insisted on smuggling across the border. It didn't take long for the local prefect to recognize this strange creature as a Curie, at which point he insisted that the family join him at his home until France was again free.

The man who had married Irène, meanwhile, went underground in Paris not long after his family left, assuming the guise of an electrician named Jean-Pierre. He slept at the home of a different friend each night, and spent his days skulking along the Seine with a fishing pole, posing as a peasant and rendezvousing with other resistance fighters. They traded rumors about Allied invasions and German crackdowns, and passed around documents with designs for home-made grenades and crude antitank missiles, in preparation for the upcoming battle for Paris.

They didn't have to wait long. By August, with the Allied army approaching Paris, the city was simmering like Mount Etna, and it finally exploded on August 19. No longer needing to hide, Joliot ran to the local police headquarters, dragging with him two suitcases full of chemicals. He recruited three men to help, and they stripped off their shirts in the heat and got to work making Molotov cocktails. Molotovs are normally crude weapons, bottles of gasoline with lit rags for fuses. As a man of science, Joliot devised something more sophisticated. Instead of straight gasoline, he had his assistants mix in sulfuric acid and potassium chlorate for added punch; they also wrapped each bottle with a rag soaked in more potassium chlorate. Luckily, the gluttonous head of the Parisian police under the Nazis had kept a huge stock of champagne in the basement of the head-quarters, which gave them access to plenty of bottles. The shirtless men began pouring it all out, filling the bottles with their own pun-gent liquor and hauling the bombs up to the roof.

They finally deployed them a few days later, when shooting broke out nearby. Several thousand French policemen rushed to the head-quarters, barricading themselves inside. And when three German tanks rolled up on August 23, the police let them have it, pelting the tanks with Joliot's cocktails. Some reports claim that Joliot himself was in the thick of things, hurling bottles down; other reports have him running back and forth through the streets, pistol in hand, to grab more supplies from his lab. "I saw the boches [German soldiers] fall like puppets in a traveling fair!" someone heard him cry. What-ever his role, the Nobel laureate did his country proud in this guerrilla fight, and when the smoke cleared the police had held off the tanks with little more than fancy chemistry.

Joliot's adventures were not over, however. Because after the battle ended and he'd returned to his lab, thrilled and exhausted, he found a strange message waiting for him. Who on earth was Boris Pash? And why was he coming after Joliot?

CHAPTER 46

Lightning-A

Boris Pash despised Samuel Goudsmit the first time he met him. Samuel Goudsmit was merely terrified of Boris Pash.

In September 1943 Goudsmit was sent to England to work out some kinks with American radar equipment there. He'd just turned forty-one and was growing gray and paunchy; he had the same dopey smile as before, but was no longer the svelte, mop-topped hotshot who'd discovered quantum spin. Still, he liked hobnobbing with the young soldiers abroad, helping them fix their radar sets by day and knocking back brewskies with them by night.

After six months in England and a brief stint back at MIT, Goudsmit was summoned to Washington in April 1944 for a new assignment—he suspected it was a top-secret radar project. The army put him up in a hotel, which impressed Goudsmit, since rooms in D.C. were scarce and expensive during the war. He waited a week or so for his briefing, and finally called to ask what the holdup was. The Pentagon blew him off. (*You can't imagine how busy it is down here.*) So he waited another week, and still nothing. Then another. He called again, crankier, and was fed more lines. (*We'll be in touch at our convenience, Sam.*) He found the delay humiliating—did they think he had nothing better to do? He wished he did.

The delay was no accident, of course. Military intelligence had placed Goudsmit under surveillance, tapping his phone at the hotel and tailing him around town to see if he met with any suspicious characters. (Moe Berg had likely been under similar surveillance during his wait in Washington.) Goudsmit apparently passed the test, and General Groves's office finally brought him to the Pentagon in May.

Goudsmit still assumed he'd be working on radar, but Groves's deputy quickly disabused him. His assignment was far more serious—nuclear espionage. Goudsmit first learned of the Alsos mission at the same time he found out he'd be running it. The deputy then made him swear an oath of allegiance to the United States.

Goudsmit later claimed to have no idea why Alsos chose him as scientific chief. This was a bit disingenuous, given his research background and familiarity with Europe, but he certainly wasn't the mission's first choice. Due to a clerical mix-up, Goudsmit accidentally saw a copy of his personnel file, which contained a blunt assessment of his pros and cons. Pro: he knew nuclear physics. Con: "his name adds nothing to the prestige of the mission." Pro: his friendship with German scientists might help solicit information. Con: he was grouchy and "tactless," both liabilities in intelligence work. The other pros were either borderline insulting ("Dr. Goudsmit has been recommended principally because he is available") or downright alarming (because he knew nothing of the Manhattan Project, no one could torture any secrets out of him). Such revelations didn't exactly swell his confidence.

Nor did meeting Boris Pash. Pash flew from Europe to Washington specifically to meet Goudsmit, but he didn't hide his skepticism about adding academic scientists to his mission. Pash called them "longhairs," radicals, and expressed doubt that they'd hold up at the front. Goudsmit agreed wholeheartedly: he didn't think he'd hold up, either. As Pash remembered it, Goudsmit confessed a preference "for the comforts of the civilized world and the quiet of a peaceful laboratory. Braving enemy fire or parachuting behind enemy lines was not his idea of recreation." Pash just laughed: "I promised that if any jumping were involved, I would go first and prepare a soft landing."

Goudsmit nevertheless swallowed his qualms and shipped out for London on D-Day, joining a small Alsos support office there. He spent the next few months gathering dossiers on French and German physicists—scientific rap sheets, essentially—and imbibing the latest intelligence reports. He also packed for his upcoming transfer to the front, using that ominous checklist—the helmet and woolen underdrawers, the gas mask and extra life insurance. Most memorably, British agents showed him photographs of the giant concrete bunkers in northern France, assuring him that the Nazis wouldn't go to so much trouble just to launch conventional warheads; this had to be for something special, possibly nuclear. Goudsmit got to test this theory when the V-1 barrage began a few days after his arrival. He spent many a harrowing hour crawling down into craters with a Geiger counter, ready to scramble back out if it started chattering. Welcome to the war, longhair.

Boris Pash, meanwhile, was preparing to invade France. General Patton's recent breakthrough there had knocked the German army back and opened up the French coast near L'Arcouest/Port Science, where Frédéric Joliot was rumored to be hiding. In early August the Pentagon sent Pash an urgent pink radiogram with orders to get there on the double and hunt down the scientist. "Still ringing in my ears," Pash recalled as he landed on Omaha Beach, "were the words I had heard [back] in Washington: 'Any slight delay in reaching your targets might cost us tremendous losses, or even the war.'" Pash and a sidekick rustled up a jeep, tied an army bedroll across the hood to protect their engine from snipers, and took off. They traveled day and night, snatching sleep in orchards or fields and weaving around the smoldering husks of tanks on the roadways.

L'Arcouest turned out to be little more than a single street, with a stone church and cemetery on one side and shops and a wine cellar on the other. The Nazis, however, hadn't quite abandoned the place; they were still fighting in the woods near Joliot's cottage. Pash decided to risk a visit anyway and commandeered a local man with a scruffy beard—a relative of Joliot's, it turned out—to show them the way. The man led them to the edge of town, to a trail overgrown with weeds. "Voilà," he said, but refused to go a step farther. The Nazis had mined the woods, and seven French resistance fighters had already died there. Still, the man did offer to do his part. "I shall return to the village," he vowed, "and drink to your health and safety." Ta-ta.

Alone now, Pash and his sidekick began creeping through the underbrush, checking every centimeter for mines and booby traps. Halfway there, someone started shooting—they couldn't tell who—and they dove for cover, scarcely daring to breathe. But they started moving again before the gunfire died down completely, not wanting to delay their mission by even a minute.

The Curie family "cottage" was a handsome two-story stone house whose façade was dominated by tall windows. But when Pash reached the clearing around *Maison Curie* that day, his rifle at the ready, he could see the front door ajar. And when he nudged it open and peeked inside, his heart sank. The place was bare: "There was not a stick of furniture or anything else to be found." His lieutenant searched the rooms anyway, but there was no point. Joliot was gone.

At least the gunfire outside had died down, so Pash motioned for them to leave. As soon as he appeared in the doorway, however, several bullets smashed into the frame near his skull. They both flung themselves to the dirt and began crawling on their bellies. Unlike the house, the woods were far from empty, and machine guns and rifles traded retorts in the air above them. The duo made it back to the village unharmed, but this was scant consolation. For all the fuss Alsos had caused in Italy, the gains had been paltry, just a few second-rate nuclear scientists. And Pash hadn't even gotten much information out of them—Moe Berg had. On his recent trips to Washington, Pash had sensed forces aligning against Alsos, with some people pushing to eighty-six it altogether. And however unfair, he knew the failure to capture Joliot would only increase the clamor.

A few days later Pash received orders to resume the hunt for Joliot as soon as Paris fell. He got no other encouragement, and while no one spelled anything out, he sensed that this was Alsos's last shot. "If we miss the boat again in Paris, I'll shoot myself or go over to the Germans," he wrote a colleague. "Or what is worse, I'll join the Russians." For a proud veteran of the White Russian army, there was no keener cry of despair than that.

✗

The Allies expected to conquer Paris in early September, and Pash planned his mission accordingly. So when the city actually started falling on August 23, he had to scramble. He cadged two jeeps for himself and three companions from the army headquarters in Rennes, then took off with little more than a map and a canteen. One of the quartet wasn't even part of Alsos; he'd simply been bored with his desk job and was looking for adventure. Pash liked the cut of his jib and recruited him on a whim. The man had recently adopted a stray puppy, a munchkin with a white body and a black head, so they named him Alsos and took him along as a mascot.

The quartet approached Paris from the south. They tried being discreet, but the delirious ovations they received made that impossible. At every farmhouse they passed, French peasants hung out of windows and waved red, white, and blue bunting. In every village, crowds squeezed in so close that their jeeps could barely pass. Men tossed bottles of wine to them and women leapt onto the running boards to kiss their cheeks. Still, the trip was not without hazard. Two thousand Nazi troops were roaming the roads leading into

Paris, and thousands more were battling partisans within the city. The biggest threat to the mission, however, was the French military. France had had a humiliating war so far—Maginot, Vichy, Nazis goose-stepping down the Champs-Élysées. French troops desperately wanted to redeem themselves, and in meetings with Allied commanders their generals insisted that the French army enter Paris first, as its liberators. Unfortunately, constant defeat had made the French army skittish about confronting the Germans. Even while brave civilians were fighting the Wehrmacht inside Paris, the troops dithered in the countryside, wasting day after day.

Boris Pash did not dither. Failure to apprehend Joliot could mean the death of Alsos. So when Pash came upon a French barricade a few miles outside of Paris and found the road closed to him, he resorted to duplicity. He located the officer in charge, a mustachioed French major, and pulled him aside. In hushed tones he confided that—*sacré bleu!*—some rogue American tanks had already made a break for Paris. Pash said he had orders to stop them from entering first, and could easily do so. But if monsieur insisted on blocking his path…

Non, non! the major cried. *You will go—you will stop zhese bad Americans!* He ordered the barricade dragged aside. Pash bid him au revoir, and roared off cackling.

His first stop was Joliot and Irène's home in the suburbs. Finding them absent, he had a servant telephone Joliot's lab. Joliot wasn't there, either, so Pash left a message that Alsos was coming.

As they drew close to Paris, the Alsos crew saw thick smoke hanging in the sky, framing the Eiffel tower in the distance. Along the way their jeeps came under fire several times, although nothing they considered serious. And amid the gunfire the French people continued to cheer the Americans. Indeed, the city folk were even more intoxicated than their country brethren. *"Vive les Americains!"* they screamed, and threw so many flowers at the jeeps, Pash recalled, "that we resembled the prize floats in the Pasadena Rose Bowl Parade." Their little mutt leapt onto the hood and barked in joy.

The Alsos quartet arrived at the nearest entrance to the city, the Porte d'Orléans, at 8:55 a.m. on August 25. The Porte bordered a square lined with balconies and cafés, and the roar that accompanied the sight of their jeeps—the first Allied troops to enter Paris—was so loud they couldn't hear each other talk. People rushed forward to grab the flowers on the jeeps as souvenirs, quickly stripping them bare. When the roar subsided, an American pilot who'd been shot down over France ran up, grinning to see some Yanks. He informed them that Nazi tanks were still prowling the Luxembourg Gardens near Joliot's lab. They could in fact hear tank fire in the distance.

Pash now had a decision to make. Counting himself, he had four men in two open-top jeeps—no match for tanks. He also suspected that the Allied command would "hang and quarter" him if he cut the line and beat the French into Paris. By all rights, he should hang back.

On the other hand, screw the French—the bastard unit had a mission. Pash gave orders to roll out, and as the jeeps took off, a mob of jubilant Parisians fell in behind them, cheering and waving rifles in the air. It proved a short-lived adventure. A few blocks on, German snipers began peppering the ragtag troops. Pash's crew tried to fight their way past them, but they simply couldn't push any farther in jeeps. As much as he hated to, Pash retreated to the Porte d'Orléans to wait for the French.

He didn't have to wait long. After asking around about the jeeps, the mustachioed major at the roadblock realized he'd been duped and informed his superiors. They were furious—those dirty Americans! They had no choice but to start marching. In a roundabout way, then, Pash's trick spurred the liberation of Paris.

Back at the Porte, Pash could hear the French coming now: the cheers were beyond deafening this time, seismic in intensity. French tanks started rolling through the square a minute later, and Pash let three pass by him before slipping his jeep in behind.

His team hewed close to the tanks for several hours, letting them clear out the main pockets of resistance. Then, as soon as they spotted an opening, the Alsos crew peeled off and began weaving toward Joliot's lab. Four separate times, gunmen on nearby rooftops opened fire, pinning them down. At one point the quartet had to abandon the jeeps entirely, taking cover in doorways and behind trees. (The puppy presumably just laid low.) Some brave French militiamen ran up to help, and with Pash's men orchestrating a counterattack, they fought their way forward house by house.

Near the Luxembourg Gardens, another group of French partisans flagged Pash down and warned him that the Germans had posted an antitank gun at the intersection just ahead. An antitank shell would of course vaporize a jeep, but Pash refused to turn around, deciding to make a run for it instead. Without any preliminaries he and the other driver revved their engines and gunned them, blowing through the intersection at top speed. Halfway across, they heard a *boom*. They'd caught the Germans off guard, and the shell exploded harmlessly behind them. Pash once again cackled.

At 4:30 p.m. they reached the courtyard that fronted Joliot's lab. Far from celebrating, however, they took a good, hard look around. Most intelligence reports had painted Joliot as a collaborator—all those Germans in his lab. Were they holed up in there now? Had Alsos dodged all that rifle and tank fire just to get killed here? Sure enough, someone opened up on them as soon as they dismounted from the jeeps. But the shots were coming from behind them, from a church belfry, and after Pash's crew returned fire and chased the gunmen away, the neighborhood fell silent. A minute later an assistant of Joliot's descended the steps of the lab and greeted them.

Not quite believing his luck yet, Pash made his way inside to Joliot's office, where the great scientist shook his hand, every bit as happy to see the Americans as the Americans were to see him. Joliot knew full well about his reputation as a collaborator and was worried about reprisals. "I am afraid for my life," he admitted. "I shall be grateful if you can give me protection." Just like that, Alsos had snagged one of the top nuclear physicists in the world.

Pash and his men were overjoyed. "Lightning-A has struck Paris!" one of them crowed, referring to the Alsos logo, the white α with the red lightning bolt. Pash liked the boast so much that he adopted "Lightning-A" as the nom de guerre of his advance scouting team for the rest of the war.

To celebrate their triumph, Lightning-A held a feast that night. As the guest of honor, Joliot generously turned his chemistry lab into a kitchen, allowing the Americans to cook food over Bunsen burners and drink champagne from beakers until the wee hours of the morning. It's hard to imagine Irène tolerating such an invasion, but Joliot had always been more of a bon vivant than the woman he married.

The next day turned out to be one of those glorious Paris afternoons that make you wonder why anyone would ever live elsewhere. Pash recalled "the golden-red leaves on the trees, the aroma of roasting chestnuts... [the] pretty girls on bicycles, riding along the Champs." Cafés were blasting the American dance music forbidden during the occupation, and the Alsos crew got so swept up in the excitement that they partied right through a second night. Pash later admitted that when he finally reported to the new army headquarters in Paris on August 27, he was still pretty "befogged."

But the message waiting at HQ—"a shock more disturbing than a bomb blast," he said—sobered him right up. His orders? To stop hunting foreign scientists and start hunting one of his own men. To his astonishment, "Samuel Goudsmit was now Alsos's target."

α

A few days earlier Goudsmit had flown from London to northern France, planning to meet Pash there and accompany him to Paris. In his tactless way, he groused about every leg of the trip. His plane was fogged in for hours at the airport, and he got so hungry that he had to beg sandwiches from Red Cross workers. The plane's cold metal bucket seats reminded him of "toilets in kindergarten." Cherbourg, his first stop, was little more than "tents and barracks and mud."

He was supposed to meet Pash in Cherbourg, but Boris was nowhere to be found. So after several hours—and plenty of begging—Goudsmit and a few companions threw their sixty-five-pound duffel bags into a truck and rode the hundred-plus miles to army headquarters near Rennes, to resume searching there. In some ways it was a pleasant ride—the weather was gorgeous, and French people flashed V-signs and shouted *"Vive l'Amerique!"* the whole way. At the same time, the truck kept having to swerve around land mines and bombed-out vehicles. Goudsmit then saw several dead bodies being pulled out of rubbled homes; the sight of one breathlessly beautiful corpse on a stretcher particularly haunted him.

Many more hours later than it should have taken, they arrived at Rennes—at which point Goudsmit realized that his duffel bag was missing. Either it had been bucked off somewhere along the road, or some cagey French thief had pinched it. He was mortified. Here it was, his first day on the front, and he'd already misplaced everything. Typical longhair. Goudsmit filed a report about the bag, then resumed looking for Pash. He still hadn't found him by dusk, when he was shunted to a cot in a dilapidated girls' school. He spent the night listening to the moans of refugees and trying to ignore the stink of the overflowing toilets.

Fearing that Pash had abandoned him, the next morning Goudsmit talked an officer into lending him a jeep and driver for a last-ditch search of the sprawling headquarters and surrounding towns. Sure enough, he heard rumors that Pash was already inside Paris and had probably snagged Joliot. Although miffed, Goudsmit could do nothing but hurry to Paris himself. But when he asked about transportation, the replies were curt: a scientist didn't rate a ride anywhere, much less to Paris. Once again he learned that he was just about the last priority in the army.

Like Pash before him, Goudsmit now had a decision to make. Pash was skeptical that academic scientists could contribute much to Alsos—they were simply too slow and timid. He'd in fact recently grumbled in a letter to a friend that "we may miss the boat in Paris if Sam doesn't get off the dime"—that is, stop fretting and show some damn initiative. Goudsmit didn't know about the letter, but he'd no doubt sensed the colonel's irritation and wanted to prove him wrong. Moreover, Goudsmit needed to prove something to himself. Driving around that day, he later recalled, "I was getting more and more excited and mad, thinking of the responsibility I had." He'd wanted so badly to contribute to the war, to fight Hitler—yet he'd been bucked off and left behind yet again, abandoned like a duffel bag on the side of the road. He *had* to get to Paris.

Still unsure of his plan, he grabbed a road map of France and some extra rations. He then approached the driver who'd been tooling him around. He knew they weren't supposed to leave the area, but Goudsmit had a few tricks up his sleeve—or had at least read about them in detective novels. He asked the driver, What were the exact orders your officer gave you?

"He told me to take you wherever you want to go."

This was just what Goudsmit hoped to hear—ambiguity. "Fine," he said. "Paris!"

"Yes, sir." If the officer hadn't been smart enough to spell out his orders, that wasn't Goudsmit's fault. They left at dawn the next morning.

When the officer realized that the longhair had hijacked his jeep, he reported it stolen and put out an APB to arrest one Samuel Goudsmit on sight. But with the map he'd pinched, Goudsmit could steer the driver around every checkpoint, and with the extra rations, they didn't need to stop for food. Goudsmit later called the caper "my first evil deed."

As they entered Paris, Goudsmit tensed: he could hear gunfire coming from every direction. He'd last seen the city in 1938, before the war started, before his parents disappeared. He'd long wondered whether he'd ever see it again, and the churn of emotions as they rolled through different neighborhoods was overwhelming. He had to fight back tears upon seeing the Sorbonne, where he'd once lectured about quantum spin—back when he was still a scientist with promise. The city was looking ragged in places, yes, with ugly scars from the war. But it somehow retained that old magic—it was still beautiful, still Paris. After a few more blocks he gave in and wept openly.

The tears stopped abruptly at Joliot's laboratory. Upon arrival he realized that Pash knew all about the stolen car. It turned out that the jeep belonged to one of the top colonels at Rennes, a big swinging dick who was royally pissed off. Goudsmit tried to play down the incident. "I guess we can settle down to business right away," he said hopefully. Pash countered by asking him whether he thought he could do his job from inside a jail cell.

Goudsmit scrambled to explain himself—until he saw Pash grinning. After all, Pash had pulled an equally dirty trick in breaking through to Paris, and no one hated empty-headed administrators more than the chief of Alsos. He practically slapped Goudsmit on the back. *Welcome to the war, longhair.* Goudsmit just about melted with relief, and no doubt flashed that dopey smile of his. He'd proved himself at last. And just as he'd hoped, it was now time to settle down to business—to start hunting for the Nazi nuclear bomb.

CHAPTER 47

Zootsuit Black

Although an army guy, Lieutenant Colonel Roy Forrest liked Joe Kennedy. The navy's Project Anvil and the army's Project Aphrodite were basically the same mission, but interservice rivalries had driven a wedge between the two. Instead of working together and trading tips, army and navy pilots had formed cliques on their miserable air base and would bark insults at each other whenever they crossed paths. A navy sentinel even pulled a gun on some army fellows for snooping around the navy planes. And Joe Kennedy—rich, pious, an ambassador's son—seemed a special magnet for hostility.

But Forrest didn't mind Kennedy. The boy knew his stuff and handled the slurs against him with grace. So as a goodwill gesture, Forrest invited Kennedy over occasionally for a nip of whiskey before dinner. In return, Kennedy did Forrest a solid a few weeks later. Forrest had been complaining one night about the ruinous cost of liquor during the war—$20 ($280 today) for a fifth of bourbon. The next day Kennedy knocked on his office door and mentioned that he could get booze cheap in London, $1.40 for a fifth. He just needed transport.

Forrest rolled his eyes. "Sure, Joe. And you'll throw in the Tower of London if I act fast, right?"

But Kennedy insisted: he had connections at the embassy and could get liquor at cost. When Forrest realized Kennedy was serious, he called in his assistant. "Shag ass over to the flight line," he told him, "and get one of those unloaded babies and fly it over to London for a new set of plugs." Then he turned to Kennedy. "Lieutenant, I just learned of a flight leaving right away for London. Maybe you can catch a ride." Six hours later Kennedy knocked again and cooed, "Delivery boy!" He handed Forrest a case of scotch, along with gin, crème de menthe, and two cases of Pabst Blue Ribbon. Total cost: $16.80 for the scotch, with the rest on the house. Yes, Roy Forrest liked this Kennedy boy just fine.

Shortly afterward, Aphrodite got grounded and Kennedy was tapped as the first Anvil pilot. Out of both professional and fatherly interest, Forrest wandered over one afternoon to check out Kennedy's plane, *Zootsuit Black*. The exterior looked a little silly, frankly. Stripped of all its guns, *Zootsuit Black* was helpless in the air, so a carpenter had mounted two black broomsticks on the underside to pass as machine guns in case they encountered enemy fighters. The interior was far more impressive. Forrest couldn't help but whistle at all the advanced electronics that engineer Bud Willy had rigged up. Willy had even installed television cameras on the nose cone, which allowed the mother ship to see exactly what *Zootsuit* saw in real time. Pretty slick.

Forrest was less impressed with the arming panel, which controlled the explosives. In fact, when he got up close and scrutinized the circuits, he was horrified. The navy had been bragging up and down about its ability to arm the explosives remotely, via radio signals, but to Forrest the panel looked like a hack job, thrown together in just two weeks and full of bad soldering and poorly grounded wires. "This looks like something you'd make with a No. 2 Erector set and Lincoln Logs," he remembered thinking. Did they really trust this thing?

That night Forrest voiced his concerns to his superior, a colonel. They were sitting at the bar on base, and when Forrest finished explaining the problem, the colonel shrugged. That's the navy's problem, he said. But that panel could kill him, Forrest countered. Besides, weren't they all fighting the same war?

The colonel raised his martini and toasted Forrest's "wild imagination." "When the navy blows up its drone," he said, "give me a ring. Nobody wanted the navy here in the first place."

If only Forrest had known that a trio of navy electricians shared his fear about the kludgy arming panel. During test flights near Philadelphia the detonation circuits had kept popping on accidentally,

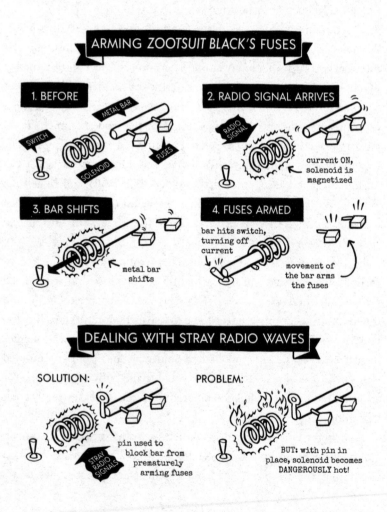

326

possibly due to radio interference. In response, Bud Willy had added a safety measure. The details get a little gnarly, but the explosives were controlled by two distinct circuits. The first armed the explosives, like cocking a gun; the second pulled the trigger. One substep of the arming process involved a coil of wire called a solenoid. Solenoids become magnetic when an electric current flows through them, and in this case the magnetized solenoid would tug on a metal bar. That bar was attached to a few nearby fuses, and when the bar moved, the fuses were armed. In essence, then, moving the bar primed the payload to explode.

During normal operation that movement wouldn't happen until the mother plane ordered it. But let's say a stray radio signal came in. This would start current flowing through the solenoid too early. This would in turn shift the bar prematurely and cock the gun when the pilots were still inside, an unacceptable risk. So for safety's sake, Willy physically blocked the bar with a metal pin to prevent premature movement. The pilot simply had to pull out the pin before jumping. Problem solved.

Or was it? After poring over the circuit diagrams, the electricians noticed something. Adding the metal pin would prevent the bar from shifting too soon, no question. But it introduced another, potentially worse problem. That's because the metal bar actually had two purposes: beyond arming the fuses, the movement of the bar also knocked into a nearby switch, which cut off the flow of current through the solenoid. That was necessary because anything with current flowing through it warms up over time. So if a stray radio signal did in fact start current flowing, but the metal pin prevented the bar from moving, then the solenoid would start to get hot. And the hotter it got, the higher the chances it would accidentally trigger one of the nearby fuses. This wouldn't happen instantly, but tests indicated that after a few minutes—150 seconds—the solenoid would reach a dangerous temperature. They decided to alert Willy.

This proved to be a bad idea. Willy was an engineering genius, but the stress of running a combat mission was eating at him. Instead of stating his wishes calmly but firmly, as an officer should, he took to shouting at his men like a drill sergeant, often ending commands by barking, "And that's an order!" Even worse, his judgment had grown erratic. He'd gotten hopelessly turned around on a practice flight a week earlier (ironically, given all its fancy electronics, his plane had a screwy compass), and rather than admit he was lost, he'd blundered into both an antiaircraft battery, which fired on the plane, and an armada of barrage balloons, which nearly sheared the wings off. Two officers more or less mutinied after the second near miss and took over the plane, and at least one man refused to fly with Willy ever again. People seriously questioned his mental health.

Sure enough, when one of the electricians knocked on Willy's door and explained his fears about the arming panel, Willy got agitated. He hadn't designed the arming components himself, but he felt defensive about every aspect of "his" plane. He insisted there was nothing wrong: "I didn't get to be executive officer of this outfit by being stupid." When the electrician pressed his case and explained why the metal pin was dangerous, Willy snapped, "Stop playing your games! That's an order." The pin would stay.

The trio of electricians debated what to do now. Two of them tried explaining the problem to Kennedy; as the pilot, he could scrub the mission if he felt unsafe. As enlisted men, though, they felt uncomfortable telling an ambassador's son what to do, and Joe didn't really grasp the technical details anyway. Kennedy also had enough responsibilities that week (practice flights, briefings, selecting a flight engineer) without fretting over electronics. My motto at this point, he told one electrician, is "keep quiet and obey orders." He didn't want anything mucking up his shot at glory.

Running out of options, the trio considered sabotaging the electronics—sneaking onto *Zootsuit Black* and snipping a certain wire, which they calculated would eliminate the threat of early detonation. But Bud Willy had threatened to court-martial them if they altered anything, and he seemed serious. So the electricians swallowed their misgivings and did nothing.

While this drama played out behind his back, Kennedy kept flying practice runs and overseeing other tasks, including the loading of explosives. Rather than napalm, the mission now planned to use Torpex, a new type of explosive that mixed TNT with aluminum powder, which extended the length and intensity of the blast. Torpex came in blocks that looked like rose-colored butter, and Joe watched a few dozen navy grunts load 347 crates of this deadly colloid onto *Zootsuit Black*; each box was lined with a beeswax cushion to reduce the likelihood of turbulence setting it off accidentally. With 21,170 pounds of Torpex inside, the plane carried the same firepower as a dozen V-1 rockets.

Kennedy's most important task involved selecting his flight engineer, who would help him set up the electronics midflight. He had two bunkmates begging for the job, but he rather coldly cut them out and picked none other than Bud Willy. Willy argued that he was the best man for the job, since he'd designed and tested the remote control and arming/detonation circuits. The engineer's erratic behavior notwithstanding, Kennedy agreed.

Kennedy and Willy spent the days before the mission waiting, as usual, for the weather to clear. To distract himself, Joe wrote letters home to his family, assuring them that he wasn't going to "risk [his] fine neck…in any crazy venture." He was more honest in a phone call with Lorelle Hearst, a family friend (and the wife of publisher William Randolph Hearst) who was stationed in England as a war correspondent. "I'm about to go into my act," he told her. "If I don't come back, tell my dad—despite our differences—that I love him very much." In between letters, Joe kept listening to the BBC and brooding over General Patton.

The gloomy weather finally broke on August 12—total CAVU, with sunshine and a glorious blue vault of sky overhead. Kennedy's buddy Roy Forrest took his plane up that afternoon for a look-see at the bunkers across the Channel and came back smiling. "The Krauts are out sunning themselves on the terrace," he reported. Everything was finally ready.

Although it was probably unnecessary—everyone knew the mission cold by then—Kennedy endured one last briefing that afternoon. It took place in a room with a gigantic 3-D model of the Mimoyecques "atomic bunker" mounted on a Ping-Pong table. The attack would take place at sunset, to blind the German antiaircraft crews to the incoming drone. They were aiming for the bunker's only point of vulnerability, a doorway roughly sixteen feet square—shorter and narrower than the fuselage of the plane.

While Kennedy was briefed, a few engineers checked over *Zootsuit Black,* wriggling around among the explosives inside. Finding nothing amiss, they signaled for Joe and Willy to climb aboard. As they did so, Joe solemnly informed them that if anything happened to him up there, they could have his most precious possession: the crate of fresh eggs he'd been hoarding in his locker. Everyone laughed. Joe and Willy then slithered in through a hatch near the nose wheel. It was the same door they'd jump out of an hour later.

Two mother ships took off at 5:55 and 5:56 p.m. They were followed by a weather plane, two reconnaissance planes with photographers, a plane to mark where the pilot and engineer landed after parachuting out, and five fighter jets, in case of a Nazi sneak attack; in all, a dozen escorts would join Joe and Willy's baby in the sky. While these aircraft circled above, pilot and engineer spent ten minutes working through various checklists. Then they gave a thumbs-up and taxied to the runway. Joe began rumbling down the strip near the beets and turnips at 6:07, then lifted off. By design he had enough fuel in the tank only for the outbound trip, not the return. One electrician called it "the most beautiful takeoff I've ever seen."

Zootsuit Black spent the next fifteen minutes banking in a lazy triangle while Joe and Willy made sure the remote-control system worked. It did, and Joe finally spoke the code phrase—"spade flush"—for the mother to take over. This step had failed several times during the Aphrodite flights, but the navy handoff couldn't have gone more smoothly. Anvil really knew its stuff.

Kennedy and Willy set about some other tasks now, such as calibrating the autopilot and altimeter. Meanwhile, high above, someone in the mother plane noticed that the baby was a tad off course; he nudged a joystick left to correct its heading. A moment later he felt a rumbling beneath him. Confused, he glanced at the television monitor showing the camera feed from the nose of *Zootsuit Black*. He saw only static.

CHAPTER 48

Catching Pretty Well

As Moe Berg might have put it, he worked like a Trojan the summer of 1944. Pictures show him looking slimmer than he had in years, closer to his playing weight, and no wonder. Every day in Rome brought new scientists to interview, new documents to translate, new reports to write—on radar and radio, torpedoes and fuses, altimeters and remote-control planes. Berg wrote several drafts of each, then relayed them in breathless cables to Washington that ran up to seventeen single-spaced pages. OSS appreciated the enthusiasm but eventually asked Berg to back off a little. The torrent of dispatches— every last one labeled TRIPLE PRIORITY—was clogging up their channels of communication. Berg ignored the hint and kept churning.

That said, Berg did find plenty of time to indulge himself in the Eternal City. OSS put a stylish black car and chauffeur at his disposal, and he stayed at the most luxurious hotel in town, the Excelsior, where he read six Italian newspapers each morning, along with *Stars and Stripes.* He had an audience with the Pope one day, and gave guided tours of the Vatican art collection to visiting officers. Overall Berg called Rome—which had suffered surprisingly little damage during the occupation—"a ray of sunshine, with no sign of war."

He also took field trips for OSS. He poked around several local fishing holes, asking people if they were getting nibbles these days. The ponds sat near hydroelectric plants that OSS suspected might be used for uranium processing; if so, runoff might be killing local fish. Berg then snuck up to Florence to infiltrate an optical plant whose equipment could be repurposed for plutonium experiments. Germany still controlled Florence at that point, so Berg had to travel incognito. As he crossed the famous Ponte Vecchio, he could hear shells exploding in the distance. He did manage to find a hotel that served high tea and had a string quartet playing in the lobby, but for once this level of luxury discomfited him, given that food and even water were scarce citywide.

Berg's biggest coup that summer involved tracking down Antonio Ferri, an aerodynamics expert who'd built the *galleria ultrasonica,* the world's most advanced wind tunnel. A year earlier, in September 1943, the German military had overrun Ferri's laboratory. Unlike Frédéric Joliot, Ferri refused to work with the Nazis in any capacity, so he lied his way back inside the lab one day, trashed his own equipment, and fled with the most vital documents. He ended up in the Apennine mountains to the north, where he and his brother (a history professor) organized a band of guerrillas they called Il Spartaco, to resist the Nazi occupation. The two scholars eventually commanded a force of several hundred tatterdemalion troops, who blew bridges, ambushed unwary patrols, and even captured a staff car and several prisoners whom they kept in caves. Tired of this harassment, and eager to retrieve the wind tunnel documents, the Germans offered a handsome ransom for Ferri's head. No one ever claimed it, and Ferri eluded capture month after month.

But where the Third Reich had failed, Moe Berg succeeded. He started by wooing Ferri's mother-in-law in Rome, flattering and charming her, and when the scientist later snuck back into the city, she helped Berg make contact. At their first meeting Berg found Ferri depressed, claiming he wanted to chuck science altogether and become a fireman. Berg soothed and consoled him, then put in time with

Ferri's family to win their trust; he taught Ferri's children to play baseball. Seeing this kindness, Ferri eventually opened up, and Berg coaxed enough technical details out of him to produce a twelve-page report on wind tunnels for OSS in August. Shortly afterward Berg persuaded Ferri to sign a three-month contract to work in the United States while Italy settled down politically. Due to immigration restrictions, OSS needed President Roosevelt to approve Ferri's visit. Upon receiving the request, FDR chuckled and said, "I see Berg is still catching pretty well."

(Once Ferri landed in New York, of course, American officials refused to let him leave, making excuse after excuse to detain him in the United States. Resigned to his fate, Ferri eventually became a U.S. citizen and went on to make fundamental contributions to American aviation. Most importantly, he helped design the aircraft that allowed Chuck Yeager to break the sound barrier a few years later.)

Even while snagging top scientists, however, Berg—being Berg— still managed to exasperate his bosses, mostly by disappearing for weeks at a time. The most egregious flake-out occurred in late August, when OSS decided to send him to Paris to make contact with Frédéric Joliot. Officials dispatched cables to every outpost between Casablanca and Florence, asking if anyone had seen their spy. No one had.

In truth, it was probably for the best that Berg didn't interfere with Boris Pash again—there might have been a homicide. Besides, OSS was already plotting another, darker mission for its star catcher.

CHAPTER 49

"I'll Be Seeing You"

On the evening of August 12, 1944, a forty-four-year-old woman named Ada Westgate was chatting with a neighbor outside her home near New Delight Wood, in extreme eastern England, when a plane appeared in the sky. The neighbors were enjoying the end of a lovely summer day, and while they surely noticed the plane—it was flying quite low, making an awful racket—there were always bombers coming and going in those days, and this one probably didn't make much of an impression. Not until it exploded.

A gigantic burst of light filled the sky, and the blast that followed must have been staggering—less a noise than a physical blow. Her mind suddenly blurry, Westgate assumed that a V-1 rocket had struck. She lived with a cousin who'd recently fled London after a V-1 destroyed his home there, and she felt sick thinking that he might have been killed out here, in the "safe" part of England. She needed to check on him, and began stumbling toward her front door—only to realize that the door was missing. It had been blown off its hinges into the yard. Lying on the ground beneath it was her cousin.

He wasn't hurt, thank God, so after dusting him off, Westgate asked her niece (who also lived with her) to fetch her coat so she could check on some neighbors down the road. The niece shot through the empty doorway and scrambled upstairs, only to hurry back down a moment later. Turns out their upstairs was missing, too. "Aunty," she cried, "the ceiling's all down!"

Until the Trinity nuclear-weapon test eleven months later, *Zootsuit Black* was the largest bomb explosion in history. A gargantuan green-yellow fireball filled the sky, and shrapnel began streaking out a moment later, including one full engine with its propellers still turning in midair. The resulting shock wave blew the roofs off several houses and shattered windows up to nine miles distant. When the fireball subsided, a gigantic octopus of smoke right out of Revelation replaced it, a thick black cloud with tentacles radiating outward wherever the shrapnel had sailed. Over the following week, local villagers found bits of the plane up to a mile away from the epicenter. No one ever found a speck of human remains; the closest they came was some parachute silk tangled in a tree.

Given who Joe Kennedy's father was, the disaster resulted in what one witness called "the most intensive investigation I have ever seen." The navy prepared two separate reports, hired electronics experts from RCA and CBS to write another, and then farmed each one out for comments. No one will ever really know what caused the explosion, but the reports did eliminate some theories, including sabotage, gasoline leaks, sparks of static electricity from nearby clouds, and stray bullets from the ground. Instead, they mostly zeroed in on the arming panel. The British had been trying to jam German communication lines with radio waves, so a stray signal might well have activated the arming circuit. And just as the electricians feared, the solenoid near the metal pin had likely heated up and triggered the detonator. Why the explosion happened when it did—just after the mother ship nudged *Zootsuit Black* to the left—remains unknown; perhaps it was coincidence.

The Kennedy family learned none of this. On August 13 they were staying at their compound in Hyannis Port, the same place where Joe had broken down in tears of frustration over his brother's heroism. Bing Crosby's "I'll Be Seeing You" was playing on the phonograph when Joe's mother, Rose, saw a dark sedan pull up. Two navy chaplains stepped out. She ran upstairs to get Kennedy Senior, who was napping. He came down, heard the news from the chaplains, and ran right back upstairs, already sobbing. Everyone else started breaking down, too. The only one who kept his head was Jack. He grabbed his twelve-year-old brother Teddy by the hand and said, "Joe wouldn't want us sitting here crying. He would want us to go sailing." Teddy would always remember that kindness.

Just as he'd promised in his final letters, Joe got a medal for the mission, the Navy Cross. He'd finally equaled his kid brother. But for security reasons, and perhaps to mask their mistakes, the navy told the Kennedys nothing about Joe's mission or why it had failed, only that he'd volunteered for it and died a hero. But Kennedy Senior's fear in 1939 had proved prescient: the war he'd never wanted had taken away his favorite son.

Incredibly, Project Anvil continued after Joe Kennedy and Bud Willy died. But every single subsequent flight failed as well, and Anvil's final record looked as bleak as Aphrodite's: combined, the two missions went 0 for 18 with the flying bombs. As historians have noted, the missions did more damage to the English countryside than to the concrete bunkers in France.

Meanwhile, the battle to stop the V-weapons raged on. As noted, the Germans fired V-1s from several sites across northern France, and Allied troops finally overran most of them in late summer 1944. This quieted the skies above England, and by September 7, Churchill's son-in-law Duncan Sandys felt confident enough to announce that "except possibly for a few last shots," the V-rocket threat was over.

Famous last words. The V-2 barrage started the next night. The first one landed in West London at 6:30 p.m. Hundreds more followed over the next few months, and the British quickly realized just how dire these new weapons were. Because they held more explosives, they packed a bigger punch, and because they approached at such incredible speeds (around 3,500 miles per hour), no one could hear them coming, let alone shoot them down. Worst of all, because the Germans fired them from mobile launch pads, they were all but impossible to eradicate. Despite the Allies' best efforts, the Reich continued to fire V-2s almost until the end of the war, killing 134 people as late as March 1945. One exploded just a half mile from General von Thoma's cushy POW camp at Trent Park. As he'd predicted, England suddenly had more "fun" than she could handle.

And what of the V-3—Hitler's high-pressure "Busy Lizzie" gun, which the Allies feared might be atomic? At almost the same time Sandys made his announcement, Patton's armies overran Mimoyecques and made a sickening discovery. The place was a sham. Beneath the impressive façade the Americans found not a single working rocket. It turned out that the earthquake bombs and other explosives had done far more damage than the Allies realized, battering delicate equipment and filling underground shafts with rubble. Nazi officials decided to abandon the site, but Hitler had kept a skeleton crew in place to fool reconnaissance planes and divert Allied attention: every bomb that fell there, he figured, meant fewer bombs falling on Berlin. In short, he outfoxed the Allies. Given the low likelihood of the Busy Lizzie ever working anyway, it's hard to avoid the conclusion that Joe Kennedy died on a futile mission.

Futile, at least, in a military sense. From a different perspective, the mission was quite consequential. Before the war the Kennedy family had the stink of defeat on them; Joseph Kennedy Sr. was the appeaser-in-chief, an American Neville Chamberlain. Jack's heroics with PT-109 counterbalanced that view, and Joe's death washed away the stigma completely. No one could question the family's commitment to fighting Hitler now. Kennedy Senior all but withdrew from public life after Joe's death, rarely writing letters or going out. But his sons emerged as heroes, and an American dynasty was born.

The type of plane that Joe Kennedy Jr. flew, a PB4Y-1 Liberator. *(Photo courtesy U.S. National Archives and Records Administration)*

Last known picture of Joe Kennedy Jr., from August 12, 1944. *(author Earl P. Olsen)*

The Curie family
cottage in L'Arcouest
(a.k.a., Port Science).
*(Photo courtesy
Musée Curie, ACJC
collection)*

Parisians tossed so many flowers at the Lightning-A advanced scout team
that Boris Pash (far left) said their cars resembled floats in the Rose Bowl
parade. *(Photo courtesy U.S. National Archives and Records Administration)*

The Lightning-A advanced scout team adopted this puppy, named Alsos, as their "mascot" on the dash into Paris. *(Photo courtesy U.S. National Archives and Records Administration)*

Doppelte Abwehr gegen die Feinde Ihrer Zähne!

A brochure advertising Doramad radioactive toothpaste. It promises "double defense against the enemies of teeth." *(Photo courtesy Oak Ridge Associated Universities, Inc.)*

Samuel Goudsmit in an armored car near Stadtilm, Germany. *(Photo courtesy AIP Emilio Segrè Visual Archives, Goudsmit Collection)*

Exterior of the "atom cellar" cave in Haigerloch that housed the final Nazi Uranium Machine. *(Photo courtesy U.S. National Archives and Records Administration)*

Fearing booby traps, Alsos members had German prisoners start to dig up the Uranium Club's cache of uranium outside of Haigerloch. Inset: the uranium cubes. *(Photo courtesy U.S. National Archives and Records Administration)*

Alsos members dismantle the Nazi reactor core in the "atom cellar" in Haigerloch, Germany. *(Photo courtesy U.S. National Archives and Records Administration)*

Cartoon of the "Nazi donkey" that kicked Boris Pash and broke his ribs during the pursuit of Werner Heisenberg. *(Photo courtesy Hoover Institute, Boris Pash papers)*

Boris Pash dancing up a storm during a party to celebrate the end of the war in Europe. *(Photo courtesy U.S. National Archives and Records Administration)*

CHAPTER 50

The Quisling Zoo

After Paris quieted down, Alsos set up headquarters in a grand hotel near the Arc de Triomphe. The Royal Monceau had fallen on hard times recently, but the staff was determined to put on a good show for the Americans. The waiters served meals in tuxedos and dished up army K-rations, said Samuel Goudsmit, "as if [they] were filet mignon aux champignons." The chefs even turned a detestable army dessert, the prune bar, into a delectable treat they dubbed "whip-a-fluffy." Still, Goudsmit couldn't help but notice that the waiters were mostly teenagers, and that the tuxedos hung loose on their adolescent frames. Apparently the men who'd once owned the suits, the original waiters, had all died at the front. It was a small, unforgettable reminder of the sadness of war.

Alsos had no time to grieve, however—there was too much work to do. Goudsmit's team began sifting through the mountain of documents that the Germans in Joliot's lab had left behind, searching for clues about progress on the atomic bomb. Boris Pash, meanwhile, turned to a sexier assignment—hunting down German uranium. In conquering Belgium, Germany had acquired the largest stockpile of uranium in Europe. So when Allied troops began liberating Belgium in early September, Pash and his Lightning-A team hopped into their jeeps—the license plates emblazoned with alphas now—and took off.

The two-day journey to Antwerp was forgettable, but once Lightning-A arrived, Pash encountered one of the most shocking scenes of his life. Over the previous year, the starving people of Antwerp had eaten most of the animals at the municipal zoo, a common occurrence during the war. What shocked Pash was that the cages hadn't remained empty: the people of Antwerp had filled them back up with Nazis. Few if any were German. Rather, these were Belgian collaborators, the men and women who'd sold their country's soul to Hitler.

"We walked into the lion house," Pash recalled, "[and] in each cage were from six to twelve prisoners.... Some looked defiantly at the tormentors, but most had the appearance of frightened, cornered animals. Men and women moved up to the bars to spit at the 'animals' and shout insults. Some poked canes or sticks into the cage, trying to jab the prisoners." Pash considered intervening, but several Belgian soldiers with tommy guns warned him off, and he left the "animals" to their fate.

Pash soon traced the Nazi uranium—much of it in the form of yellowcake, a powdered mineral—to a refinery town twenty-eight miles southeast of Antwerp. The refinery was surrounded on three sides by brick walls and on the fourth by a canal. Because the Germans still controlled the land across the canal, the Alsos crew had to dodge machine-gun fire as they approached. They zipped past one stronghold by gunning their jeeps and using an elevated railroad crossing as a ramp; it sent them "sailing through the air like water-skiers," Pash recalled. Entering the refinery was even more dangerous. When the Germans saw American troops poking around, they began sending shells over, forcing Lightning-A to dive for cover. They resumed the search on their hands and knees.

Eventually the crew located the uranium, or at least part of it. Records indicated that nearly two million pounds (a thousand tons) had been shipped via railroad to France, but sixty-eight tons remained somewhere in town. Pash's crew spent the next few weeks searching and finally unearthed the ore in an abandoned warehouse. It was packed into barrels, and as the Germans continued to shell them, they loaded the barrels into trucks. By late September Alsos had moved all sixty-eight tons beyond the reach of the Reich.

As for the yellowcake shipped to France, Alsos traced several tons of it to Toulouse by tracking the serial numbers of railroad cars. (This was the uranium that, during the takeover of his lab in 1940, Joliot had claimed was in North Africa.) Its presence in France should have been good news for Alsos—France was an American ally, after all—but French troops gave Pash almost as much trouble as the Nazis had. At the plant where the barrels of yellowcake were stored, Pash had to threaten the manager with a jeep-mounted machine gun before he surrendered any of it. Then, after loading the eighty-one barrels onto trucks and taking off for Marseilles, Pash's team got stuck behind a streetcar that refused to pull over and let the convoy pass; the streetcar would in fact speed up to block the trucks if they tried. (In contrast to Parisians, the people here resented American troops.) A fed-up Pash finally gunned his jeep, swerved in front of the streetcar, and slammed on his brakes. Both vehicles suffered heavy damage in the wreck, but the convoy got past.

Pash ran into another sticky situation at the Marseilles harbor. To load the barrels onto a ship, he recruited some American soldiers, who of course were told nothing about the top-secret material they were handling. They just knew that it was back-breakingly heavy and that handling the barrels left a metallic yellow film on their hands. They quickly deduced that they were stealing French gold—a rumor the French readily accepted. This didn't reflect well on America, but Pash judged it better to let them believe that than to violate security and reveal the truth.

The thirty tons of Marseilles yellowcake was shipped to Boston and then Oak Ridge, and the uranium-235 inside them ended up in the bomb that later devastated Hiroshima. The sixty-eight tons of

Belgian ore met the same fate. That added up to roughly a hundred tons total—which in most circumstances would have been a fantastic haul. But records in Belgium had spoken of a thousand tons of uranium. Where was the rest? As long as it remained at large, the Allies couldn't shake the fear of waking up one morning to a mushroom cloud.

α

That autumn also witnessed the notorious Rhine wine fiasco. It all started with a half-cocked idea of Robert Oppenheimer's. The Manhattan Project had recently built a reactor in Washington State that used water from the Columbia River for cooling. Oppenheimer reasoned that a German reactor would likely use river water, too, and he proposed monitoring the Rhine River for clues. A lack of winter ice near an industrial plant, for instance, would indicate that something inside was pumping out vast amounts of heat. Even in the summertime, the river might be contaminated with radioactive isotopes, and he suggested securing some water to test it.

Given his other responsibilities, Oppenheimer likely forgot about the idea, but General Groves's office didn't. The Rhine flows north through Germany before emptying into Holland, so as the Allied army pushed through Belgium and entered the Netherlands, the Alsos mission received an urgent pink radiogram to beat cheeks over there and take a sample. Problem was, the Allies had conquered only one side of the Rhine at that point. The Reich still controlled the far bank, with the bridges between considered no-man's-land. So when some poor Alsos sap showed up with a bucket and a rope one day and asked an American captain for permission to crawl out onto a bridge, the captain said that if he was stupid enough to go out there, no one would stop him. Indeed, a crowd soon gathered to watch— on both sides of the river. The German rubberneckers even decided to get some target practice in and began taking potshots. But the

Alsos soldier kept his head down and crept out far enough to lower the bucket into the water; he then scrambled to safety without spilling too much.

Back at Alsos headquarters, scientists sealed the water in bottles and boxed it up for Washington. At the last minute, as a joke, someone included a bottle of Roussillon, a local red wine, with a note to test that for "activity," too. Unfortunately the officers under Groves lacked comic sensibility. Upon receiving the package, they dutifully poured the wine into test tubes and broke out the Geiger counters. To their dismay, they discovered radioactive isotopes in it.

A dispatch soon reached Alsos headquarters in Paris: "Water negative, wine positive. Send more." Samuel Goudsmit and company had a good laugh at that. (*You know, those military folks aren't so bad.*) Goudsmit crumpled the message up and got back to work. But a second telegram arrived a few days later, demanding to know where the wine was. A confused Goudsmit tried to explain the joke, but to no avail. An order's an order, Washington said. Send more wine. So to his disgust, Goudsmit had to pull a scientist away from examining German documents—actually valuable work—and send him on a ridiculous ten-day errand to gather bottles from around the region.

It turned out that grapevines there were probably sucking up naturally radioactive atoms from the soil and concentrating them. (Tobacco plants do something similar.) Knowing that it was futile to explain this, Alsos gathered an entire case of red for Washington and shipped it off. Goudsmit did make the most of things, however. He insisted that his errand boy grab two bottles of every vintage he came across—one for Washington, and a "file copy" for Alsos. They too would test the wine for activity, the old-fashioned way.

In addition to humorless officers, Goudsmit had something else to worry over as the Allies pushed into Holland—the fate of his parents.

Before the war, Goudsmit had feared he would never see his home-land again, and when he finally entered Holland on September 29, he found it in a grim state. The tidal battle between advancing and retreating armies had laid waste to huge areas, and thousands of people had no food or water. But there were moments of hope, too, like when a small boy gave him an orange bow to wear. "It makes me feel as proud as if it was a real campaign ribbon," he wrote his wife. He pinned it to a picture of her.

Goudsmit was traveling to Holland to examine the records of an electronics plant that had once supplied Peenemünde with vacuum tubes; he suspected the Uranium Club had ordered parts, too. The plant was located just eighty miles from Goudsmit's boyhood home in The Hague, and while he couldn't make it all the way there (the Germans controlled the land in between), he hoped that someone in the area had heard some news.

Sure enough, he ran into a young Dutch physicist at the plant, a student of his friend Dirk Coster. The student confirmed that Goud-smit's parents had indeed been deported; the postmark on the last letter Goudsmit had received, from the concentration camp in Czecho-slovakia, had been accurate. Still, Coster hadn't given up. He'd been sending them care packages of food in the camp, to help them get by. Even better, the student mentioned that Coster had appealed to "influential German colleagues" to intervene on their behalf.

It's not clear whether the student mentioned that Werner Heisen-berg was one of these colleagues, but it seems likely. If so, Goudsmit must have felt a pang. Earlier in the war he'd suggested kidnapping Heisenberg, and now that Joliot was in custody, Heisenberg had become Alsos's top target. Nevertheless, this tidbit must have given him hope. After all, he and Heisenberg had been good friends once—Heisenberg had even dined at his parents' home. Surely, surely he'd help if he could.

CHAPTER 51

Healthy Rays, Healthy Teeth, Healthy Paranoia

Goudsmit's team of technical sleuths consisted of thirty or so scientists ranging in age from grad student to retired. Most wore military uniforms on the job, although they didn't have to drill or salute anyone. Misfits abounded. Several of them took advantage of being separated from their wives to experiment with growing mustaches.

They spent most days examining documents, which they raided from labs and industrial firms around Paris. Technical reports were the most valuable items, but they also paid close attention to personal papers and office detritus, which helped them piece together the social networks of scientists. If you knew Fritz worked on fission, the thinking went, and Fritz had taken a train to visit Dieter, then Dieter probably worked on fission as well. No piece of paper was too trivial to escape scrutiny: over the next few months, Goudsmit's team gleaned clues from desk calendars, appointment books, bills of lading, streetcar stubs, office seating charts, and bookstore receipts. Goudsmit was especially giddy to find a sheaf of used carbon-paper inserts for duplicating typewritten letters. He'd read a trick once in a detective novel, and sure enough, by holding the sheets up to the sun, he could read the original messages. (They looked like photograph negatives, white text on black paper.) Watching the scientists work, Boris Pash would chuckle to himself as they gathered around some catalogue or address book, reading it with all the zeal, he said, of "a spicy French novel." Sometimes they got so caught up in work they'd forget to eat, letting their fancy K-ration dishes grow cold.

Occasionally what they uncovered made them chuckle. In one letter the wife of a German scientist ordered him to bring some Chanel No. 5 home from Paris—or face her wrath. They also got a kick out of German research into invisible inks and amphibious cars (which looked like VW Bugs crossed with motorboats). The U.S. military provided plenty of mirth as well, albeit inadvertently. Analysts in Washington were always sending psychological profiles of German physicists in which they'd emphasize how so-and-so "engage[d] in beer-drinking jousts" or had "an atrophied right testicle"—as if those traits held the key to hunting them down. Other reports were straight-up phrenology, instructing Goudsmit on how to unlock people's true characters by examining the warts and knobs on their faces. This was what passed for military intelligence.

Still, after years of feeling useless, the scientific sleuthing reinvigorated Goudsmit, and he broke his first big case in mid-October 1944. From seized documents, he learned that a Nazi-run company had stolen a huge cache of thorium in France and hidden it deep inside the Reich. Thorium did have industrial uses in making gasoline and ceramics, but only in tiny quantities; the amount stolen would have fulfilled any one firm's demand for decades. To Alsos, that left only one explanation. The main isotope of this element is thorium-232. When exposed to neutrons, thorium-232 becomes thorium-233, which undergoes two beta decays and becomes uranium-233. Uranium-233 was every bit as fissionable as uranium-235 and therefore seemed promising as bomb material. And unlike the uranium-235 in natural ores (which is mixed with other uranium), the uranium-233 would be easy to separate chemically from thorium. Accumulating a large store of thorium, then, was a quick way to produce bomb cores.

The firm that stole the thorium had an office in Paris, so Goudsmit's team raced over to raid it, taking a new staff car. (One of Pash's less scrupulous officers had illegally commandeered several civilian cars and forged dummy registration tags for them, so Goudsmit didn't have to suffer the humiliation of begging rides everywhere.) Alas, the Germans had cleaned out the office before abandoning it—a sign they were hiding something big.

Given the stakes, Goudsmit observed, "the thorium mystery became an obsession." Among the few papers that remained in the office, Goudsmit found references to a chemist named Jansen and his secretary, Ilse Hermanns. Cross-checking a telephone log revealed that Hermanns had made several long-distance calls on Jansen's behalf. Considering the expense of such calls during wartime, Jansen was clearly someone important. What's more, by examining lists of correspondence, Goudsmit determined that Jansen had sent a registered letter to Hermanns in Eupen, a town in Belgium, just before the fall of Paris. Perhaps she was still there. So as soon as the Allies secured control of Eupen, in early November, the Lightning-A crew hopped into jeeps and took off.

When they reached the address on the registered letter, Boris Pash knocked sharply on the door. To his surprise, Frau Hermanns answered. Pash bullied his way inside and discovered that she lived there with her parents, upper-middle-class folks with a grand piano and handsome furniture. Hermanns's father objected to Lightning-A searching the house but was helpless to stop Pash, who uncovered a suitcase full of documents addressed to Jansen. Even better, they came across a locked closet door upstairs, which the father refused to open. Instead he pulled Pash aside and, man to man, appealed to his sense of decency. A special admirer of his daughter's was in there, he confided. Her beau. Couldn't Pash just leave things be and spare the couple some embarrassment?

Pash could not. Raising his voice, he announced that either they could open the closet door or he'd shoot the lock off. As he hoped, the man inside heard this and began fumbling with the knob. The door swung open to reveal none other than Jansen himself. Hermanns, evidently, was more than his secretary.

Pash arrested Jansen and hauled him back to Paris for interrogation—a big moment for Goudsmit. One of his first jobs for Alsos had been to interrogate Frédéric Joliot several weeks earlier, and he'd botched it royally. Alsos had learned a few nuggets from the Frenchman, especially about the shocking success of Kurt Diebner's experiments. (They'd never heard of Diebner before.) By feigning ignorance, though, Joliot had successfully drawn Goudsmit out on certain points, and he ended up learning far more from Alsos about American nuclear science than Alsos had learned from Joliot. Boris Pash had not been pleased.

Goudsmit hoped to redeem himself with Jansen. To increase the intimidation factor, he and Pash broke out their crispest uniforms, with Pash adding several ribbons and medals. Jansen proved a tough nut, however, revealing almost nothing under questioning. All they squeezed out of him was that his firm had shipped some of the missing thorium to Belgium for safekeeping, and in the chaos of the war, several train cars' worth had disappeared. Jansen claimed he'd gone to Belgium to track them down, and when he found himself in the neighborhood of Eupen, he'd decided to swing by his secretary's home for a little hanky-panky. Alas, the German defensive lines had collapsed during his visit, and he'd hidden at her parents' house for safety. That's why Pash had found him there. As for what the thorium was for, Jansen claimed to have no idea.

For Goudsmit this story didn't add up. Yet he couldn't press Jansen about anything related to atomic weapons, lest he tip the German off about American interest in them. His only hope now lay with the suitcase of documents Pash had confiscated. So after striking out with Jensen, Goudsmit sat down that night and began paging through them for clues.

A cold snap had descended on Paris that week, so Goudsmit

crawled into bed with the bundle, burrowing under his blanket for warmth. The first few dozen pages were tedious and irrelevant; he must have been on the verge of nodding off. But he soon came across something, he said, that made him "almost fall out of bed from excitement."

It was a hotel bill from Hechingen, a village in the Black Forest region of southern Germany. Moe Berg and other sources had already traced Uranium Club members to the area, and here was evidence that Jansen—whose firm was smuggling thorium—had visited, too. What's more, a personal letter in the suitcase referred to Hechingen as a "restricted" area, closed to outsiders. What were the Germans hiding there?

When confronted with the hotel bill, Jansen responded that he'd been visiting his mother, who lived nearby. *Bullroar,* Goudsmit thought. As for the "restricted" status, Jansen claimed it was a misunderstanding. That simply meant that no refugees from bombed-out cities up north could resettle there. *Likely story.* With Jansen stonewalling, Goudsmit kept digging through the documents and finally discovered where the thorium had ended up, in a medieval castle just two miles south of Hechingen. In addition, another scrap of paper in the suitcase contained the address of a woman named Carmen. When asked about her, Jansen at first denied all knowledge. He then broke down and claimed she was a prostitute who'd cheated him. Goudsmit knew better: Carmen was clearly a secret agent, "a Nazi Mata Hari."

Every detail of the Jansen case fit together beautifully. A highly radioactive element had been stolen from France and rerouted to secret lairs in southern Germany, lairs where the world's top nuclear scientists just happened to be stationed. It was scary and exhilarating all at once, and Samuel Goudsmit was having a grand old time piecing it all together—right up until the bottom fell out of everything.

One set of documents he came across confirmed the transfer of the thorium to the Hechingen castle. But after scrutinizing some shipping papers, and cross-checking with other scraps, he finally learned the real reason the firm had stolen it: to make toothpaste. Jansen's bosses had grown rich manufacturing gas masks and searchlight filaments for the Wehrmacht, but with the end of the war looming, they wanted to diversify and shift into cosmetics. One promising product in their portfolio was a thorium-infused toothpaste called Doramad, whose radioactive properties supposedly whitened and brightened teeth. (A Doramad ad from the time declared, "My rays massage your gums. Healthy rays—healthy teeth!") The thorium heist had simply been a dirty trick to corner the postwar market. As for the hotel bill, Jansen really had been visiting his mother.

In a way this was great news: the Nazis hadn't developed a short-cut for making nuclear bombs. (Research later revealed that thorium wasn't much use for bombs anyway, because making uranium-233 produces by-products that cause chain reactions to fizzle.) Still, Goudsmit's crew felt pretty sheepish for whipping themselves into a frenzy over toothpaste. Even worse, they'd still learned nothing useful about the Nazi atomic bomb.

CHAPTER 52

The Deadliest Hombre

In 1944 the much-discussed plan to kidnap Werner Heisenberg suddenly gained momentum and took off at a gallop. Spies had finally traced the physicist to Germany's Black Forest, a name right out of Grimm's, and with their top scientific target in sight, General Groves and Wild Bill Donovan began plotting to lay hands on him.

The soldier they chose to lead the mission was Carl Eifler, once described as "without a doubt the toughest, deadliest hombre in the whole OSS menagerie." Eifler, who'd joined the army at age fifteen, had all the pugnaciousness of Boris Pash and none of the scruples. "Consider yourself a criminal," he once advised OSS recruits. "Break every law that was ever made." He stood well over six feet tall and weighed 280 pounds, and his size and demeanor regularly drew comparisons to a grizzly bear. His hobbies included shooting shot glasses off the tops of people's heads while drunk.

Eifler made his reputation in 1943 while battling the Japanese in the jungles of Burma. Despite starting with just a few dozen soldiers (including film director John Ford, who recorded several live ambushes), Eifler soon commanded an army of 10,000 angry Burmese natives, and dollar for dollar he ran one of the most successful campaigns of the war. Using hit-and-run tactics, the Burmese irregulars destroyed dozens of airfields and railroad stations across 10,000 square miles of jungle; they also killed and wounded 15,000 Japanese troops, compared to 85 deaths on the Burmese/American side. (One favorite trick involved planting bamboo spikes in the weeds alongside a road and ambushing Japanese soldiers, who impaled themselves when they dove for cover.) Many Burmese natives kept shriveled brown ears from Japanese corpses as souvenirs, which they carried in bamboo tubes slung around their necks. Some had several dozen.

In his most famous exploit, Eifler and a few men set out to raid a Japanese base along the Burma coast one day in March 1943. But choppy seas prevented their boats from landing, and they risked being dashed on nearby rocks or swept out to sea. Just before they lost control, Eifler grabbed a towline, leapt into the water, and started thrashing for shore. The chop slammed his head against the rocks several times, and the undertow almost dragged him down for good, but several minutes later he managed to stagger ashore, bloody and woozy. He then used the towline to drag all five boats to safety before passing out. In terms of sheer bravado, it surpassed even JFK's exploits.

Although he saved the mission, Eifler woke up with ringing in his ears and an excruciating headache. (In smashing against the rocks, he'd probably suffered several concussions, if not outright brain damage. He would later experience seizures and other neurological problems.) Over the next few weeks his thinking grew foggy, and he had so much trouble getting to sleep that he had to scarf painkillers and gulp bourbon each night in order to black out. This didn't exactly improve his mental state, and he sometimes burst out crying for no reason. Rumors of his erratic behavior finally reached Wild Bill Donovan, who had little choice but strip him of his command, a humiliating development.

The demotion pained Donovan, too: as a fellow reckless warrior, he sympathized with Eifler's plight. Moreover, Eifler was too valuable an asset to let sit idle. So at some point over the next few months Donovan mentioned another adventure that he thought might appeal to the deadly hombre—a little lark to hunt down a certain scientist in Germany.

After some much-needed rest, Eifler began meeting with OSS officials in early 1944 to discuss the mission, which was so secret it never received a code name. Accounts of the meetings differ, but Eifler first sat down with one of Groves's deputies, who briefed him on Heisenberg and nuclear bombs. Eifler had no idea what "fission" meant and even less of a clue who Heisenberg was, but he understood dirty warfare. He finally interrupted, "So you want me to bump him off?"

Groves's man winced. "By no means." Instead Eifler should "deny the enemy his brain," the euphemism they'd settled on for abduction. "Do you think you can kidnap this man and bring him to us?"

Eifler didn't hesitate. "When do I start?"

"My God," Groves's man marveled, "we finally got someone to say yes."

A few days later Eifler met with Donovan and other officials. By this time he'd conjured up a cover story, a tale meant to fool the Allies as much as the Germans. He planned to launch the mission from Switzerland. He'd pose as an American customs agent there, claiming that he wanted to study how neutral countries handled border control during wartime. "This gives me the opportunity to observe their borders," he explained, "and figure out how to violate them." For the actual kidnapping, he suggested using a dozen commandos to raid Heisenberg's new lab, located just fifty miles from the Swiss border. After bludgeoning the physicist into submission, they'd smuggle him back to Switzerland, steal a plane, and fly to England. No sweat.

Donovan immediately objected—not to the manhandling of Heisenberg or the gross violation of Swiss neutrality, but to landing in England. Under no circumstances did they want the bloody British involved. Eifler accepted this limitation. He then began thinking out loud and came up with an even wackier scheme. Instead of steering the stolen plane for England, they could fly toward the Mediterranean. Then maybe they could ditch the plane over the water and parachute into the freezing sea. Oh, and then rendezvous with a submarine in the darkness, which would whisk them through the phalanx of U-boats there to safety.

Lacking any limeys, this plan suited Donovan just fine. Ironically, it was Eifler who started thinking it through more, and pointed out some potential difficulties. What if the submarine got delayed? Or what if storms prevented the rendezvous? Donovan merely chuckled. Carl Eifler, he said, you're the last man on earth to be worrying about undue risks. That was probably true, Eifler conceded. But he did have one more question: What if we get caught?

The answer sounded familiar: "Deny Germany the use of his brain." But the meaning of the euphemism had shifted, taking on a darker cast. Eifler now had permission to shoot Heisenberg rather than risk him falling back into German hands.

Eifler nodded. "So I bump him off and get arrested for murder. Now what?"

"We deny *you*." Deny they'd ever heard of him.

Okey-doke, said Eifler. It was the answer he'd expected, and it didn't trouble him. Time to get to work.

For once, though, OSS had second thoughts about a nutty scheme. Donovan admired Eifler's gifts, such as they were, but the man was simply too gung-ho for a mission like this. Germany wasn't a jungle where you could butcher your way out of a tight spot. Kidnapping Heisenberg would require tact and subtlety—less smash and more dash. So in the summer of 1944 Donovan once again had to yank command from his top warrior, and this second demotion proved even more gut-wrenching than the first. For privacy, they met on the balcony of OSS headquarters in Algiers, where Eifler was training. Donovan gave some bullshit excuse about how the Manhattan Project had "cracked the atom," which supposedly rendered the kidnapping moot.

Still unclear about the physics, all Eifler understood was that he'd been relieved of command again. He managed to keep himself together in front of Donovan, but a few days later, during a conversation with a friend, he lost it. The embarrassment, the stress, the continuing mental turmoil from his injuries in Burma—it proved too much, and he broke down sobbing.

Eifler might have been even more upset if he'd realized something else: that contrary to what Donovan had told him, the plot against Heisenberg had not been canceled. Groves and Donovan had simply reworked it, enlisting other, less volatile hombres to carry it out.

One was Paul Scherrer. Code-named Flute, he worked as a physicist at the prestigious Swiss Federal Institute of Technology (ETH) in Zurich, Albert Einstein's alma mater. It was the ideal cover for an atomic spy. Switzerland had remained neutral during the war, so Axis and Allied citizens alike could travel there freely. Switzerland also bordered France, Germany, and Italy, which made it a convenient central location. Zurich became the epicenter of wartime espionage as a result, a hotbed of spies and double agents. Only a fool trusted the strangers he met in bars and cafés there.

Flute and Heisenberg had once been friends, and Heisenberg still trusted him completely. For Flute, things were more complicated. The war—and especially Heisenberg's refusal to denounce German aggression—had driven a wedge between them. Flute didn't cut off his friend entirely, but he did feel obligated to feed information about his whereabouts to the Allies, including his occasional trips to Switzerland. Two years earlier, in fact, in November 1942, Flute had invited Heisenberg to give a lecture at ETH on something called S-matrix theory. (It described how elementary particles collided; S stood for scattering.) This lecture had actually been the impetus for Samuel Goudsmit's friends to suggest kidnapping Heisenberg in the first place. That plot had died of inattention, but as luck would have it, Flute invited Heisenberg back to ETH in the fall of 1944—and this time Groves and OSS pounced. Apprehending Heisenberg in neutral Zurich would eliminate the need for an Eifler-type raid into hostile territory.

Besides Flute, the other agents enlisted in this plot were Moe Berg and Samuel Goudsmit, who'd formed an unlikely friendship. The two probably first met in Paris. Although Alsos folks and OSS folks normally kept their distance—Berg and Boris Pash were outright enemies, for instance—Berg quickly warmed to the urbane, multilingual physicist, and Goudsmit more than reciprocated this feeling for the dapper catcher. "I was intrigued," Goudsmit remembered, "that a professional baseball player would be so effective in this totally different sphere of activity [i.e., espionage]." Once again, Berg's athletic prowess left an intellectual swooning.

The new plot to seize Heisenberg called for Berg to slip into Zurich and make first contact. Goudsmit, who knew Heisenberg personally, would follow a few days later. Beyond this, details were sketchy. They talked sometimes of merely grabbing and interrogating Heisenberg before ultimately letting him go. Other times, they planned to spirit him away to Allied territory. In either case, they'd need more muscle: Berg and Goudsmit were hardly seasoned bounty hunters. And the plot would violate nearly every international law in existence about wartime neutrality. The Swiss tolerated spying unofficially, but you had to be discreet; you couldn't shanghai Nobel laureates on the street. Indeed, one top OSS official in Switzerland (Allen Dulles) vehemently protested the kidnapping on these grounds, fearing that the Swiss would sever diplomatic relations with the United States and thereby endanger the entire U.S. intelligence apparatus abroad. But as usual, fear of an atomic Hitler overwhelmed every other consideration.

In preparation for the mission, Berg mostly behaved himself and kept OSS informed of his whereabouts. After finally leaving Italy in September 1944, he'd split most of his time between London and Paris. It probably helped his mood that OSS advanced him more than $3,000 ($41,000 today) for hotel rooms and meals during those months, in addition to a 20 percent raise in salary, to $4,600, in anticipation of his dangerous assignment. To further placate the catcher, OSS officials did their utmost to keep him updated, via telegraph, on Major League Baseball standings.

The mission came together smoothly at first. Heisenberg accepted Flute's invitation to lecture and they settled on a date in mid-December. Berg would attend the lecture and listen for clues about Germany's progress with nuclear fission. Flute would then arrange a meeting between Berg and Heisenberg to establish contact, with Goudsmit joining later. Flute of course knew that Berg was working for American intelligence and wanted to feel Heisenberg out about nuclear bombs. But he assumed the meeting was a straightforward interview. He would have been horrified to learn that he was abetting a kidnapping. Suspecting as much, OSS kept him in the dark.

For his part, Heisenberg accepted Flute's invitation for several reasons, none of them related to physics. He was feeling more isolated than ever from the world scientific community and longed to reconnect with an old friend. Visiting in mid-December would also give him a chance to pick up new winter clothes and better Christmas presents for his wife and children. As a neutral country, Switzerland faced no shipping embargo, and Swiss manufacturers (unlike German ones) didn't have to divert all their energy into producing war goods, which meant that toys and chocolates and other treats were readily available in Zurich. In short, Heisenberg had visions of sugarplums dancing in his head, and had no inkling he was walking into a trap.

But the trap soon experienced a snag. In addition to inviting Heisenberg, Flute had also invited the diplomat's son Carl von Weizsäcker to speak in Zurich on November 30. Weizsäcker knew that his visit would be controversial, given his father's status, so he tried to mitigate his presence by lecturing on an innocuous topic, the evolution of the solar system. The ploy didn't work. A mob of ETH students turned out to protest the Nazi's son, and a riot nearly broke out. ETH officials had to lock the lecture hall for safety, and Weizsäcker returned to Germany shaken.

Hearing this, Heisenberg put strict stipulations on his appearance: the crowd had to be small, so he could speak freely, and only other physicists could attend. This presented difficulties for Berg, who as a six-foot-one, 200-pound nonphysicist was likely to attract scrutiny at an intimate gathering. Still, Flute had no choice but to give in to Heisenberg's demands.

Then, just days before the mission started, the plot endured another, more serious blow. For security reasons, Samuel Goudsmit had to be cut out.

CHAPTER 53

Nazi U

Among the spicy appointment books and desk ledgers, the Alsos crew in Paris recovered a university course catalogue, a paperback booklet whose cover showed the handsome cathedral and stout medieval walls of the city of Strasbourg. The Third Reich had opened a university there in 1941 as a propaganda outfit to spread fascism in France—Nazi U, essentially. It had quickly grown to 3,500 students.

Now, a course catalogue might seem mundane, but it did list the professors teaching each class at Nazi U, and a few names jumped out immediately—including that of Carl von Weizsäcker. Given that he belonged to the Uranium Club, Alsos probably should have dropped everything else to pursue him when the city fell in mid-November 1944. But at that point Goudsmit and Pash were distracted with the thorium-toothpaste scare, and by the time Lightning-A reached Strasbourg—on November 25, ten days after Patton's armies had rolled through—the diplomat's son had fled.

All wasn't lost, though. Through various ruses, Pash tracked down several other nuclear physicists in the city. (In one case, when a woman claimed that the scientist who lived next door to her had fled weeks before, Pash tripped her up by pointing out the fresh eggs and soft, fresh bread in the man's kitchen.) Pash also ransacked Weizsäcker's apartment and his office at the university. The apartment proved a disappointment. It was as bare as Joliot's cottage had been, containing nothing but a potbelly stove and some ashes—the remnants of all the papers he'd burned. Moving on to his office, they found the door locked, so several stout fellows lowered their shoulders and slammed into it. It didn't budge, so they started kicking. When that failed, they grabbed an axe. Only after smashing through the wood did they realize that the door opened outward and hadn't been locked at all. Luckily, Weizsäcker hadn't had time to clean out his office, and piles of documents remained. Pash sent word to Goudsmit to hurry to Strasbourg.

Goudsmit did so, albeit reluctantly. He took an open-top jeep there, and the weather on the two-day, three-hundred-mile journey was so bone-chilling that he resorted to wearing his pajamas beneath his uniform in an effort to keep warm. It didn't help, and the trip put him in a foul mood. "A jeep is not a proper method of travel for… over-draft-age, desk-type, and blackboard scientists," he griped. But honestly, cold weather wasn't the real problem. The disappearance of his parents was still festering in his mind. He'd also grown acutely homesick. Amid the jubilation in Paris he couldn't help but notice, with a pang, all the families sharing meals and laughing together in the cafés, while he ate alone. In letters home he began begging his wife and daughter to write more often: "Mail," he told them, "is more important than sleep and food." (His daughter complied; his wife did not.) Going to Strasbourg only dragged him deeper into the war and farther away from his family. He nevertheless sucked it up and went, and when he arrived at Nazi U, he began going through the documents from Weizsäcker's office.

Because the electricity in Strasbourg was out, everyone in Alsos huddled together in the same room around a few candles that night. While the scientists read, the soldiers played one of their endless games of what Pash called "applied mathematics"—poker. (They preferred playing with inflation-prone French money because the piles of cash looked more impressive.) It was a sleepy evening overall, although shells continued to fall and dogfights occasionally erupted overhead. Halfway through the stack of papers, however, Goudsmit and a colleague both yelped and jumped to their feet. This startled the soldiers, who dropped their cards and whipped their rifles around. A little sheepishly, Goudsmit told them to stand down, but he was overjoyed. He'd discovered a cache of letters between Weizsäcker and Heisenberg on nuclear fission—their first real lead on that front.

In a huge security lapse, the letterhead on one page provided not only Heisenberg's new address in southern Germany, Weiherstrasse 1, #405, but his telephone exchange. (Goudsmit suggested, only half facetiously, sneaking into Switzerland and calling up ol' Werner.) More importantly, the letters included several pages of reactor calculations and references to a *"spezialmetall,"* obviously uranium. One especially valuable letter had been discovered in Weizsäcker's trash can, torn into little pieces. Weizsäcker had taken a harsh tone in it, and had apparently thought better of sending it. When reconstructed, the scraps contained valuable clues about the state of German research and how the Uranium Club was proceeding.

Goudsmit studied these letters by candlelight until his eyes ached, and continued studying them for the next three days. At the end of which he jumped to a dramatic conclusion. The Nazi atomic bomb project, he announced, was a sham—a crude, poorly funded effort that would never produce a nuclear weapon. You could see it in the calculations, he said: the Nazis were years behind the Manhattan Project. Feeling triumphant, he summarized his conclusions in a memo to General Groves, then rewarded himself with a little cognac.

If Goudsmit was expecting praise, he obviously didn't know Leslie Groves. In fact, Groves didn't believe Goudsmit, and picked apart every one of his conclusions. The torn-up letter in the trash can, for instance, seemed like an obvious plant. Why would someone who'd scrupulously burned every document at home leave behind that one vital piece of evidence, unless he wanted it discovered? In fact, how did Goudsmit know that *any* of the documents were genuine?

Perhaps they were part of a misinformation campaign designed to lull them into complacency.

Goudsmit didn't have a good answer, and it soon became clear that he'd been sloppy and hasty, glossing over several bits of evidence that might argue *for* a Nazi nuclear threat. One document spoke of "large scale" fission experiments being conducted just fifteen miles from Berlin. Another revealed that German officials had already briefed Hitler about atomic weapons. Perhaps not coincidentally, reports began trickling in to intelligence officials that very week about something Hitler had said at a military conference: "Up to this day I can answer for all the acts I've committed before God and my compatriots. But for what I'm going to order in the near future, I will no longer be able to justify before God." What else could he be referring to?

Groves also had access to reports that Goudsmit didn't — including one report he called "the biggest scare to date" in the war. A few days after Alsos arrived in Strasbourg, reconnaissance planes had snapped photos of several mysterious buildings in a valley not far from the Black Forest. They were clearly industrial sites — they had huge chimneys and grids of pipes, with railroad spurs and slave quarters. What most alarmed Groves was the speed of their construction. The first recon sorties spotted three buildings; a few weeks later there were fourteen, spread over twenty miles. Was this a uranium enrichment facility, the German Oak Ridge? Even the unflappable British began trembling.

Ten days of crisis followed in Washington and London before the truth emerged. Someone realized that all the plants lay on the same geological contour, and a trip to the library revealed that this contour was mostly shale. Shale often contains uranium, but in this case it contained oil as well. The Allies knew that Germany was running low on petroleum, and more likely than not, the site in the valley was a new refinery.

Groves bombed the hell out of the place anyway, just to be safe.

And even after he destroyed it, the specter of it continued to haunt him. However bruised and battered, Germany could still throw together huge industrial projects with terrifying speed. The Allies had been lucky to spot the buildings—Germany was shifting most manufacturing underground at that point. So who knew what else they'd missed? Robert Oppenheimer had once warned Groves that the Manhattan Project's facilities for enriching uranium, including the sprawling Oak Ridge site, might be anomalies. Nuclear science was still in its infancy, and Oppenheimer had enough humility to admit that some clever German "might come up with a way to [enrich uranium] in his kitchen sink." In which case the Allies would never notice a thing, until it was too late.

Rather bravely, Goudsmit continued to defy Groves on this matter, insisting that the Nazis would never succeed in making an atomic bomb. But he was destined to lose this power struggle with the general. With hundreds of tons of uranium still missing, and Heisenberg and Weizsäcker still at large, Groves simply couldn't take a chance otherwise.

$$\alpha$$

Aside from losing this fight with Groves, Goudsmit found Strasbourg demoralizing for another, darker reason. As noted, Alsos scientists were gathering intelligence in fields beyond nuclear physics. In particular, they'd been tracking rumors about Nazi doctors experimenting on prisoners, and what they found in Strasbourg confirmed those tales in the most awful way.

It started with a vaccine biologist named Eugen von Haagen, who ran the "Hygiene Institute" at Nazi U. Some of the labs there seemed innocent enough: bays of rats, mice, goats, and other livestock on which to test serums. But Alsos soon discovered a second facility at an isolated fort outside of town. According to his records, von Haagen had been infecting prisoners there with spotted fever and other diseases, then sacrificing them at fixed intervals to monitor the decay of their organs. He'd even had the audacity to complain to prison officials that they were sending him weak, inferior "material" (i.e., human beings) to study. To understand the true effects of these diseases, he argued, he needed to kill healthy men and women. How else could he save people's lives?

Strasbourg was also home to Heinrich Himmler's personal anthropology institute, Das Ahnenerbe (The Legacy of the Ancestors), dedicated to proving the racial superiority of Aryans. Much of the "research" there focused on anatomy, and the staff kept an extensive reliquary of human skulls to study. Himmler had in fact supported Germany's disastrous invasion of the Soviet Union in 1941 in part to gain access to the skulls of "Jewish-Bolsheviks" and other degenerate types. While virtually every other Nazi prosecuting the war on the Eastern Front was focused on, say, defeating the Russians, Himmler and his minions were dispatching high-priority messages about where to find new "specimens," along with detailed instructions for preserving their bodies in "specially constructed airtight tin containers filled with preservative." The Alsos team saw the strange fruits of this labor in Strasbourg: half-dissected limbs lying about, and giant tanks full of alcohol with human corpses floating inside. These turned out to be prisoners from Auschwitz. Many were sickeningly emaciated and had numerical IDs tattooed on their arms. Most had been gassed and had arrived in Strasbourg with their bloodshot eyes wide open.

News of other sadistic experiments would emerge over the next several months as Allied troops liberated concentration camps throughout the Reich. But even this glimpse of the atrocities proved too much for Goudsmit. By awful coincidence, he was billeted in Eugen von Haagen's lavish house in Strasbourg, an unnerving experience. Worse still, he had to sleep in the bedroom of von Haagen's son. "All his toys were still there," Goudsmit remembered. "An electric train, a movie projector, an old microscope of his father's, an aquarium with snails, books, tools. But also a lot of Hitler Jugend [Hitler Youth] insignia....I was thinking whether he missed his toys now."

Had Goudsmit hardened his heart and thought only of von Haagen and his crimes, he might have gotten through the weeks there. But he was pining for his own daughter, and as soon as he allowed himself a little sympathy for the von Haagen child, he broke down and began sobbing. The images he'd seen over the past few days—tanks of bodies, shelves of skulls—flashed through his mind and snapped the last frayed strands of emotional restraint he had left. This most timid of physicists stormed out of the child's room and began rampaging around the house, stomping his feet and screaming. The soldiers staying with him were stunned. "He just went off his rocker," recalled one. "He was furious at the Germans, weeping and thrashing around." It was a full-fledged mental crack-up, and it took a full half hour to calm him down.

Goudsmit never mentioned the incident in letters home. But shortly afterward, while writing yet another unanswered missive to his wife, he lamented, "I fear I am too soft for this [espionage] game." The army was inclined to agree. After his breakdown Alsos put him on rest leave, and officials quietly arranged for him to return to the States to see his family. According to some accounts, he visited a psychiatrist there—a drastic act back then, something only desperate people did.

During his absence the army debated cutting him out of Alsos altogether. The Pentagon eventually allowed him to remain on duty, but it did veto his participation in another mission. Throughout his time in Strasbourg, Goudsmit had been fielding cables from OSS about his trip to Zurich to seize Werner Heisenberg. His nervous breakdown eliminated him from that operation. Someone liable to crack-ups clearly couldn't be sent undercover to a foreign country.

Several other members of the Zurich kidnapping team had to drop out as well, due to travel delays and other foul-ups. And given the timing here—Goudsmit fell apart in early December, just two weeks shy of Heisenberg's lecture—intelligence officials didn't have time to find substitutes. Moe Berg would have to stalk Werner Heisenberg alone.

CHAPTER 54

Uncertainty, Principles

Paradoxically, going after Heisenberg with the gentle Moe Berg rather than the brutal Carl Eifler all but ensured that the mission would turn violent. To be sure, ever since OSS first enlisted Eifler, the plot had had an undertone of violence. Eifler, though, simply had permission to kill Heisenberg if things went south; his death had never been the point of the mission. But with Berg going solo, OSS had to abandon all hopes of a kidnapping: Berg would never be able to manhandle the physicist alone, and Samuel Goudsmit wouldn't be there to interrogate him anyway. As a result, assassinating the physicist seemed the only viable option. Moe Berg would have to become a deadly hombre.

Although too unstable to accompany Berg, Goudsmit did help brief him in Paris on December 13, passing along final instructions from Groves. Berg was to take his pistol to Heisenberg's lecture and listen closely to determine how much progress the Reich had made toward a nuclear bomb. It seemed unlikely that Heisenberg would blatantly threaten the allies, of course, but he might drop clues here and there, hints. If nothing else, the laddish Heisenberg might well brag in front of his peers. And if he did so, Berg had to render him "hors de combat," as Goudsmit put it. That was more than a flourish of French between two polyglots. Literally, the phrase means "outside the fight," or more colloquially, "out of action." It usually refers to battlefield casualties, and as one historian noted, "There is a very narrow range of ways in which a gun may be used to take an opponent out of battle." Five years earlier Goudsmit had welcomed Heisenberg into his home in Michigan. Three years after that he'd proposed kidnapping him. Now he was telling another friend to shoot the man, for trying to build something he didn't believe the Germans even could. Perhaps Goudsmit wasn't as "soft" as he claimed.

Heisenberg arrived in Zurich by train on December 16. Riding alongside him was Carl von Weizsäcker, who served as his official escort. Beyond that, Heisenberg had no security with him. What had he to fear in Switzerland?

In fact, news from the front lines soon put Heisenberg in fine fettle. The day he arrived in Zurich, the Third Reich launched its last major offensive of the war, now known as the Battle of the Bulge, in the dense forests of Belgium. To everyone's surprise the Wehrmacht still had plenty of fight left, and the Germans knocked the Allies in the teeth and sent them staggering backward. (The Alsos crew at Nazi U had to abandon their headquarters and flee thirty miles west.) The German press was ecstatic, and a few delirious reporters hinted that the Nazis might deploy atomic weapons soon, driving the Allies off the continent forever.

Heisenberg finally gave his lecture on the eighteenth, a week before Christmas. Despite the abundance of consumer goods there, Switzerland did ration fuel during the war, and the first-floor lecture hall at ETH was chattering cold. Weizsäcker had no doubt warned Heisenberg about the near riot during his talk three weeks earlier, but thanks to Heisenberg's insistence that the talk not be publicized, only twenty physicists showed up.

Them, and a pair of spies. Berg entered the freezing lecture hall with a fully loaded Beretta—the one he'd fumbled on the plane—concealed beneath his suit coat. Although just three months younger than Heisenberg, he was posing as a Swiss graduate student learning the intricacies of quantum mechanics. At Berg's side was an OSS agent named Leo, sent to escort Berg and presumably help him escape after the deed. And if Leo failed, the catcher had his lethal cyanide L-pill in his jacket pocket. One sharp bite down, and he'd render himself hors de combat, too.

Berg took a seat in the second row and produced a small notebook and pencil, as if to take notes on the lecture. In fact he drew a map of the room and began taking notes on the other attendees. At one point he also tried out his German and offered his coat to a man seated in front of him who seemed chilly. Carl Friedrich Freiherr von Weizsäcker turned his deep-set eyes on the stranger and curtly told him no. Berg jotted "Nazi" in his notebook and fingered him as Heisenberg's minder.

At 4:15 p.m. Berg finally laid eyes on the man he'd spent months obsessing over. Heisenberg emerged onstage in a dark suit, and after having some trouble cranking the blackboard into place, he wrote out several equations. While he did so, Berg took notes on his manner and appearance. It was a superfluous act—it's hard to imagine Carl Eifler bothering—but Berg wanted to size up the man he'd emotionally prepared himself to kill. He described Heisenberg as looking "Irish," with an oversized head, ruddy hair, and a bald spot on the crown. He wore a wedding band on his ring finger, and his furry eyebrows couldn't quite conceal two "sinister eyes."

His equations ready, Heisenberg began speaking, blithely unaware that his survival depended on what he'd say over the next few hours. He'd decided to lecture on developments in S-matrix theory, the scattering theory he'd first outlined at ETH two years earlier. Berg could not have been pleased with this choice of topic. Again, he didn't expect Heisenberg to sketch a bomb on the blackboard and start cackling, but he must have hoped for something at least related to fission—a lecture on reactors, perhaps. Instead Heisenberg wanted to talk pure theory, especially his hope that S-matrix theory could reconcile quantum mechanics and general relativity.

He began by outlining a history of the topic, but a colleague interrupted. Don't bother, he said, we know all this. Heisenberg was one of those easygoing lecturers who didn't mind interruptions, so he shrugged and skipped ahead. Berg noted that he paced as he talked, his left hand thrust in his jacket pocket. And despite the esoteric topic, Berg strained for any hint that Heisenberg might be betraying more than he realized. Do those equations relate to fission somehow? Is scattering important for chain reactions? At one point Heisenberg's gaze lingered on the unibrowed stranger for several seconds; they might well have locked eyes. "H. likes my interest in his lecture," Berg wrote.

No matter how hard he strained, though, Berg couldn't decipher the equations. It all seemed like innocuous physics, but how could he be sure? Was he missing something? Doubt began gnawing at him, and his mind inevitably circled back to the most famous discovery of the man now pacing the stage. Berg scribbled in his notebook, "As I listen, I am uncertain—see: Heisenberg's uncertainty principle—what to do."

Meanwhile, the physicists in the room remained oblivious to Berg's torment, focused on the equations. "Discussing math while Rome burns," Berg wrote. "If they [only] knew what I'm thinking." In truth, Berg himself didn't know what to think. Failing to act could hand Hitler the Bomb, and Europe with it. (*So do I shoot and potentially save the world?*) Then again, could he really shoot a man without hard evidence—especially knowing that he'd sacrifice himself in the process? He could feel the gun heavy in his pocket, but as Heisenberg droned on, that other piece of OSS gear, the L-pill, must have weighed more and more heavily on his mind.

This private torture continued for two and a half hours. And in the end, uncertainty made up his mind for him. When the lecture ended, he still couldn't bring himself to shoot.

Afterward, the score of physicists broke into small groups to chat, and a few rushed the stage to talk to Heisenberg. Berg took the opportunity to introduce himself to Flute, reciting a prearranged code phrase, "Doctor Suits sends his regards from Schenectady." Berg also passed along a gift from Allied intelligence—a vial of heavy water. The two spies quietly arranged to meet later that night in Flute's office.

Berg then crept close to the scrum around Heisenberg to eavesdrop, pretending to study the equations on the blackboard. Might Heisenberg let his guard down now, brag about something? No. After a bit of chitchat, a few old chums whisked Heisenberg off to dinner at the famous Kronenhalle café, leaving Berg behind in the freezing lecture hall. Having nothing to do, he skulked off to meet Flute—emotionally wrung out and still uncertain whether he'd done the right thing.

Heisenberg, meanwhile, was in high spirits at dinner. The lecture had gone well, he was surrounded by friends, and a newspaper story about the new German offensive left him so thrilled that he read it aloud at the table. "They're coming on now!" he marveled. Oblivious as ever, he failed to notice that his hosts were mortified.

Later that week Berg got a second chance. Flute was holding a small dinner party in Heisenberg's honor, and Berg hoped that, away from the seminar room, in a relaxed atmosphere with wine and food, some unguarded remark would betray the status of the German nuclear bomb program.

As with his lecture, Heisenberg put stipulations on the party, telling Flute that politics and the war were verboten topics of conversation. He had good reason for doing this. In the few days since he'd crowed over the newspaper article, the Battle of the Bulge had turned against Germany. The Wehrmacht certainly hadn't been crushed, but the fighting had deteriorated into a yard-by-yard struggle amid snow-drifts and icy streams—exactly the sort of long slog that a depleted Germany could never win. Heisenberg's grand, improbable dream—for a stalemate that would somehow both discredit the Nazis and keep the Allies out of his homeland—seemed less and less likely. Moreover, in the nest of spies and informants that was Zurich, he didn't want to talk politics among strangers. Even at this late stage of the war, "defeatist remarks" could get you shot in Germany.

Flute agreed to Heisenberg's demands, but as soon as the party started, Heisenberg realized the futility of this promise. Even if Flute stayed quiet, he had no control over his guests, several of whom cornered the physicist to pelt him with questions. Berg sidled up to listen, and it was ugly from the start. No one wanted to hear Heisenberg's rationalizations, and when he began whining about how the world was demonizing the good people of Germany, the other guests jumped down his throat, reminding him who exactly had started this war. They also scoffed when Heisenberg claimed to know nothing of Jews and other undesirables disappearing from Germany in large numbers. (In truth he almost certainly knew that his precious uranium came from processing plants that employed female slaves.) The guests backed down only when he stoked their fears about the Soviet Union: he argued that Germany was the one bulwark between civilized Europe and the hordes of Red barbarians eager to overrun it, a scenario that frightened your average Swiss even more than a German invasion. Heisenberg being Heisenberg, however, he managed to swallow his foot one last time near the end of the interrogation. Someone said, "You have to admit the war is lost." Heisenberg sighed and said, "But it would have been so good if we had won."

Heisenberg no doubt left the party exhausted and alone; all the residual joy from the lecture a few days earlier had dissipated. But as he donned his coat and stepped outside to walk home, someone joined him, someone he recognized from the lecture: the Swiss physics student with the heavy eyebrows. They were going the same way, it turned out, so he and Moe Berg—gun in pocket, pill in pocket—slipped off together.

As they walked, Berg pestered the physicist with questions. Drawing on his lawyerly training, he made several leading statements as well, trying to draw Heisenberg out. He complained about how boring Zurich was, saying that he'd give anything to be in Germany right now, where you could really fight the enemy. Heisenberg muttered that he disagreed but didn't elaborate.

As they stalked the dark streets of Zurich, Berg continued to press and Heisenberg continued to parry: years of living under Hitler had conditioned him to guard his opinions, and he answered the "student's" questions as vaguely as he could without being rude. Still, he had no inkling that this man was prepared to shoot him; even a jest, an ironic comment taken the wrong way, could have fatal consequences. Berg, meanwhile, had a perfect opportunity to carry out the execution. They were walking alone, at night; he easily could have ditched the gun and fled. So why not shoot Heisenberg, just to be safe?

In the end, uncertainty triumphed again: Berg simply couldn't do it. The two men parted at Heisenberg's hotel, and when Heisenberg turned his back one last time, Berg made himself walk away. Heisenberg entered the lobby and put the encounter out of his mind. Berg never could.

Heisenberg left Zurich the next day to spend Christmas with his family in Germany; he had toys for his children and skin cream and a sweater for his wife. To cut down on smuggling, Germany had banned the importation of certain goods from Switzerland, so Heisenberg had to pull the woman's sweater over his shirt at the border and pretend it was his.

Berg continued to poke around Zurich for the next week, and he did pick up some choice bits of intelligence from Flute. These included claims of a "supercyclotron" in Germany that could separate fissile isotopes much faster than any previous method—exactly the sort of "kitchen sink" apparatus Oppenheimer had warned about. Berg also confirmed earlier reports on the whereabouts of Heisenberg's new lab, as well as his family cottage south of Munich.

Despite the praise these reports won him, Berg still felt tortured by doubt as 1944 came to a close. Fate had thrown him two chances to take out Germany's top nuclear scientist, and he'd watched both pitches go by. Would he come to regret his prudence? Would the world?

If Berg couldn't act, though, his sometime nemesis Boris Pash would have no such qualms—no such uncertainty—about the need to render Werner Heisenberg hors de combat.

PART VI

1945

CHAPTER 55

Operation Big

The spring of 1945 saw the last days of World War II in Europe. But for a roving unit like Alsos, the chaotic end of the war was in many ways the most dangerous time.

Allied armies poured into Germany after the Nazis lost the Battle of the Bulge, and Alsos finally entered the country in March. In contrast to the jubilant Parisians, German citizens met them with hard stares and tried to undermine their progress. Sometimes this was merely annoying. Boris Pash once asked two German villagers which road to take to get to Heidelberg. Each man pointed—in a different direction. Pash growled that he was going to seize both of them, take each road in turn, and hang the one who'd lied. They blanched and pointed to a third road, muttering that they'd misunderstood the question. Other times Germans came within a hair of murdering an Alsos member. A sniper put a hole in the windshield of one vehicle, and a few weeks later Samuel Goudsmit—who was standing up in a jeep doing 50 miles an hour at the time—was nearly scalped when his head struck a wire strung across the road. Only his steel helmet saved him.

Despite these dangers Pash pushed his troops relentlessly. Stories abound of Alsos folks jumping onto jeeps and racing off to some recently liberated lab with nary a toothbrush or a change of socks.

When one fellow ran out of skivvies, rather than waste time returning to base, he liberated a pink pair of women's long underwear from a nearby home and trimmed them strategically to fit him. (This earned him some snickers, but they had to be more comfortable than the army's standard-issue long underdrawers, which according to one account were "so stiff that you didn't have to hang them up; you just stood them up on the floor beside your cot.")

Thanks to relentless Allied bombing campaigns, large swaths of Germany lay in ruins. At some labs the Alsos crews were more archaeologists than anything, digging documents and equipment out of rubble. In one cratered-out city a physicist named Smyth recalled driving past a few gutted buildings and seeing water pipes "almost tied into knots by the violence of the explosions." In another city he noticed roses blooming unaccountably early that spring on a grassy strip of median. It seemed a poetic moment—renewed life amid the destruction—until he realized that the heat of a few smoldering buildings nearby had simply tricked the bushes into thinking it was summer. So much for metaphor.

On April 11, Lightning-A reached the central German town of Stadtilm, where Pash encountered the rowdiest mob he'd seen since the Antwerp zoo. The town had virtually no electricity, and refugees and escaped prisoners were running amok. One gang found a railroad car full of industrial alcohol and tapped it. Two people died and one more went blind before Pash prevailed on the local burgermeister to drain it. After the chaos subsided, Pash learned that Uranium Club founder Kurt Diebner had been stationed in town but had slipped away just hours before.

Eventually Alsos established a new headquarters in Heidelberg, taking over a picturesque estate. From their ragtag start in Italy, the mission had grown into a force of one hundred plus troops—enough for Coach Pash to organize nine-on-nine baseball games on the grounds, even hurling a few innings himself. Amid the games, however, Alsos had some serious decisions to make. The United States,

United Kingdom, Soviet Union, and France had agreed in February 1945 to partition postwar Germany into four zones, each one occupied by a different conquering power. Unfortunately, one top Alsos target lay inside the Soviet zone—a huge stash of uranium at Stassfurt, ninety miles southwest of Berlin. Pash wanted nothing more than to raid the stash and stick it to the Russians, but he'd recently been feeling some heat from Washington, having been threatened with three separate court-martials by then for his freewheelin' ways. So before he risked angering the Soviets and causing a diplomatic incident, he sent a message to General Omar Bradley soliciting his thoughts. Bradley's response? "To hell with the Russians." That meant yes, and Lightning-A rolled out in mid-April.

A few months earlier, in Belgium and Toulouse, Pash had seized a few dozen tons of uranium. In Stassfurt he found more than a thousand tons, 2.2 million pounds total. The problem was, most of it was stored in small barrels that had cracked or split, spilling radioactive ore all over. Even worse, the Russians were already on the march and were making straight for the area.

Refusing to concede defeat, Pash summoned two military trucking units to the site. Then he rushed over to a nearby factory that made barrels and heavy-duty bags for packaging fruit and persuaded the workers to fire up their machines. They cranked out several thousand containers over the next few days, then pitched in to help stuff uranium inside. Despite occasional shells and gunfire, they managed to fill 260 vehicles and clear out every last ounce of ore in less than a week. General Groves later called the heist one of the biggest reliefs of the war for him: the vast majority of the missing Nazi uranium was now accounted for.

The trucking troops didn't know what they were hauling away, of course, only that the barrels were unaccountably heavy. Because one of the Alsos officers was named Calvert, some speculated they were transporting whiskey. Others swore they were stealing Nazi gold. The Soviets had no such illusions. When technical crews arrived in

Stassfurt a few weeks later, they discovered the heist and were furious. Boris Pash's little caper, in fact, served as one of the opening salvos of the Cold War.

α

With the uranium ore secure, Alsos turned its attention to hunting down the Uranium Club—an urgent task, since the Russians were also eager to lay hands on German scientists. In Heidelberg, one Alsos unit had already snagged the lovelorn Walther Bothe, whom Goudsmit (recently returned from rest leave) interrogated at his lab. Bothe was the first German prisoner Goudsmit knew personally, and they shook hands upon greeting, despite the U.S. army's rules against fraternizing with the enemy. With boyish pride Bothe showed off his cyclotron, which at long last had sputtered to life. Goudsmit tried to smile, but he found the scene pathetic: the United States had a score of working cyclotrons by then, any number of which were far more powerful than Bothe's toy.

Most club members proved harder to track down. One complication was that the Black Forest region where they'd fled to happened to lie within the French zone of occupation, so politics once again came into play. Pash at first suggested a deadly-hombre mission that involved parachuting into the area ahead of the French army and seizing all the scientists. Nothing ever came of this, but with the German resistance crumbling and the French army making good progress for once, Alsos had to hurry. For a while American officials seriously debated carpet-bombing the advancing French troops to slow them down. Eisenhower's chief of staff finally had to put the kibosh on that: "We cannot bomb the French," he said with a sigh, "much as I would like to." Only one option remained: Lightning-A would have to race south and sweep in ahead of the French troops. They called the plan Operation Big.

Luckily for Pash, many of the advancing French battalions had higher priorities than defeating Germany. Stuffing their pockets with loot, for instance, or hunting down Vichy traitors. A few avant-garde units were fast approaching the Black Forest, however, and Pash had to dip into his bag of ruses to get around them. At one point on the way south, Lightning-A encountered a formidable German barricade—thirty massive logs strewn across the road, with several more driven into the ground like pikes. They had no choice but stop and clear them. Meanwhile, a French unit had skirted the barrier by going off-road, dropping down into a nearby streambed. With their head start, they stood a good chance of opening up an insurmountable lead on Alsos.

Unfortunately for the French, the warm spring weather had started to thaw the streambed; Pash called it some of the swampiest, ooziest ground he'd ever seen outside an Alaskan bog. The French vehicles could still make their way along, provided they didn't slow down. So Pash waited until the convoy had reached the soupiest part, then yelled out for the leader of the column, a major, to dismount and stand at attention. The major ignored him, so Pash scrambled down the slope from the road, screaming in French, "When a colonel speaks to you, you assume attention!" Soldierly instinct won out over common sense, and the major stopped his car and obeyed. Splashing up to him, Pash introduced himself and was the soul of courtesy thereafter, asking all sorts of thoughtful questions about the man's trip so far. The major answered Pash—he had no choice—but his eyes kept darting up to the road, where Alsos engineers were making short work of the barricade. Meanwhile the major's car had sunk up to its axles in muck, and a similar fate befell every vehicle stopped behind him. Pash continued to blather until Alsos cleared the road-block. Then he wished the major a *bonne journée*, and raced off cackling.

This put Alsos among the new avant-garde, an enviable but dangerous position. In most villages in the region, the inhabitants had already given up, hanging pillowcases or sheets out their windows to signal surrender. But those white cloths sometimes concealed nests of SS troops, who had no compunctions about faking a surrender and opening fire. Moreover, bands of Werwolfen—self-proclaimed Nazi "werewolves"—continued to fight for their Führer. Even with the war lost, they wanted to ambush and kill as many enemy soldiers as possible.

Lightning-A's top priority in the region was reaching Haigerloch, the village with the cave where Heisenberg was building his final Uranium Machine. One historian compared the town to "the setting of some extravagant Wagner or Weber opera," with cobblestone streets and houses dating back to the 1100s; it was nestled against an eighty-foot limestone cliff with a stone castle and white baroque cathedral at the top. The landscape surrounding the town was just as magical, with rolling farmlands and fields; it was famous in springtime for its lilacs.

In the spring of 1945, lilacs were scarce. Several villages around Haigerloch had been reduced to piles of stones, and dust from the rubble blew through the streets like Wild West ghost towns. The landscape seemed almost too quiet, in fact, and Lightning-A rolled into Haigerloch warily, spooked at the lack of resistance.

They located Heisenberg's cave across the street from a few Bavarian-style homes with white walls and wooden shutters. It was a natural cave, with a jagged entrance roughly eight feet high. The cathedral atop the cliff had once stored its sacramental wine there, and in taking it over, Heisenberg and his assistants had code-named it the Speleological Research Institute. Pash's crew dubbed it the "atom cellar." The huge metal door covering the entrance was padlocked, so Pash hunted down the burgher with the keys and ordered him to open it. When he protested, Pash turned to his lieutenant: "Shoot the lock off. If he gets in the way, shoot him." The keymaster was far more accommodating after that.

The cave stretched twenty-five yards deep into the limestone, its walls cool and damp. The candles that Pash and his sidekicks held aloft could barely penetrate the gloom, and they stepped forward cautiously, wary of booby traps. After a few yards they came to a huge pit in the floor, ten feet across, with a thick metal lid and an aluminum cylinder embedded inside. Above the pit, anchored to the low ceiling, was an overhead crane that drew power from a diesel generator in the *Bierstube* (pub) across the street. Wires for hanging uranium cubes lay nearby, and pipes and electrical cables snaked all around. It was the sanctum sanctorum, the Nazi nuclear reactor. On a blackboard to the side, someone had written a cryptic message in German: "Let rest be holy to mankind. Only crazy people are in a hurry."

To their disappointment, Alsos found no uranium or heavy water in the cave; the Germans had apparently removed everything. But Pash was determined to deal the Nazi nuclear project a mortal blow. He put the whole town on lockdown, forcing everyone indoors. Then, after hauling out some valuable-looking equipment, he ordered his men to pack the cave with dynamite. He planned to bring down the whole cliff and caboodle above it, including the church, burying the atom cellar under thousands of tons of rock.

At the last minute, however, according to town lore, the local priest came scampering down a set of stairs carved into the cliff and begged Pash to spare the cathedral. Pash had orders to the contrary, but as the son of a bishop, he couldn't deny a man of the cloth. So Alsos set off a much smaller charge, one that simply collapsed the roof of the cave. It was still enough to destroy the last lingering remnants of the Nazi atomic bomb project.

❖

With the atom cellar buried, the hunt for the Uranium Club members began in earnest. Werner Heisenberg had already fled the area, reportedly to his family's cabin in the Bavarian Alps. Other choice targets remained, however. Alsos found Otto Hahn working quietly in a nearby village, his briefcase already packed. Hahn detested the war—his son Hanno, Lise Meitner's godson, had lost his arm on the Eastern Front fighting against Russia—and when an Alsos soldier entered his office, Hahn looked up and said in English, "I have been expecting you." He went along quietly. Carl von Weizsäcker showed a little more attitude. Although relieved that the Americans and not the Russians had picked him up, he nevertheless looked down his nose at his captors as uncivilized apes. (To be fair, it couldn't have helped his mood that the Alsos team, upon taking him into custody, had looted his wine cellar.)

In interrogating them, Pash and Goudsmit asked Weizsäcker and Hahn three things. Where's the uranium from the atom cellar? Where's the heavy water? And where are your technical reports? Just like at Strasbourg, Weizsäcker answered that he'd burned the papers— a disappointing but unsurprising answer. As for the three thousand pounds of heavy water, the Germans had siphoned it into oil drums, which they hid in a grist mill three miles distant. It was quickly recovered. Similarly, the Germans had buried two tons of uranium in a field atop a nearby hill. Liberating this took some work. Fearing booby traps, Pash forced a group of German prisoners to dig it up. And because the uranium had been cut into several hundred two-inch cubes, the Alsos team had to transfer it to waiting trucks by passing the cubes hand to hand down the hill, bucket brigade–style. A few scientists, including Goudsmit, pocketed one as a souvenir.

After securing the uranium and heavy water, Pash thought it prudent to scoot before the French caught on to his pilfering. His jeeps were headed out of town, in fact, when Weizsäcker spoke up and confessed something. He'd been listening to their questions over the past few days and had determined that Alsos wasn't what he'd feared—a trophy brigade out for loot. (His wine notwithstanding.) They were "intelligent people," he decided, people he could "talk sense" with. He therefore admitted that he hadn't burned the German technical reports after all. He'd hidden them.

The jeeps screeched to a halt. Where? That was the rub. To ensure their safety, he'd sealed the papers in a drum and buried them where no one would ever look: in a latrine, beneath several years' accumulation of human waste.

Perhaps Weizsäcker didn't believe they'd actually go after the drum, but if so, he badly underestimated Alsos. Goudsmit dispatched two soldiers—each wielding a long pole with a hook—to go fishing for it. As a joke, Goudsmit declined to tell them what they had in store, only that it was a "very important top-secret assignment." Both expressed gratitude that he'd singled them out for this honor. When they found out about the cesspool, they were awfully sore with him, but they got their revenge soon enough. After returning to the Alsos base they dumped the filth-streaked drum beneath the window of Goudsmit's room. There was no need to alert him that they'd found their target—he could smell that for himself. After hosing the drum down, Goudsmit pried the lid off and discovered virtually the entire archive of the Nazi atomic bomb project. Boris Pash sent a cable to Washington announcing that "Alsos has hit the jackpot."

The only sour note to Operation Big was that Werner Heisenberg had eluded capture. Alsos nevertheless managed to find his office before evacuating the region. It was located in an old woolen mill a few miles away. Few documents remained when Pash broke in, certainly nothing of value. But one item there remained burned into his memory forever. In the middle of the desk, in a place of pride, sat a framed photograph of Heisenberg and Goudsmit. They'd taken it together in Michigan in 1939, and for whatever reason Heisenberg had wanted it with him when he moved his lab south. Perhaps it reminded him of the last peaceful time in his life—before the war, before nuclear bombs, before all the trouble began.

Sorry to say, though, trouble had not stopped stalking Werner Heisenberg.

CHAPTER 56

The Lonely Organist

In January 1945, a few weeks after returning home from his lecture in Switzerland, Werner Heisenberg found himself in trouble with the Gestapo again. It turned out that an informant had infiltrated the dinner party in Zurich and had overheard him making "defeatist" remarks. Heisenberg instantly recalled the heavy-browed "Swiss physics student" who'd followed him to his hotel—clearly a Nazi agent. Who the real informant was, no one knows, but it took Heisenberg's superiors every ounce of persuasion they had to spare him.

The month only went downhill from there. After losing the Battle of the Bulge, Germany revoked the military exemptions of most scientists, and Heisenberg was duly drafted into the Volkssturm—the People's Militia, organized for a suicidal last defense of the Reich. He had to waste every Sunday training with them now instead of doing science. Not that his science was going much better. He'd recently moved his lab to the atom cellar in Haigerloch, where he planned to build his biggest Uranium Machine yet—one he hoped would go critical. He simply needed a few hundred gallons of heavy water, currently stored in Berlin, shipped south to Haigerloch.

But on February 1, the administrative head of the atomic bomb project, who was escorting the heavy water, called to tell Heisenberg that he'd made an impromptu decision: he was diverting the entire stock to — it was outrageous! — Kurt Diebner, who was working in a schoolhouse basement in a different town. From an impartial perspective, the decision made sense. However pathetic, Diebner had proved himself a dynamic nuclear scientist, and if anyone could throw together a working reactor at this stage, it was he. Heisenberg, though, wouldn't stand for it. That was *his* heavy water; it was a matter of scientific honor. So on February 5, Heisenberg and Weizsäcker set out to take the D_2O back, making a dash north that might have turned even Boris Pash's hair white.

The duo started before dawn on bicycles, then hopped onto a train. When the track ahead of them got obliterated in an air strike, they arranged for a car to cover the remaining distance. But another air raid swept in while they were waiting, and they spent the next few hours huddled in a cellar listening to a cello sonata on the radio, with the bombs above adding an unwelcome bass line. Then things got really hairy. When they finally hit the road again, they discovered that cars were irresistible targets for gunners in planes. (Germany had zero air defense left, so the Allies could strafe at will.) Every time a plane appeared overhead, the two physicists and their driver had to screech to a stop and leap into the weeds alongside the road to keep from being blown to bits. It was long after dark when Heisenberg finally reached his boss. The heavy water probably still should have gone to Diebner, but the boss could hardly deny his star scientist after all he'd been through that day. Heisenberg returned triumphant with the canisters, his scientific honor restored.

He spent the next two months setting up the final Uranium Machine in the atom cellar. They were productive months, but lonely; to pass the time between experiments, he'd ascend to the cathedral on the cliff and play Bach fugues on the organ. Finally, in late March, the machine was ready. The business end of it looked like a Calder mobile — 664 uranium cubes dangling from wires, eight or nine per strand. These were lowered into an aluminum vat in the floor filled with heavy water. To start things cooking, someone would shove a neutron source into a shaft leading into the center of the vat. The setup was dangerous—like dropping a grenade through the chimney of a gunpowder plant—but Heisenberg's crew didn't have time to set up safety shields. All they had was a lump of cadmium to cram into the chimney if things got out of hand.

In the end the experiment proved both a triumph and a dead end. Heisenberg's crew managed to produce a neutron multiplication factor of 670 percent—a huge leap toward a self-sustaining chain reaction. Not even Diebner had approached such numbers. (No one in the world had, as far as Heisenberg knew.) Still, without more heavy water or uranium, he simply couldn't goose any more neutrons out of the setup and achieve criticality. He and his team did keep tinkering—they had nothing better to do. But for all intents and purposes, this was the last gasp of the Nazi Manhattan Project.

By mid-April, the residents of Haigerloch could hear enemy tanks firing in the distance. Even this fairy-tale town, it seemed, was not exempt from the ravages of modern war. Heisenberg soon announced plans to evacuate and dismantle the Uranium Machine. He and Weizsäcker then hid the cubes in the field and the heavy water in the mill and the documents in the latrine.

Heisenberg was finally ready to leave town on the evening of April 20, but practically on his way out the door, he heard someone knock. It was Weizsäcker's wife, visibly distraught. She said that her husband had left their home several hours ago on his bicycle to pick up some equipment at his lab and hadn't returned. Had Heisenberg seen him?

No, he said. You're sure he's not somewhere else? She was sure, so he invited her in to wait. They spent the next hour drinking wine and halfheartedly reassuring each other that her husband was fine. Yet with mobs of soldiers and Nazi werewolves roaming about—one historian described the final weeks of the Third Reich as "a complete breakdown of military and civilian order"—they eventually stopped believing their own words. With every minute that passed, the odds increased that something awful had befallen the diplomat's son.

Weizsäcker finally returned just after midnight. He was unharmed and had no idea he'd caused his wife and friend such distress. Although relieved, Heisenberg said a hasty goodbye to the Weizsäckers and left town at 3 a.m.

With no other means of transportation, he had to pedal a bicycle 150 miles east to his family's cabin, the Eagle's Nest. By this point in the war, Allied planes were gunning down even bicyclists, so he traveled at night. He covered roughly fifty miles between each sunset and sunrise, and spent each day concealed in hedgerows, both hands gripping his bicycle while he slept, lest someone steal it. He tried not to talk to anyone, scrounging food from farms and orchards like a criminal on the lam, and he gave a wide berth to the roving bands of foreign troops in the area. He also encountered a wandering platoon of fifteen-year-old German soldiers, lost and hungry and crying.

Despite his caution, an SS guard caught Heisenberg on the final night of his hegira. The guard demanded to know why an able-bodied man like him wasn't with his Volksturm unit, defending the Reich. He then asked for Heisenberg's papers. With a sinking feeling Heisenberg dug through his pockets and produced them. The documents excused him from military duty, but they were crude forgeries—no better than a schoolboy's "doctor note," and the SS man knew it.

This all but sealed Heisenberg's fate: he would be shot for deserting. But before the SS man laid hands on him, Heisenberg decided to gamble. He would bribe the soldier, using the one currency that anyone in this whole crazy war would accept, no questions asked— American cigarettes. "I'm sure you haven't smoked a good cigarette in a while," he said, removing a crumpled pack of Pall Malls from his pocket. He held out an unsteady hand in the dark.

The SS guard had his duty to carry out: Heisenberg had betrayed the Reich, and the very notion of a bribe was offensive. But the prospect of smooth American tobacco proved overpowering. Despising Heisenberg, and himself even more, he grabbed the pack and waved the physicist on.

Heisenberg arrived at the Eagle's Nest filthy and exhausted. Yet however glad she was to see him, Elisabeth Heisenberg had little sympathy for Werner. While he'd been gallivanting around Europe the past few months, eating fancy dinners in Zurich and tinkering with Uranium Machines, she'd been stuck with their six children in the tiny cabin, which despite its grand name was rather dilapidated. The previous winter the roof had collapsed from excess snowfall, and they'd had to scavenge tiles from their old home in Leipzig to patch it up. Worse, she struggled to gather firewood in the mountains, and she could neither grow much food in the poor soil nor buy it from stingy local farmers. Elisabeth and the children suffered from constant bouts of illness as a result—and not just sniffles, but serious ailments like scarlet fever.

Chastened, Heisenberg did his best to protect his family now that he was there. He and Elisabeth stacked sandbags in front of the cellar windows for protection from blasts and set about buying as much food as they could afford. Heisenberg had also moved his elderly mother into a cottage in the area, and he made several trips to check on her. Then the Heisenbergs hunkered down to wait out the endgame of the war. Stray bullets still whizzed by the cabin sometimes, smashing into trees, and a local SS unit hanged sixteen men for desertion. Hundreds of soldiers nevertheless tried to escape through the woods, and more than once Elisabeth found her children playing with discarded guns.

The only pleasurable moment that week occurred on May 1, when the Heisenbergs heard the news of Adolf Hitler's suicide. The previous summer, Heisenberg had despaired while listening to the radio at the cabin, convinced he'd be rounded up in the Wednesday Club conspiracy. Now he could relax at last, and he and Elisabeth celebrated with a bottle of wine they'd been saving for one of their children's baptisms. After that, they could only wait for what came—and hope like hell that the Americans got to him before the Germans or Russians.

CHAPTER 57

Triumph and Loss

Werewolves—of course it was werewolves. Lightning-A had spent the past few days tracking the scent of Werner Heisenberg across southern Germany. Their jeeps and armored cars had crossed hundreds of miles of war-torn land, weaving around pockets of Nazi troops and careening over hill and dale. Even a freak spring blizzard in the foothills of the Bavarian Alps hadn't slowed them down. Now, just miles short of their destination, they were stuck. A pack of teenage Werwolfen had blown a bridge across a gorge, cutting off the only road forward.

Pash sent orders down the mountain to summon a team of engineers to rebuild it. But it would take at least a day to fix, and waiting around all day wasn't Boris Pash's style. So he grabbed a handful of doughty men and, taking a page from the Norwegian heavy-water commandos, scaled the gorge in front of him. He then plunged into the mountains to hunt Werner Heisenberg on foot.

According to Pash, his crew looked less like traditional alpine mountaineers and more like "Pancho Villa bandits, with ammunition tucked in pockets and cargo bandoliers slung across our shoulders." Though it was May, the snow reached halfway to their knees, and every so often they had to flush out pockets of enemy soldiers. Because Pash had neither the time nor the personnel to handle prisoners, he simply slit the waistbands of their trousers so they couldn't run or fight and ordered them down the mountain to surrender. One man endured even deeper humiliation. Despite reports of Hitler's suicide, rumors were swirling that the Führer had faked his death and fled into the mountains. Sure enough, Lightning-A captured one German who was the spitting image of Adolf. But upon stripping the man down to his skivvies, they turned him loose. Hitler, everyone agreed, would never wear such ratty underwear.

After five miles of trudging, Lightning-A reached the town of Urfeld, an idyllic mountain village overlooking a cold, clear lake. An innkeeper waving a white tablecloth pointed them toward Heisenberg's cabin up the road. Pash assigned several men to guard the town while he pursued the physicist. Somewhat rashly, he then divided his forces further, thanks to a hot tip. A Nazi propagandist named Colin Ross—who had traveled widely in the United States, and whom Goebbels had employed to urge Americans to embrace the master race—had also fled to the area with his family, and Pash decided it would be a feather in Alsos's cap to arrest him. He sent three men in that direction and went to look for Heisenberg with just one companion.

Although less nerve-racking than the approach to Joliot-Curie's cottage, the trek to the Eagle's Nest was more arduous—a long uphill slog through snowdrifts. But at 4:30 p.m. on May 2, Pash finally arrived. The six Heisenberg children gawked at the Americans, and a bemused Elisabeth informed Pash that her husband had gone to visit his mother. She telephoned Werner and told him to hurry back.

A few minutes later Pash finally laid eyes on the man he'd been stalking for months. Far from looking like the most dangerous scientist in the world, Heisenberg seemed haggard, already defeated. Pash was cordial but firm in placing him under arrest. "At that moment, I took a deep breath," he said. "Alsos was about to close the book on one of the most successful intelligence missions of the war—or so I thought." That qualifier was necessary, because as soon as he and Heisenberg sat down to discuss what would happen now, gunfire erupted in the distance.

Not wanting to risk Heisenberg's life, Pash left him behind and raced back to Urfeld. He found several bloody bodies lying in the village square. A group of German soldiers had attacked the Alsos guards there, but the Nazis got the worst of it, with two killed, three hurt, and fifteen taken prisoner. The remaining Germans had escaped into the forest to regroup—and possibly rally more troops.

Coincidentally, the trio Pash had sent to Colin Ross's cabin returned to Urfeld around this time. They were ashen, shaken. The propagandist had remained loyal to his Führer to the very last, they reported: after hearing of Hitler's death, he'd poisoned his wife and young son before shooting himself in the head.

As if the little stage of Urfeld needed more drama, a German general marched into town at this point and told Pash—in excellent English—that he had seven hundred soldiers nearby and wished to surrender. Not wanting to reveal how few troops he had, Pash accepted the surrender but said he'd have to wait until the next day to round up the Germans, given all his other responsibilities. In another bluff, Pash then turned to his lieutenant and told him to double the number of men on watch duty. The lieutenant, a little slow on the uptake, blurted out, "But we only have seven men, Colonel." Pash groaned—the general was standing right there. It wasn't clear whether he'd caught the blunder, though, so Pash feigned calm. He took the lieutenant by the arm, squeezing so hard he left a bruise, and repeated his order before shoving him off.

After dismissing the general, Pash ran to a local hotel, grabbed a phone, and demanded to be put through to a nearby town where American troops were stationed. He needed reinforcements on the double. But during the wait, the telephone wires got crossed, and a familiar voice with a German accent came on the line: "Ve know zere are not many uff you...Ve die for der Führer. Heil Hitler!" That answered that: the general had caught the blunder. Pash had to get the hell out of town, and as much as it killed him, he couldn't risk taking Heisenberg along.

Furious, Pash hurried back to the blown bridge. If he could just get a few armored vehicles across, he might be able to dart over and nab

Heisenberg. The engineers had arrived by then, and Pash drove them mercilessly, begging them to erect something, anything. It only had to hold for a few hours.

While they worked, a scout returned with ominous news. Back in Urfeld, a hundred SS troops had stormed the village square a few hours ago to hunt down the Americans. Finding none, they'd instead butchered some local townsfolk who'd cooperated with Alsos earlier that day, accusing them of collaborating with the enemy. Then they dumped the bodies in the lake and disappeared.

The scout had no idea where the Germans were now—perhaps lying in ambush. Pash didn't hesitate. The engineers got a makeshift bridge erected around 4 a.m., and practically before the last boards were in place, Lightning-A rattled across in a convoy.

They crept along the road for several miles in the dark, stopping regularly to check for booby traps. But upon reaching Urfeld they got aggressive, making a show of force by sending in several armored vehicles with mounted machine guns. The bravado worked, and shortly after dawn the rogue German units in the forest began dribbling into the village to surrender.

Alas, this included a mountaineering unit with an ornery mule. No one could calm it down, and with the village square growing more chaotic by the minute, it finally reached its limit. Rearing up and kicking, it lashed out at the most convenient target in the vicinity—Boris Pash, who happened to be standing nearby. "The two hoofs came at me like pistons," he recalled, catching him square in the back. "I landed ten feet farther away than I could have jumped under my own power." The blow broke three ribs, and a gasping Pash was dragged to a local inn for treatment. Somewhat anticlimactically, then, the colonel did not make the final trip up the mountain to take Werner Heisenberg into custody.

Still, the pain couldn't dampen Pash's triumph, not after all he'd been through—not only that day, but stretching back to those long, futile months in Italy when Alsos seemed destined to be disbanded. He knew that Kurt Diebner had been arrested near Munich the previous day, and when his men returned an hour later with Heisenberg in tow, he felt a wave of satisfaction: every last member of the original Uranium Club was now in American hands. Lightning-A had struck Germany.

Noting the number of troops and vehicles the Americans had risked to capture this one man, Heisenberg's neighbors in the village were awed. "Even Stalin could not have been better escorted," one marveled. He wasn't the only one who was impressed. As the retreating convoy left Urfeld, Heisenberg asked one of his American captors what he thought of this place. The soldier turned and took in the view—the forest, the mountains, the cold, clear lake nestled in the valley. He'd been all over the world, he finally said, but this was the most beautiful spot he'd ever seen.

α

The war in Europe ended five days later with Germany's surrender. Alsos celebrated by getting stinking drunk with the wine they'd looted from Weizsäcker's basement—all of excellent vintage. Having shot the moon with the Uranium Club, the bastard brigade could afford to be magnanimous to the Soviets, too, and invited a nearby platoon to their party. A thoroughly lubricated Boris Pash—his ribs apparently healing well—broke into a traditional dance "in the Russian manner," squatting and kicking up a storm.

Not every member of Alsos, however, was feeling so cheerful. Samuel Goudsmit could be grouchy at the best of times, and he'd had an especially rough go of things that week. Two days earlier he'd had to interrogate his onetime idol Werner Heisenberg, an unpleasant but unavoidable duty. Despite Germany's defeat, Heisenberg remained as arrogant and oblivious as ever, and when Goudsmit began asking questions, he launched into an enthusiastic account of all he'd achieved with his Uranium Machines, especially the 670 percent increase in neutrons. He then asked Goudsmit if the Americans had dared attempt such sophisticated research. For security reasons Goudsmit couldn't tell Heisenberg the truth—that the Manhattan Project had lapped the

Uranium Club, that scientists in Los Alamos would laugh at his puny 670. He nevertheless hesitated to lie to his old friend. He finally muttered that "certain features of the German experiments were new" to him and let Heisenberg draw his own conclusion. Heisenberg did—the most flattering one possible. He told Goudsmit not to worry: he'd catch the world up on this reactor business soon enough. The soul of generosity, Heisenberg even offered to give the Americans a tour of his atom cellar—the one Pash had dynamited.

This "sad and ironic" encounter left Goudsmit feeling queasy. And yet, perhaps trying to rekindle their old friendship, Goudsmit offered Heisenberg a job in the United States at one point, just as he had in Michigan six years earlier. "Wouldn't you want to come to America now and work with us?" he said.

Heisenberg refused. "Germany needs me," he insisted. Goudsmit could only sigh. It was the exact response Heisenberg had given back in Ann Arbor, word for word. After six years of war, Werner Heisenberg hadn't changed in the slightest.

Goudsmit had another bitter pill to swallow that week. A few days before the Heisenberg interview, while still in Haigerloch, he'd had a poignant conversation with Max von Laue, one of the nuclear physicists arrested there. Von Laue had long been friends with Dirk Coster, the Dutch physicist who was trying to save Goudsmit's parents. As with Heisenberg, in fact, Coster had written von Laue a letter begging him to intercede with German authorities on their behalf. Goudsmit knew this, and he once again violated the rules for fraternizing with the enemy by pulling von Laue aside one day to ask, sotto voce, if he'd heard any definite news about Isaac and Marianne.

Von Laue had. I'm sorry, he said.

Goudsmit later learned that they'd both died in Auschwitz on February 11, 1943—his father's seventieth birthday. Although he'd long suspected the truth, hearing von Laue confirm it tore the scab off his heart again. The pain hadn't dissipated days later, and prevented him from joining fully in the revelry of V-E Day.

Months later, Goudsmit heard something else that made the week even more painful—poisoned it in retrospect. For he finally heard the full backstory of the letter Dirk Coster had sent to Heisenberg, begging for help on behalf of his parents. The exact date remains unknown, but Heisenberg received the letter either in late 1942 or very early 1943, and despite the urgency of the plea, he did nothing. He dithered, he dallied, letting several weeks pass before replying to Coster in a short, tepid note. He began it by praising Goudsmit's scientific work. Then, in a calculated fib, he declared that Goudsmit had always championed German scientists in America. At last Heisenberg came to the crux of the matter, writing that it would be "highly deplorable if [Goudsmit's] parents, for reasons unknown to me, experienced difficulties in Holland." The note was dated February 16, five days after the Goudsmits died.

When Goudsmit finally learned of the note, he was dumbstruck. *Difficulties? For reasons unknown to me?* Was that what Heisenberg thought the Holocaust was, some bureaucratic blunder? It was mealymouthed, cowardly bullshit. Moreover, his old friend hadn't even tried to contact German authorities. He'd merely penned a private letter to Coster and washed his hands of the matter.

It's hard not to compare Heisenberg's behavior here to his vigorous defense of himself during the "Jew physics" debacle. In that case he'd appealed straight to Himmler, enduring a full SS investigation and risking his family's freedom and reputation to preserve his honor. With his friend's parents' lives at stake, he could barely be moved to put pen to paper. Goudsmit was worldly enough to know that Heisenberg probably couldn't have saved his parents no matter what. But he never forgave his friend for not trying. As Goudsmit later said, Heisenberg tried to save "'Jewish physics' with more vigor and success than he tried to save Jewish lives."

CHAPTER 58

Goimany

Moe Berg stuck around Switzerland after the aborted assassination attempt, and soon became fast friends with his fellow undercover agent Flute. They took long swims and bicycle rides together and spent whole days devouring newspapers in the cafés of Zurich. Flute's wife and children adored the big lug, too, nicknaming him "Bushie" and inviting him on skiing holidays. Berg in turn wrote musical ditties about the family, which he belted out with tone-deaf bravado. Ted Williams once said of his Red Sox teammate, "I don't ever remember seeing him laugh," but one of Flute's children remembered a different Berg: "We never saw him sad."

Beneath the revelry, however, Berg was still working as a spy in Zurich. He was preparing to infiltrate Germany, in fact, and in the meantime began stalking some pro-Nazi physicists around town. In one case, Berg powdered his hair white, followed a scientist into a library, and snooped on what books he was reading. When it became clear that the man was sketching plans for a cyclotron, Berg somehow swiped copies of his blueprints and shipped them to Washington.

Eventually, though, the old wanderlust got to Berg. And when General Leslie Groves abruptly canceled the Germany mission, the catcher abandoned his post in Zurich and began drifting around Europe. Some of his stops had potential intelligence value: he happened to be visiting Buchenwald the day the concentration camp there was liberated. Other stops didn't: at one point he dropped in on a linguist whose work on Celtic and Roman place-names happened to amuse him. Most of the time Berg simply went AWOL, slipping into various cities to see friends by night and disappearing on the morning wind.

During these months Berg indulged in a little chicanery to keep himself traveling in style. In each new city he'd find the local OSS branch and ask the special funds office for an advance, explaining that they could transfer the debt to his account in Washington. He regularly walked away with $100 or $200, and occasionally up to two grand ($30,000 today). He'd also have people check out of hotels for him and send those bills to Washington, knowing that the poorly managed OSS would never catch on.

Berg did consent to do some actual work during these months, including couriering a sheaf of espionage documents to the United States in April. But after returning to Europe in May, he resumed bouncing around the continent—London, Paris, Zurich, Marseille, Rome, Florence, Salzburg, Munich, Frankfurt. (Of those last few places, he gloated, "Germany is beaten, suffering, and how I love it.") He had no real agenda, just a vague hope that OSS or Groves would think of something for him to do. He felt cast aside.

Still, a famous baseball player was hardly anonymous in Europe then, given the hordes of American soldiers in every city. More than one stranger rushed up to him on the street and brayed, "Ain't you Moe Berg?" He usually either ignored them or feigned surprise and muttered something in the local French or Italian dialect before scurrying away. Every so often, though, Berg couldn't resist a little mischief. One time he came across a group of soldiers organizing a ball game. They needed a catcher to fill out their roster, and they asked the tall, dark stranger hovering nearby if he could play that position. Berg said he'd do his best. Pretty soon some cocky GI on first decided to test the old man's arm by stealing a base. He took off with the next pitch. Berg snapped up and smoked one across the diamond; the runner was out by ten feet. It finally dawned on another player, a lad from Brooklyn, who this mystery man was. "Jesus Christ, Moe Boig!" he yelled. "And here you are in Goimany."

In June 1945, Groves finally scrounged up a job for Moe: spying on the Russkies, America's new enemy. Groves was worried about German scientists defecting to the USSR and turning against the United States. So Berg crept into Rome to investigate rumors that Edoardo Amaldi and Gian-Carlo Wick were entertaining offers. He also snuck into Sweden to check on Lise Meitner, the physicist who'd fled Berlin in 1938 and later convinced Otto Hahn of the reality of fission. Meitner had endured seven miserable years in Stockholm, and even the conclusion of the war brought her little relief: upon hearing about the horrors of Buchenwald and Bergen-Belsen on the radio, she broke down sobbing and couldn't sleep. Despite her age (sixty-six), she was still one of the foremost minds in Europe, and Groves feared that Russia would scoop her up. He dispatched Berg to sound her out.

Berg arrived at Meitner's door with a letter of introduction from Flute, and spent several hours chatting with her over tea to win her trust. He returned the next day and only then proceeded to ask some questions. Meitner probably would have talked to a tree stump at that point—she was desperately lonely—but that old Berg magic had her bouncing like a schoolgirl. She autographed a paper of hers for him ("With many thanks and kindest regards, L.M., Stockholm") and in a letter to Flute afterward, she gushed, "Dr. [sic] Berg was extremely friendly. This does one so much good after the long years of isolation." She also entrusted Berg with a letter for Otto Hahn, then in custody. Berg promised to deliver it.

He didn't. Instead he tore it open the moment he'd turned the corner. It was an emotional and highly personal note, the product of several years' resentment. Among other things, Meitner accused Hahn of de facto collaboration with the Nazis for not standing up to them more forcefully. ("Your pacifism made you an accessory.")

Although it made for salacious reading, it had zero intelligence value. Berg nevertheless forwarded it to Allied officials, and it ended up on the desk of General Groves.

☢

As summer gave way to fall, Berg changed employers, albeit not by choice. In September 1945, President Harry Truman abolished OSS. The agency had never been popular with the American public, who feared the rise of a domestic Gestapo, or U.S. intelligence analysts, who hated its footloose ways. (Washington eventually founded a more professional spook service, the Central Intelligence Agency, which had little patience for eccentrics like Berg.) Berg got shifted over to the State Department, although he remained under the control of Leslie Groves.

The general soon attached Berg to a team of technical experts headed for Soviet-occupied Eastern Europe. They planned to investigate rumors that the Soviets were dragooning scientists there — packing them into trucks and trains emblazoned with the Soviet red star and forcing them to Moscow at gunpoint. The seizure of nuclear scientists was especially worrisome.

To get through checkpoints, Berg's team would flash phony passes or bribe hungry Russian guards with Spam and Hershey bars. When the bribes didn't work, they resorted to Pashian ruses. At one point near Prague, a band of angry Russian soldiers chased down Berg's jeep and surrounded it. Berg calmly fished a letter out of his pocket and unfolded it, tapping the bright red star at the top. The soldiers saluted this symbol of Soviet might and waved the Americans on, apparently unfamiliar with the Texaco oil company logo. Over the next few weeks Berg confirmed that the Soviets, gearing up for the Cold War, were indeed kidnapping scores of scientists.

That winter Groves sent Berg to interview Niels Bohr in Copenhagen. He also visited Lise Meitner in Stockholm, a second and far less pleasant call. A few months earlier, miffed that Otto Hahn hadn't answered her first letter, Meitner had written him another note to ask why. Hahn, confused, replied that he'd never received any letter from her. Meitner quickly deduced that the "extremely friendly" Dr. Berg had double-crossed her. So when he showed up at her home again, all smiles, she gave him the worst tongue-lashing he'd endured since Boris Pash laid into him in Rome.

Berg calmed her down the fastest way he knew how, by spinning more lies. He claimed that no one had read the first letter; authorities simply hadn't delivered it to Hahn because he was in custody. Berg then had the chutzpah to invite himself into her home, where they spent the next few hours chatting things over. At the end of which — the man had a way — Meitner entrusted him with another letter, for physicist Max Planck. Berg immediately opened this letter, too.

Berg and Meitner never met again. Incredibly, though, absorbing the brunt of her wrath stirred up some real emotions in Berg, a rare thing for him. She continued to haunt his thoughts for years, in fact, and he talked about her regularly; friends wondered whether he hadn't fallen a little in love with her, perhaps even tried to seduce her. In sum, the hulking big league catcher from Newark seemingly had a thing for a prim Austrian physicist who half hated his guts, and whose trust and confidence he repeatedly betrayed. If you ever needed further proof that Cupid is a perverse little imp, there you go.

Eventually Groves ran out of missions for Berg and cut him loose. More humiliating still, Berg discovered the hard way that OSS hadn't quite disbanded, not entirely. The agency no longer ran missions abroad, but it still had a stable of accountants in Washington tying up loose ends, and they had a few questions for Mr. Berg about the money that OSS had advanced him over the years: a total of $21,439.14 ($300,000 today). Mind you, no one ever accused Berg of theft or misappropriation (at least not publicly); the bean counters simply needed some sort of justification for the money in their ledgers. But Berg smelled a conspiracy afoot, or perhaps he decided to mask his guilt with bluster. Either way, he refused to account for a single dime, setting off a brawl that lasted for years.

Even if OSS accountants thought him a bounder, Berg still had champions in other areas of government, and in December 1946 the White House awarded the catcher the Medal for Merit, at that time the nation's highest civilian honor for wartime service. The nomination letter specifically cited his intelligence gathering in Italy, and his stalking of Werner Heisenberg in Zurich.

Berg rejected the award. When asked why, he stayed true to character and refused to explain himself. "It embarrasses me," was all he'd say. He'd already resigned from the State Department by then, and after refusing the medal he deflected all inquiries about what he'd do next. When he turned in his gear, he kept just two mementos from his days as an atomic spy: the pistol he'd carried to Heisenberg's lecture, which he'd still never fired, and the cyanide-filled rubber L-pill.

CHAPTER 59

The Bomb Drops

Alsos arrested dozens of German nuclear physicists throughout the spring of 1945 but ultimately kept just ten in custody, including Heisenberg, Hahn, Weizsäcker, and Diebner. All ten assumed, of course, that the Americans wanted to milk them for atomic secrets, based on the incredible Uranium Machines they'd built. (Did we mention the 670 percent increase in neutrons?) But the plain truth was that American scientists would learn almost nothing from their German counterparts. Alsos detained them in large part to keep them away from the Russians.

While debating where to intern this new Uranium Club, Alsos put them under the guard of some U.S. troops in Heidelberg who happened to be black—an act the Germans protested as insulting. (Apparently the attitudes of the Third Reich had rubbed off on them more than they realized.) There was talk of sending them to rural Montana to keep them isolated, until an American general proposed saving fuel and shooting them instead. Aghast, the British stepped in at this point and assumed control. They shifted the scientists to Versailles and then Belgium before finally flying them to England on July 3. It was a routine flight, but the Germans immediately grew suspicious. If the Allies wanted to wipe out the entire German nuclear physics program, they had a grand opportunity here: one plane crash would do it. All ten were no doubt white-knuckled the whole flight, flinching at every shudder of turbulence until they landed.

The British transferred them to Farm Hall, an estate north of London that was every bit as cushy as the one the German generals had enjoyed. The scientists dined on rich food and had full access to newspapers and radios. Heisenberg played Beethoven sonatas on the piano, and Hahn weeded the rose garden. A British soldier read passages from Dickens to them to improve their English. They played volleyball. The only complaints were a lack of contact with their families and boredom. One physicist read Lewis Carroll's *Alice* books several times through and couldn't stomach another journey into Wonderland.

Most of the ten got on well, enjoying the solidarity of prisoners, but Heisenberg and his clique continued to freeze out Diebner, snubbing him during activities and speaking to him as little as possible. The one time Diebner opened up to Heisenberg, the latter practically laughed in his face. Diebner had just confessed his fear that the British were monitoring their conversations, perhaps through microphones hidden around the estate. Heisenberg rolled his eyes: "They're not as cute as all that. I don't think they know the real Gestapo methods. They're a bit old-fashioned." Same old Diebner, pathetic as ever.

Turns out Diebner was right. In direct violation of the Geneva Conventions, the British had wired the whole house and were taping every conversation.

Political considerations kept the Alsos crew out of Berlin until late July 1945, a full three months after the Soviet army marched in. The clean, efficient German capital that Samuel Goudsmit remembered from before the war was gone. Mobs of grubby Russian soldiers roamed about, spreading typhoid fever and lice. Food and clothing were scarce, and a ruthless black market had popped up: everywhere

American soldiers went, people swarmed their jeeps, offering choco-late bars for 50 Reichsmarks ($70 today) or cigarettes for 100. Bombed-out cars clogged the streets, and Alsos sometimes showed up at scientific institutes to find little more than an address plate atop a pile of rubble. (In future years many university students had to help clear debris or do construction work as a condition of admission.)

While digging through one rubble pile, Goudsmit found the skull of an infant buried in ash. He also visited the old Egyptian museum in Berlin, once a favorite haunt of his, and found it still standing, albeit barely. Sadly, the old museum security guard, not knowing what else to do with himself, still reported to work each day and stood there all alone. Goudsmit shared his memories of the collec-tion, and the guard was so moved that he offered Goudsmit a mummy to take home. Although tempted—Goudsmit could ship anything he wanted to the United States, by claiming he needed to test it for radioactivity—the souvenir was too unwieldy to fit in his jeep; he settled for some painted scraps of mummy wrappings instead.

One day during the first week of August, while Goudsmit was excavating the ruins of Himmler's "race academy," a jeep screeched up beside him and an officer jumped out. You have fifteen minutes to catch a plane, he told Goudsmit. Goudsmit asked what the fuss was. The officer replied that he'd been scouring Berlin for him, and had finally seen the Alsos lightning logo on his jeep. Now he had to rush him to the airport for an emergency flight to Frankfurt, where Boris Pash was waiting. He couldn't reveal why.

After digging through debris all day, Goudsmit was tired and dirty, and asked to grab his toothbrush and pajamas first. No time, the officer said, and dragged him off. Goudsmit arrived at the air-strip to find the plane's propellers already spinning, "just like in the movies," and they started down the runway the moment he slammed the door shut. He arrived in Frankfurt to find two weeks' worth of fresh laundry waiting; clearly he wasn't going anywhere soon. But when he asked why he'd been summoned, no one would say. Even

Pash mumbled excuses. Annoyed, Goudsmit wandered off to have dinner with some old Alsos colleagues stationed nearby.

That evening everything became clear. After dinner Goudsmit chaperoned an Alsos secretary back to her hotel. In the lobby sat a drowsy sergeant listening to big-band music on the radio. Suddenly the broadcast cut out for a news flash—the United States had dropped a devastating new bomb on the Japanese city of Hiroshima. The report riveted Goudsmit: he couldn't believe how much technical detail the announcer was revealing. After years of intense secrecy, the public now knew almost as much about nuclear weapons as he did.

As the broadcast wound down, Goudsmit felt a wave of foreboding. One day a few months earlier, after realizing how far behind the German bomb project was lagging, he'd turned to one of Groves's deputies and said, "Isn't it wonderful that the Germans have no atom bomb? Now we won't have to use ours." The deputy just looked at him: "You understand, Sam, that if we have such a weapon we're going to use it." The prophecy had now come true. Atomic bombs had originated in fear, as defensive weapons to counteract the threat of a nuclear Reich. But as the German threat receded, so too had the notion of merely playing defense. Imperceptibly but inevitably, the Bomb had been transmuted into something else—the most blatantly offensive weapon in history. And the world, Goudsmit sensed, would never again be the same.

Unlike Goudsmit, Otto Hahn received a courtesy warning about Hiroshima. A British officer at Farm Hall pulled the chemist aside before dinner on August 6 and broke the news. Hahn was devastated. As codiscoverer of fission, he felt morally responsible for the destruction, and all the old horrors clutched at his heart. When the soldier offered him gin to steady his nerves, he took several gulps.

Hahn told the other nine German scientists at dinner around 7:45 p.m. that evening. There was an immediate uproar: none of them believed him. If Germany hadn't succeeded in building a nuclear bomb, then no other country could have, either. Heisenberg was especially adamant. This was a trick, he insisted, propaganda. The Allies had just slapped the name "nuclear" on some large conventional explosive. Besides, he'd already asked his friend Samuel Goudsmit about an American bomb, and Goudsmit would have told him if they had one.

Their protests continued right up until the BBC broadcast at 9 p.m.—which hushed every mouth in the room. The announcer went into a convincing amount of technical detail, and even mentioned uranium fission. Each physicist there suddenly felt as devastated as Hahn. As Goudsmit later commented, "That was the first time they really felt that Germany had lost the war. Up to that time, they had believed that Germany at least had won the war of the laboratory." No more.

Each scientist dealt with his grief differently. Diebner sat silent. Heisenberg kept asking how on earth the Americans, of all people, had beaten them. Hahn, perhaps tipsy, loosed his wicked tongue and began taunting people: "If the Americans have a uranium bomb, then you're all second-raters. Poor old Heisenberg." Weizsäcker, naturally, started making political calculations. "If the Americans and the British were good imperialists," he declared, "they would attack Stalin with the thing tomorrow, but they won't do that."

Weizsäcker also began rallying his comrades and—in a move that's still causing acrimony today—laying out ways for German nuclear scientists to save face. He emphasized two stratagems. First, they should blame their failure on a lack of resources and the stringent German wartime economy. Clearly they *could* have succeeded, if only they'd had the manpower and materials. At the same time, they should emphasize that they hadn't *wanted* to build a nuclear bomb—that they'd taken a moral stand to keep it out of Hitler's hands. As if trying the line out that evening, Weizsäcker said at one point, "The physicists didn't want to do it, on principle. If we had all wanted Germany to win the war, we would have succeeded." Hahn immediately objected to this ("I don't believe that"), as did another scientist. But you have to hand it to Weizsäcker: the strategy was clever. It would convince the Allies that they'd opposed Hitler, since they supposedly hadn't wanted to build it; convince their fellow Germans that they hadn't traitorously sabotaged the war effort, since building a bomb was economically impossible; and perhaps most importantly, convince the scientific world that they could have built a bomb if they'd been as rich and well supported as the Americans.

After the discussion broke up that night, well past 1 a.m., the hidden microphones at Farm Hall captured some poignant scenes. The administrative head of the German bomb project locked himself in his room and wept. When Hahn retired, two other scientists rendezvoused in the hallway and began whispering. Would Hahn kill himself that night? They feared so, and cracked open his bedroom door to peek. They waited and waited, and were immensely relieved to see him drift off at last.

Heisenberg, too, likely slept very little that night. His thoughts churning, he began working out a complete theory of bomb design the next morning. He did so all by himself, without access to any books or technical data—and in less than a week, he managed to reproduce most aspects of the top-secret American bombs. Why Heisenberg hadn't done these calculations in, say, 1939 remains unknown. But this wizardry showed exactly why the Allies considered him the most dangerous scientist in the world.

Weizsäcker, meanwhile, kept crafting the German political response, and he pushed his fellow captives to release a statement to exonerate them in the eyes of the world. Heisenberg duly wrote something up, and all ten Farm Hall scientists signed it, albeit some reluctantly.

Hahn was one of the reluctants, and over the next few months he continued to agonize. How had his simple chemistry experiments—a purely scientific venture—mutated into something so monstrous? Things got even more complicated when he learned, after reading a story about himself in the newspaper, that he'd won the 1945 Nobel Prize in Chemistry for discovering fission—surely the only scientific laureate to receive the news while incarcerated. It later emerged that one of the first scientists to nominate him for the fission work, way back in 1941, had been Samuel Goudsmit.

Fearing that Russian agents in Berlin would snatch him up and drag him to Moscow, the American army tried to detain Goudsmit in Frankfurt for most of August. Goudsmit refused, returning to the German capital as soon as he could—he was tired of the army jerking him around. And while a Russian agent did approach him there a few days later, it was only to buy his watch. With the flourishing black market, Goudsmit stood to make a fortune on it, $250 ($3,500 today). Except it had stopped ticking that very morning, and he got nothing.

In late August he had his final Alsos adventure. Unlike Goudsmit, the Russians really did want to kidnap the publisher-spy Paul Rosbaud (the Griffin), who knew the inside scoop on everything from the Peenemünde rockets to the Black Forest Uranium Machines. So when the Soviets lured the Griffin to a hotel one day, supposedly to meet with a famous Russian physicist, he and Goudsmit smelled a trap, and Goudsmit ordered two armored jeeps to follow Rosbaud. Sure enough, Soviet soldiers seized him and tried to drag him off. American troops had to wrestle him free in the street. Goudsmit then helped smuggle Rosbaud out of Berlin in an army uniform, taking the autobahn west to the American quadrant of Germany.

After Rosbaud's rescue, Goudsmit's work for Alsos effectively came to an end. He could now return to the United States and pick up the life he'd left behind. But before leaving Europe, he had one last mission, a personal one, to attend to.

His hometown of The Hague had been liberated in May 1945, and he finally found time to visit one windy day in September. He was surprised at how cramped and narrow the streets in his old neighborhood felt, and was saddened to see that his mother's hat shop had been gutted. Otherwise, things looked fairly normal. The familiar smell of the sea filled his nostrils, and as he approached his boyhood home, "I dreamed that I would find my aged parents...waiting for me," he said, "just as I had last seen them." For one fleeting moment he could even believe this—the house with its high front porch was still standing. But as he parked his jeep, he saw that all the windows had been smashed out; curtains on the third floor were flapping naked in the breeze. He climbed inside through a broken window and found the house empty, more than empty. The previous winter had been brutal in Holland, and every burnable item inside—doors, ceiling panels, stairs, all his father's handcrafted furniture—had been scavenged for fuel.

"Climbing into the little [bed]room where I had spent so many hours of my life," he later wrote, "I found a few scattered papers, among them my high-school report cards that my parents had saved." He wandered back downstairs and peeked outside. "The little garden in the back of the house looked sadly neglected. Only the lilac tree was still standing." His mother's breakfast nook was vacant, as was the corner where they'd kept the piano; the bookcase was bereft of books.

Until then, he'd merely felt furious with the Nazis for murdering his parents. But as he wandered the vacant house that day, a sense of guilt washed over him. Why hadn't he acted sooner to get Isaac and Marianne to safety? "If I had hurried a little more," he remembered thinking, "if I had written those necessary letters a little faster, surely I could have rescued them." It was the Nazis' final cruelty: to make the victims punish themselves. Alone in the empty house, he began weeping.

"I have learned since," Goudsmit added, "that mine was an emotion shared by many who lost their nearest and dearest to the Nazis." He could only thank God, once again, that things had turned out as they had—that the most vicious regime in modern history, despite having all the initial advantages in scientific talent and industrial might, had somehow lost out on the race to build the most awful weapons the world had ever known.

411

EPILOGUES

1946 and Beyond

When they first heard the news about Hiroshima, Irène and Frédéric Joliot-Curie were waiting out the end of the war at—where else?—their little family cottage in Port Science/L'Arcouest. Irène later said, repeatedly, that she thanked God her mother never lived to see her beloved radioactive elements turned into weapons.

Despite their heroics during the war, Irène and Joliot found themselves increasingly marginalized afterward, mostly for their unabashed support of communism. Irène was refused entry to the United States in March 1948 for her political beliefs and spent a long night in an Ellis Island detention center darning old socks. Joliot was evicted from a hotel in Stockholm in 1950 because the owner despised Reds—the same hotel he'd stayed at fifteen years earlier when collecting his Nobel Prize. At times even France proved unwelcoming, especially to Irène. Despite all her accomplishments, the hidebound French Academy of Science refused to admit her (or any other woman) as a member, just as they'd denied the honor to Marie. "At least they're consistent," Irène deadpanned.

Meanwhile, Irène's health continued to deteriorate. "To breathe, to eat—the most elemental functions are painful to me," she told a friend. Little wonder that by 1955 she was rapidly losing weight again. To revive her spirits, the family took an extended vacation that summer to L'Arcouest. While the trip did buoy her for a while, she kept repeating an ominous phrase: "How tired I am." Marie had often said the same thing during her final months.

While Irène rested, Joliot would go fishing or sailing, usually alone—he still felt like an outsider there. And although her symptoms were more acute, he could feel his own grip on life weakening. One day in the summer of 1955, he decided to go hunting in the woods surrounding the family cottage, the same woods through which Boris Pash had stalked his ghost a decade earlier. Joliot had always been an avid outdoorsman, and when a bird suddenly appeared before him—a perfect shot—he snapped up his rifle. But when he saw that the bird was tending several chicks in a nest, he couldn't bring himself to pull the trigger. He'd been a hard man during the war, when the times called for it. Now, that man was slipping away. "The old hunter in him," his biographer said, "was gone."

Irène finally died of leukemia in March 1956. After thirty years of living with and loving her, Joliot was broken by the loss and died of liver damage two years later. Both their maladies were the result of decades' worth of exposure to radioactivity. Even in cause of death, the Joliot-Curies remained united.

After their release from Farm Hall in January 1946, Werner Heisenberg and Carl von Weizsäcker faced some pointed criticism about their conduct during the war—especially about doing fission research for the Third Reich. Oddly, though, what upset Heisenberg most wasn't the charge that he'd collaborated with a monstrous regime. It was a charge leveled by Goudsmit, that the Germans had failed in part because they didn't understand the physics of atomic weapons. This insult Heisenberg would not stand for, and he defended his scientific honor vehemently. Goudsmit later admitted that he'd exaggerated the scientific shortcomings of the Uranium

Club, but he nevertheless remained baffled that, as he put it, "They find it much worse to be accused of stupidity than of Nazi sympathies." Historians have since debated the failure of the Nazi atomic bomb program ad nauseam, with some accepting the Heisenberg/Weizsäcker version of events and others finding it self-serving and misleading.

Regardless, the war did change Heisenberg. The laddish physicist of before disappeared, and a middle-aged man took his place; people made whispered references to the ending of *The Picture of Dorian Gray*. And although he tried to patch things up with colleagues in other countries, the old warmth between him and his friends had evaporated: stories spread in the 1960s that, whenever Heisenberg visited the CERN particle accelerator, he ate alone at the cafeteria each day. One of his few consolations in old age was that—decades after Weizsäcker's sister Adelheid had spurned him—his son ended up marrying Adelheid's daughter, uniting the families at last.

Heisenberg visited the United States several times after the war, and even called on Samuel Goudsmit once or twice. He always stayed at hotels, however, never Goudsmit's home. And Goudsmit never could summon up the courage to ask his sometime idol about the tepid letter he'd written in defense of his parents. When Heisenberg died in 1976, Goudsmit penned a generous obituary, concluding, "In my opinion he must be considered...in some respects a victim of the Nazi regime."

When the Alsos mission ended, Boris Pash transferred to Tokyo to help the Japanese people transition to democracy. But the old White Russian in him couldn't resist pulling one over on the Reds while he was there. Shortly after arriving, he learned that the Soviets had hatched a scheme to infiltrate the Russian Orthodox Church in Japan by sending over communist agents disguised as bishops. The very idea disgusted Pash, himself the son of an Orthodox holy man, and he outmaneuvered Moscow by arranging for an American bishop to come instead. The Soviets were livid, and when the American bishop arrived to say his first Mass, rumors flared that Russian provocateurs would make trouble at the church, maybe even riot. So Pash made a show of force in response: "I entered...the House of God" that day, he reported in a letter to his father, "with a prayer in my heart and a blackjack in my pocket." The troublemakers backed down, and the new bishop began his reign in peace.

In 1946, Coach Pash accepted a teaching job at a Los Angeles high school. But after the thrill of war, teaching had lost its luster, and he eventually quit to join the nascent CIA in Washington. (Unlike Moe Berg, Pash was exactly the sort of professional spook they were looking for.) No one quite knows what Pash did there, although E. Howard Hunt of Watergate infamy later accused him of running a "wet affairs" unit that specialized in liquidations and assassinations in communist Europe. (Pash hotly denied this.) Documents also hinted that Pash was involved in efforts to overthrow the Albanian government, and he reportedly tried to develop a poisoned cigar to knock off Fidel Castro. No charges ever stuck, and after his retirement Pash died at his home in California in May 1995, age ninety-four. Given his murky dealings, both during and after Alsos, latter-day conspiracy theorists have had little trouble linking him to everything from the rise of the Illuminati to the assassination of John F. Kennedy.

With his namesake dead, Joseph Kennedy Sr. channeled his political ambitions into his second son, who won the presidency in 1960. But Joe Junior's spirit was always lurking in the White House. During the Cuban missile crisis, President Kennedy helped enforce the blockade of Cuba with a destroyer named after his brother. And when JFK pledged to put a man on the moon, NASA's top engineer on the project was none other than Wernher von Braun, the German rocket scientist whose deadly V-weapons had spooked the Allies into launching the mission that killed Joe in the first place.

Over the years the Kennedys kept in touch with the family of Bud Willy, the engineer whose faulty arming panel probably doomed the final flight of *Zootsuit Black*. Willy's death had left the family destitute, so Kennedy Senior established a college fund to educate his three children. After becoming president, JFK even met Bud Willy's widow and daughter at a breakfast one morning while visiting Texas in 1963. Dallas, Texas. It was the last breakfast of his life.

JFK's favorite ballplayer as a boy, Moe Berg, received several offers to return to the Major Leagues as a coach. He spurned every one of them. Instead he tried to get back into the intelligence game through various freelance gigs for the CIA and NATO, such as interviewing scientists in Europe about missile defense. Berg being Berg, however, he racked up huge bills in hotels and restaurants on these trips, or blew off meetings in order to, say, eat fondue with Flute in Zurich. Not surprisingly, Berg's employers didn't appreciate this lackadaisical attitude ("This operation is going down the toilet," the CIA once complained), and the freelance gigs began drying up in the 1950s. By then Berg was living with his older brother Sam in Newark, and whenever the postman came, Berg would ask, "Any mail from Washington?" Increasingly, the answer was no.

Between gigs Berg rarely worked, preferring instead to mooch off people. He treated his brother's house like a storage unit for books and "live" newspapers, piling them up by the hundreds on tables and chairs and beds. When his brother finally kicked him out, he moved in with his sister, Ethel, a borderline schizophrenic who lived a few blocks away. Berg also traveled relentlessly, buttering up train conductors for free fares and showing up unannounced at the houses of old friends, carrying little more than a razor and a toothbrush. Many a sportswriter checked into his hotel room on the road to find Berg already taking a bath there, having sweet-talked the front desk into letting him in. He accompanied one friend on his honeymoon.

Things nearly turned around for Berg in the 1960s, when a publisher offered $35,000 for a memoir about his life as an atomic spy. But at some initial meeting, an ignoramus junior editor confused Moe Berg with Moe Howard of *The Three Stooges* ("I loved all your pictures"), and Berg stomped off in a huff, refusing to write a word. In reality, Berg was probably looking for an excuse to wriggle out of the deal. He'd already tried several times to piece together a memoir, jotting down stray paragraphs on envelopes, napkins, public library slips, train schedules, and page-a-day calendar sheets. (One entry refers to Einstein's famous equation as $m = Ec^2$. Berg still needed a little tutoring, apparently.) But the scraps never added up to anything coherent, and he finally gave up. However gifted a raconteur, the lonely drudgery of writing was not for him, and it no doubt proved easier to blame a junior editor than to try again and fail.

Berg continued to mooch off friends and travel, singing for his suppers and swapping stories about Tokyo or Babe Ruth for a place to stay that week. Most of his tales were lighthearted, but among close friends he occasionally delved into darker material. He seemed especially haunted by the near assassination of Werner Heisenberg, and ruminated on those three hours in the freezing lecture hall for the remainder of his life. During one memoir draft, from 1966, he scribbled, "Would I have disposed of them?" Perhaps he didn't want to think of himself as a coldblooded, deadly-hombre killer. Or perhaps he feared the opposite—that deep down he was a coward and never could have taken Heisenberg out. Regardless, the stress continued to gnaw at him, and he never quite got over Zurich. Once he even showed his brother the cyanide-filled rubber L-capsule he'd taken to the lecture, which he kept his entire life.

Berg had always had a paranoid streak, and as he got older it became more pronounced. He started cutting off friends for no reason, either refusing to write them or sending cryptic postcards that lacked a return address. (One from Havana read "Castro fiddles while Moe burns.") He also let his once-impeccable appearance slide: several hosts caught him washing his increasingly seedy suit in their bathroom sinks. The only reliable place to find Berg in later years was at the World Series. Major League Baseball had given him an engraved, silver-plated card that granted him free lifetime admission to any ballpark, and Berg took full advantage. But whenever an old chum spotted him in the stands and waved, he'd put his finger to his lips, whisper *shhh*, and melt away.

In May 1972, at age seventy, Berg contracted a viral infection that left him weak and dizzy. A few days later he fell at his sister's home and struck a low table. He began bleeding internally—his torso looked like one gigantic bruise—and upon admission to the hospital, his heart began failing. He died shortly after. His last words, spoken to a nurse, were, "How'd the Mets do today?"

After the war in Europe, Samuel Goudsmit found the prospect of returning to Michigan dreadful—it seemed too dull, too provincial. He eventually accepted a job at the Brookhaven particle accelerator in New York, where he continued to research and became editor of *Physical Review Letters*. Beyond science, he served as an advisor for Upton Sinclair's novel *O Shepherd, Speak!*, about an Alsos-like mission in Europe. He also revived his love of ancient Egypt by writing an introductory hieroglyphics text (*Ramses's First Reader*) and by developing a scheme to search for hidden chambers inside the pyramids at Giza with cosmic rays (nothing ever came of it).

"I sometimes look back, a bad habit," he wrote his daughter in 1973, "and am surprised...that so many of my infantile childhood dreams have come true, but not the more important mature wishes."

Among the infantile dreams, he mentioned studying Egyptology and "indulg[ing] in secret intelligence operations." But he never did secure a prestigious physics post in Europe. And while his fellow physicists nominated him forty-eight times for the Nobel Prize, the award always eluded him. Most disappointing of all, he lived long enough to see himself forgotten in some circles, his worst fear. In 1977 a branch of the American Physical Society held a special meeting to celebrate the fiftieth anniversary of quantum spin, his great discovery. Yet neither Goudsmit nor his old partner George Uhlenbeck received an invitation; they weren't even mentioned during the program. No wonder that, as Goudsmit wrote his daughter, "I am very depressed about my retirement." You can't help but wonder sometimes, reading Goudsmit's letters, whether his melancholy and cynicism might be an act, a put-on. This is not one of those times.

As he got older, Goudsmit increasingly took refuge in friendships. Whenever he could, he attended the Alsos reunions that took place every few autumns in Washington, D.C. The old troops would attend a boozy party Friday night, then hit the links Saturday morning for a golf scramble; the winners claimed the coveted Pash Cup. Goudsmit also grew quite close to Moe Berg, whom he considered one of the few "real friends" he ever had, "someone with whom you can discuss everything, no matter how personal." One personal matter they discussed was Goudsmit's wife. After years of disenchantment— foreshadowed by her refusal to write him letters during the war, no matter how much he begged—Goudsmit decided to divorce her. They were already living apart, and he established part-time residency in Nevada to expedite the process. She resisted, however, and when it came time to serve her papers, she refused to reveal her address. All Goudsmit knew was that she lived on Cape Cod and that he sent a monthly check to a bank in Boston. So he paid Berg $100 to dust off his old spy tricks and track her down. Berg got friendly with a Cape Cod sheriff and postmistress and found her in no time.

But as Berg did with everyone, he eventually cut Goudsmit out of his life and stopped answering his letters. No one knew why, Goudsmit least of all. He began writing desperate messages to Berg's brother and other acquaintances, begging for news. Aware that Berg loved newspaper puzzles, Goudsmit also crafted a cryptogram and sold it to the *New York Herald*; when solved, it read "Moe Berg, where are you?" Berg never responded, and the breach left Goudsmit despondent. He'd already lost the friendship of Werner Heisenberg to the war. Now Moe Berg had slipped away, too.

It wasn't just friendships that the war upended. Goudsmit believed that physics itself had changed, and not for the better. Before the war physicists were nobodies—benignly neglected savants putzing about in the lab, blissfully ignorant of the larger world. After the war—in large part because of atomic bombs—physics was too important to leave to the physicists. Generals and politicians got involved, and annual budgets ballooned to millions of dollars. Gone were the days of "string and sealing wax," as Goudsmit put it, when experiments were cheap and rickety, and two bozo graduate students could stumble into a fundamental discovery like quantum spin.

A colleague of Goudsmit's once observed that "Sam never did recover the very light touch that he had before the war," and you could say the same of almost everyone wrapped up in atomic espionage—the Joliot-Curies, Werner Heisenberg, Moe Berg, the Kennedys. Fission was one of the seminal discoveries of twentieth-century physics, but it proved as much a social phenomenon as a scientific one. In their desperation to keep the Bomb away from a madman, Allied scientists inaugurated a new kind of madness: the madness of heavy-water raids and geological commandos, of assassinations and radioactive toothpaste—not to mention Hiroshima and Nagasaki. At every step, the men and women involved believed they were doing the right thing. But in splitting the atom, they'd riven the world.

A Thank-You and a Bonus

I hope you enjoyed *The Bastard Brigade* and its colorful characters. And if you want to learn more about these bastards—or just read some fascinating and amusing anecdotes—check out the bonus material on my website at samkean.com/books/the-bastard-brigade/ extras/notes/. There are bonus photographs as well, at samkean.com/ books/the-bastard-brigade/extras/photos/.

If so moved, you can also drop me a line at samkean.com/contact. I love hearing from readers...

Acknowledgments

This book was a departure for me, requiring much more archival and historical research than anything I've ever done. It was an intimidating world to jump into, and I never could have pulled it off without the support of several people.

First, a big thanks to all the archivists and librarians across the country who put up with my ill-formed questions and vague inquiries into obscure topics. They were the pluckiest, most helpful bunch I could have hoped for. I'm especially grateful to the staffs of the American Physical Society, Princeton University Library, the Hoover Institute, the Library of Congress, and the John F. Kennedy Presidential Library.

As always, I owe more than I can say to my friends and family. My mom and dad, Jean and Gene [*sic*], have always supported me as a writer and kept me in good spirits. My brother Ben has become a great friend as well, and I wish him and Nicole luck in their new home. My sister Becca has been a constant source of fun and love throughout the years; I look forward to revisiting Ken's Korner soon. I hope someday that my books can give as much delight to my niece Penny and my nephew Harry as they've given me. And to my friends in Washington and South Dakota and across the world—some of whom I've known for decades now—you've made my first forty years wonderful. I look forward to forty more.

This is my fifth book with Rick Broadhead, who's continued to be a wise and vigilant agent. And it's my first of hopefully many books with my editor Phil Marino, who guided the manuscript with a deft touch. Kevin Cannon delivered, on short notice, the illustrations that punch up the text. I also want to thank everyone else in and around Little, Brown who worked with me on this book and others, including Anna Goodlett, Chris Jerome, and Michael Noon.

A few words on a page aren't enough to express my gratitude, and if I've left anyone off this list, I remain thankful, if embarrassed.

Major characters

Moe Berg – An ex–Major League Baseball catcher turned atomic spy. Berg attended Princeton University and spoke (some people claimed) a dozen languages. An enigmatic man, he was once called the strangest fellow to ever play professional baseball, which is saying something.

Samuel Goudsmit – A Dutch-born American physicist with a cynical sense of humor. He peaked too early in his career and spent the rest of his life feeling like a scientific has-been. In addition to hunting down the Nazi atomic bomb, he spent his time in Europe searching for his parents, who'd been swept into a concentration camp. In many ways the emotional core of this book.

Werner Heisenberg – One of the most brilliant physicists in history, and one of the most maddening. Worked on "uranium machines" for the German atomic bomb project, which infuriated many of his dearest friends, including Samuel Goudsmit. Was the number one target of the Alsos mission.

Frédéric Joliot-Curie – Nobel Prize–winning physicist and the husband of Irène Joliot-Curie. Although a pioneer in nuclear science, he bungled several major discoveries in the 1930s. Eventually became an underground resistance fighter in Paris during the war.

Irène Joliot-Curie – Nobel Prize–winning physicist and the wife of Frédéric Joliot-Curie; also the daughter of the legendary Marie Curie. A brave and outspoken critic of the Nazis, although ill health limited her active resistance. Sometimes overshadowed by her more personable, more extroverted husband, but she was every bit his equal as a scientist.

Joe Kennedy Jr. – A navy pilot, and the older brother of future president John F. Kennedy. Joe spent most of the war trying to one-up his little brother, which led him to volunteer for several ridiculously dangerous missions, including one to wipe out a supposed atomic missile bunker on the northern coast of France.

Colonel Boris Pash – A hard-charging and somewhat reckless soldier who led the Alsos mission to capture members of the Nazi atomic bomb program. Had Russian roots and fought in the White Army against the communists during the Russian Revolution.

Minor characters

Edoardo Amaldi – Assistant to Enrico Fermi; used to race him down the hallways of their institute in Rome. Top target of the Alsos mission in Italy.

Niels Bohr – Legendary physicist and brave critic of the Nazi regime during the war. Had a major falling-out with Werner Heisenberg over the latter's work on atomic fission.

Walther Bothe – Lovelorn German physicist whose bungled graphite experiments convinced the Uranium Club to use heavy water. Later worked in Joliot's cyclotron lab in Paris.

Dirk Coster – Dutch physicist who helped smuggle Lise Meitner out of Berlin. Later helped Samuel Goudsmit track down his parents in a concentration camp.

Kurt Diebner – German military physicist whom most of the Uranium Club dismissed as a pathetic, striving loser. He nevertheless proved a dynamic nuclear scientist.

William "Wild Bill" Donovan – Head of the Office of Strategic Services, a precursor to the CIA. Equal parts inspiring and incompetent. Managed Moe Berg at OSS.

Enrico Fermi – Virtuoso theoretical and experimental physicist. Born in Italy, he emigrated to the United States in 1938 and produced the world's first self-sustaining chain reaction in Chicago.

General Leslie Groves – The blustering, bullying, and remarkably effective head of the Manhattan Project. Started Alsos as a way to spy on, and potentially kill, German nuclear scientists.

Minor characters

Otto Hahn – German chemist and scientific partner of Lise Meitner. Their work on nuclear fission alerted the world to this dangerous new force. The military implications of fission left Hahn so distraught that he considered killing himself, but did eventually work for the Uranium Club.

Lise Meitner – Austrian physicist who interpreted Otto Hahn's confusing chemistry experiments and thereby discovered nuclear fission. After being run out of Berlin, she spent the balance of the war in exile in Stockholm and never got full credit for her work.

Robert Oppenheimer – Debonair physicist who led the weapons design lab at Los Alamos. Clashed with Boris Pash over his long history of supporting communist causes.

Paul Rosbaud – Scientific editor who hurried Otto Hahn's first fission paper into print. A dedicated foe of the Nazis, he spied for the Allies in Berlin under the nom de guerre the Griffin.

Wernher von Braun – Legendary German engineer who designed the deadly V-rockets at Peenemünde. Later became the driving force behind the Apollo moon-landing program.

Carl Friedrich Freiherr von Weizsäcker – Patrician German physicist on the Nazi atomic bomb project and close friend of Werner Heisenberg. Son of a major Nazi diplomat.

Sources

In putting together *The Bastard Brigade*, I consulted several thousand books, articles, and primary source documents—far too many to list individually. The list below contains only the most important sources, followed by the archives and libraries whose materials proved especially helpful.

Books

Age of Radiance: The Epic Rise and Dramatic Fall of the Atomic Era, by Craig Nelson, Scribner, 2015

Alsos, by Samuel Goudsmit, Tomash, 1947 (1983)

The Alsos Mission, by Boris Pash, Charter, 1969 (1980)

American Prometheus: The Triumph and Tragedy of J. Robert Oppenheimer, by Kai Bird and Martin J. Sherwin, Knopf, 2005

Aphrodite: Desperate Mission, by Jack Olsen, Putnam, 1970

Before the Fallout: From Marie Curie to Hiroshima, by Diana Preston, Walker Books, 2005

Beyond Uncertainty: Heisenberg, Quantum Physics, and The Bomb, by David C. Cassidy, Bellevue Literary Press, 2009

Blood and Water: Sabotaging Hitler's Bomb, by Dan Kurzman, Henry Holt & Company, 1997

Brighter Than a Thousand Suns: A Personal History of the Atomic Scientists, by Robert Jungk, Mariner, 1970

Brotherhood of the Bomb: The Tangled Lives and Loyalties of Robert Oppenheimer, Ernest Lawrence, and Edward Teller, by Gregg Herken, Holt, 2007

Sources

The Catcher Was a Spy: The Mysterious Life of Moe Berg, by Nicholas Dawidoff, Vintage, 1995

The Civilian Bomb-Disposing Earl: Jack Howard and Bomb Disposal in WW2, by Kerin Freeman, Pen and Sword, 2015

Dark Sun: The Making of the Hydrogen Bomb, by Richard Rhodes, Simon & Schuster, 1996

The Deadliest Colonel, by Thomas N. Moon, Vantage Press, 1975

Devotion to Their Science: Pioneer Women of Radioactivity, by Marelene F. Rayner-Canham and Geoffrey W. Rayner-Canham, Chemical Heritage Foundation, 1997

Donovan: America's Master Spy, by Richard Dunlop, Skyhorse Publishing, 2014

First War of Physics: The Secret History of the Atom Bomb, 1939–1949, by Jim Baggott, Pegasus Books, 2011

The Fitzgeralds and the Kennedys: An American Saga, by Doris Kearns Goodwin, St. Martin's Press, 1991

Frédéric Joliot-Curie: A Biography, by Maurice Goldsmith, Lawrence & Wishart, 1977

Frédéric Joliot-Curie: The Man and His Theories, by Pierre Biquard and Geoffrey Strachan, Fawcett Premier, 1966

German National Socialism and the Quest for Nuclear Power, by Mark Walker, Cambridge University Press, 1993

The Giant-Killers: The Story of the Danish Resistance Movement, 1940–1945, by John Oram Thomas, Taplinger Publishing Company, 1976

Grand Obsession: Madame Curie and Her World, by Rosalynd Pflaum, Doubleday, 1989

The Griffin: The Greatest Untold Espionage Story of World War II, by Arnold Kramish, Houghton Mifflin, 1986

Heavy Water and the Wartime Race for Nuclear Energy, by Per Dahl, Institute of Physics Publishing, 1999

The Heavy-Water Raid, by Jens-Anton Poulsson, Orion Forlag AS, 2009

Heisenberg and the Nazi Atomic Bomb Project, 1939–1945, by Paul Lawrence Rose, University of California Press, 2002

Heisenberg's War: The Secret History of the German Bomb, by Thomas Powers, Knopf, 1993

Sources

Hitler's Uranium Club: The Secret Recordings at Farm Hall, by Jeremy Bernstein, Copernicus, 2000

Hollywood High, by John Blumenthal, Ballantine Books, 1988

Lise Meitner: A Life in Physics, by Ruth Lewin Sime, University of California Press, 1997

Lost Destiny: Joe Kennedy Jr. and the Doomed WWII Mission to Save London, by Alan Axelrod, St. Martin's Press, 2015

The Lost Prince: Young Joe, The Forgotten Kennedy: The Story of the Oldest Brother, by Hank Searls, World Publishing Company, 1969

The Making of the Atomic Bomb, by Richard Rhodes, Simon & Schuster, 1987

Manhattan: The Army and the Atomic Bomb, by Vincent C. Jones, U.S. Army Center of Military History, 1985

The Mare's Nest: The War Against Hitler's Secret Vengeance Weapons, by David Irving, Panther Press, 1985

Marie Curie: A Life, by Francoise Giroud, Holmes & Meier, 1986

Marie Curie and Her Daughters: The Private Lives of Science's First Family, by Shelley Emling, St. Martin's Griffin, 2013

Moe Berg: Athlete, Scholar, Spy, by Louis Kaufman, Barbara Fitzgerald, and Tom Sewell, Little, Brown, 1974

Most Secret War, by R. V. Jones, Penguin, 2009

Now It Can Be Told: The Story of the Manhattan Project, by Leslie Groves, Da Capo Press, 1983

Of Spies and Stratagems, by Stanley P. Lovell, Prentice-Hall, 1963

Operation Big: The Race to Stop Hitler's A-Bomb, by Colin Brown, Amberley Publishing, 2016

Operation Freshman: The Rjukan Heavy Water Raid, by Richard Wiggan, HarperCollins, 1986

The Oppenheimer Case: Security on Trial, by Philip M. Stern, HarperCollins, 1969

OSS Training in the National Parks and Service Abroad During World War II, by John Whiteclay Chambers II, U.S. National Park Service, 2008

Physics and Beyond: Encounters and Conversations, by Werner Heisenberg, Harper & Row, 1972

Sources

Physics and National Socialism: An Anthology of Primary Sources, by Klaus Hentschel, ed., Birkhäuser, 1996

The Quest for Absolute Security: The Failed Relations Among U.S. Intelligence Agencies, by Athan Theoharis, Ivan R. Dee, 2007

Red Spies in America: Stolen Secrets and the Dawn of the Cold War, by Katherine A. S. Sibley, University Press of Kansas, 2004

Robert Oppenheimer: A Life Inside the Center, by Ray Monk, Random House, 2012

The Secret History of the Atomic Bomb, by Anthony Cave Brown, Dial Books, 1977

She Lived For Science, by Robin McKown, Julian Messner, 1962

Spying on the Bomb: American Nuclear Intelligence from Nazi Germany to Iran and North Korea, by Jeffrey T. Richelson, W. W. Norton, 2006

Tapping Hitler's Generals: Transcripts of Secret Conversations 1942–45, by Sönke Neitzel, Frontline Books, 2007

The Virus House: Germany's Atomic Research and Allied Countermeasures, by David Irving, Kimber, 1967

Wild Bill Donovan: The Spymaster Who Created the OSS and Modern American Espionage, by Douglas Waller, Free Press, 2012

The Winter Fortress: The Epic Mission to Sabotage Hitler's Atomic Bomb, by Neal Bascomb, Mariner Books, 2017

Archives and Libraries

American Institute of Physics, Samuel Goudsmit Collection, College Park, Maryland

American Jewish Historical Society, New York, New York

California Institute of Technology Archives, Pasadena, California

Columbia University, Moe Berg Papers, New York, New York

Danish National Archives, Copenhagen, Denmark

Hoover Institute, Boris Pash Papers, Palo Alto, California

Iowa State University Archives, Ames, Iowa

John F. Kennedy Presidential Library, Boston, Massachusetts

Library of Congress, Washington, D.C.

Los Angeles Public Library Archives, Los Angeles, California

Sources

Massachusetts Institute of Technology Archives, Cambridge, Massachusetts

Military Personnel Records, National Archives and Records Administration, St. Louis, Missouri

National Archives and Records Administration, Silver Spring, Maryland

New York Public Library, New York, New York

Oregon State University Archives, Corvallis, Oregon

Princeton University, Moe Berg Papers, Princeton, New Jersey

Richard M. Nixon Presidential Library, Yorba Linda, California

Rockefeller Institute Archives, Sleepy Hollow, New York

Science History Institute (formerly the Chemical Heritage Foundation), Philadelphia, Pennsylvania

Stanford University Archives, Palo Alto, California

University of California at San Diego Archives, San Diego, California

University of Chicago Archives, Chicago, Illinois

University of Michigan, Bentley Historical Library Archives, Ann Arbor, Michigan

Index

Allier, Jacques, 107–10, 113–14
alpha particles, 28–29, 32–35, 136–37
Alsos
    as Bastard Unit, 3
    origins of, 253–54
Amagiri (Heavenly Mist) (Japanese
    destroyer), 236
Amaldi, Edoardo, 427
    and Alsos mission, 272
    Moe Berg and, 291–93
    and Fermi's departure from Italy, 56
    and Fermi's physics team, 56
    and Operation Shark, 273–75
    radiation experiments, 39, 40
    rumors of defection to USSR, 400
Ann Arbor, Michigan, 63–67
anti-submarine warfare, 276–77
antimatter, 32
appeasement, 86–87
Arvad, Inga, 234
"Aryan physics," 68–72
atomic rockets, 226, 260
atomic structure, 30
Auschwitz, 364, 396

Baja Peninsula, Mexico, 157–59
barium, 52
baseball
    Moe Berg and, 14–22, 75–81, 147,
        148, 418
    Boris Pash and, 46, 158–59
Bastard Unit, origins of name, 3
Bay of Biscay, 235, 276
Belgium, 122, 339–42
Berg, Bernard, 13–14, 20, 148
Berg, Ethel, 416

Berg, Moe, 9, 13–22, 425
    and baseball, 14–22, 75–81, 147,
        148, 418
    birth of, 14
    in Eastern Europe, 401
    espionage training, 248–49
    in Europe after cancellation of
        Germany mission, 398–400
    Flute and, 398
    Goudsmit and, 419–20
    Heisenberg kidnapping plot, 356–57,
        365–73, 368–72
    interview methods, 293
    Frédéric Joliot and, 334
    John F. Kennedy and, 144
    Medal for Merit, 403
    and OSS, 152, 156, 243–45
    Boris Pash and, 292
    postwar life, 416–18
    in Rome, 290–93, 332–34
    in Scandinavia, 401–2
    in South America, 147–51
    and State Department, 401
    swindling of OSS, 399, 402–3
    and Vemork plant, 247
    in Washington, 289
    Gian-Carlo Wick interview,
        293–95
    in Zurich after Heisenberg's
        departure, 373
Berg, Rose, 13
Berg, Sam, 416
Berkeley Faculty Club, 221
Berlin, Allied bombing of, 209–12
beryllium, 28–30, 136–37
beta decay, 98–100

435

Index

Index

Index

Fermi, Enrico *(cont.)*
and nuclear "pile" at University of
Chicago, 182–83
radiation experiments, 38–40
strontium-90 proposal, 251–52
at University of Michigan, 72, 73
and uranium fission, 55–57
Fermi, Laura, 55, 56
Ferri, Antonio, 333–34
fission, physics of, 53, 57–58, 95–96
Flute (Paul Scherrer)
Moe Berg and, 373, 398, 400
dinner party for Heisenberg, 370–72
Heisenberg kidnapping plot, 355–58
flying bombs, 278–80
Ford, John, 155–56, 351
Forrest, Roy, 324–26, 330
Franco, Francisco, 83–85
Frankfurt, Germany, 406–7
French Academy of Science, 412
French resistance, 133–35, 309–11

gamma rays, 29, 30
Garland, Judy, 46
Geiger counters, 5, 287
Gentner, Wolfgang, 121–22, 133
Gestapo
Bohr's escape from Denmark, 238–40
Heisenberg and, 386
and Joliot's involvement with French
resistance, 133–35
massacre of captives from Operation
Freshman, 177, 178
Lise Meitner and, 50
"glandular approach" to victory, 155
Goebbels, Joseph, 162–63, 216, 234,
252
Gomez, Lefty, 290
Göring, Hermann, 216, 234, 304
Goudsmit, Esther, 66
Goudsmit, Isaac, 66, 67, 124–25, 168,
396, 411
Goudsmit, Jaantje, 63, 64
Goudsmit, Marianne, 66–67, 124–25,
168, 396, 411
Goudsmit, Samuel, 61–64, 425
in Berlin, 405–6

Bothe interrogation, 380
childhood, 61–62
discovery of Heisenberg's
correspondence with Weizsäcker,
361–63
in England, 312, 314
European sabbatical (1938), 64–65
and evidence of Nazi human
experimentation, 363–65
on failure of German nuclear bomb
program, 413–14
in Frankfurt, 406–7
in Germany, 377
and Hahn's Nobel Prize, 410
at Harvard, 123–24
and Heisenberg, 74, 414
Heisenberg interrogation, 395–96
Heisenberg kidnapping plot, 167,
262, 356–58, 365
in London, 4–6
nervous breakdown, 365
and news of Hiroshima bombing,
407, 408
offers of espionage services, 167–68
parents' fate, 124–25, 168, 344, 396,
397, 411
in Paris, 322–23
Boris Pash and, 312–14, 320–23
postwar life, 418–20
radar research at MIT, 164
return to the Hague, 410–11
Rhine wine fiasco, 343
on Royal Monceau food, 339
search for evidence of German
atomic bomb project, 345–50
in Strasbourg, 360
thorium fiasco, 346–50
thwarting of Soviet kidnapping of
Rosbaud, 410
at University of Michigan, 63–67,
72, 73
in Washington, 312–13
Weizsäcker/Hahn interrogations, 384
graphite, 102–4, 180–83
Great Britain
A-bomb research partnership with
US, 128

438

Index

Index

Index

Index

Index

Index

Index

and uranium fission paper, 54
and Vemork water evacuation, 265
Ross, Colin, 392, 393
Royal Air Force, 231, 240
Royal Monceau Hotel (Paris), 339
Russian civil war, 41–42
Russian Orthodox Church, 414–15
Ruth, Babe, 18–19, 75, 76, 149

S-matrix theory, 355, 368–69
sabotage. *See* Operation Freshman
St. Luke's Hospital (Tokyo), 77–78
Sandys, Duncan, 227–29, 337
Santa Barbara, California, 157
Schardin, Hubert, 209
Scherrer, Paul. *See* Flute
Schumann, Erich, 120–22
SCRAM (safety control-rod axe man), 181–82
Securities and Exchange Commission, 86
Sharkey, Jack, 290
Sicily, 272
Sinclair, Upton, 418
slaves, 232
Solomon Islands, 234, 236
sonar, 276
the Sorbonne, 16–17, 26
South America, 147–51
Soviet Union
  Moe Berg's espionage, 400
  and Eastern Europe, 401
  espionage, 220
  and German partitioning, 379
  Heisenberg's view of, 371
  Himmler's support for Nazi invasion of, 364
  meddling in Japan after war, 414–15
  Rosbaud kidnapping attempt, 410
Spanish Civil War, 82–85, 161
Special Attack Unit No. 1, 299
Special Operations Executive, 170
Speer, Albert, 137, 219, 259
Spemann, Hans, 36
Stadtilm, Germany, 378
Stark, Johannes, 69–70
State Department, U.S., 401

Stimson, Henry, 151–52
Strasbourg, France, 359–65
strontium-90, 251–52
supercyclotron, 373
Swiss Federal Institute of Technology (ETH), 355–58, 367–70
Szilard, Leo, 59

Tag der Weisheit (The Day of Wisdom), 141–42
Theodosia, Crimea, 41–45
Theresienstadt, 168
Thoma, Wilhelm von, 214–18, 338
thorium, 48, 52, 346–50
Tinnsjö, Lake, 264, 265, 269
Tokyo, Japan, 77–79, 149, 414–15
Toledo Mud Hens, 17
toothpaste, thorium and, 350
Torpex, 329
Trampoline to Victory, 150, 271
transuranic elements, 49
Trent Park, 214–18, 338
Truman, Harry, 401
Turner, Lana, 46

U-boats, 235, 276–77
Uhlenbeck, George, 62, 63, 419
Uncertainty Principle, 69, 93, 249
University of California at Berkeley, 101, 161, 220–25
University of Chicago, 180–83
University of Michigan (Ann Arbor), 63–67
uranium, 48–49, 52
  and Alsos interrogation of Weizsäcker and Hahn, 384
  fission, 53–57
  Boris Pash's hunt for German stockpile, 339–42
  removal from Soviet zone of Germany, 379–80
uranium-233, 346, 350
uranium-235
  explosive power of, 127
  heavy water and, 105
  in Marseilles yellowcake, 341
  physics of fission, 95–98, 188

Index

Index

Wormwood Scrubs prison (England), 117–18

X-ray stations, 25

Yeager, Chuck, 334
yellowcake uranium, 340–42

Zootsuit Black, 325–31, 336, 337
Zurich, Switzerland, 164–67, 367, 371–73, 398. *See also* Swiss Federal Institute of Technology (ETH)

About the Author

Sam Kean is the *New York Times* bestselling author of *Caesar's Last Breath*, *The Tale of the Dueling Neurosurgeons*, *The Disappearing Spoon*, and *The Violinist's Thumb*. *Caesar's Last Breath* was named *The Guardian*'s top science book of 2017. *The Disappearing Spoon* was a runner-up for the Royal Society of London's book of the year for 2010, and *The Violinist's Thumb* and *The Tale of the Dueling Neurosurgeons* were nominated for the PEN/E. O. Wilson Literary Science Writing Award in 2013 and 2015, as well as the AAAS/Subaru SB&F prize. Kean edited the 2018 edition of *The Best American Nature and Science Writing*, and his work has appeared in *The New Yorker*, *The Atlantic*, the *New York Times Magazine*, *Psychology Today*, *Slate*, *Mental Floss*, and other publications. He has also been featured on NPR's *Radiolab*, *All Things Considered*, *Science Friday*, and *Fresh Air*.

Also by Sam Kean

The Disappearing Spoon: And Other True Tales of Madness,
Love, and the History of the World from the Periodic
Table of the Elements

"A nonstop parade of lively science stories."

—Janet Maslin, *New York Times*

The Violinist's Thumb: And Other Lost Tales of Love,
War, and Genius, as Written by Our Genetic Code

"A slew of intriguing tales...Sam Kean is the best science teacher
you never had." —Keith Staskiewicz, *Entertainment Weekly*

The Tale of the Dueling Neurosurgeons:
The History of the Human Brain as Revealed by
True Stories of Trauma, Madness, and Recovery

"Mesmerizing...With a razor-edged wit and intriguing narrative,
the pages are easily devoured, all while Kean explores the deepest
labyrinths of the brain."

—Mellinda Hensley, *Los Angeles Magazine*

Caesar's Last Breath: And Other True Tales of History, Science,
and the Sextillions of Molecules in the Air Around Us

"Brims with such fascinating tales of chemical history that it'll
change the very way you think about breathing."

—Chelsea Leu, *San Francisco Chronicle*

Back Bay Books